Springer Science+Business Media, LLC

Yasuyuki Horie   Lee Davison
Naresh N. Thadhani

Editors

High-Pressure Shock Compression of Solids VI

Old Paradigms and New Challenges

With 138 Illustrations

 Springer

Yasuyuki Horie
MS D413
Los Alamos
  National Laboratory
Los Alamos, NM 87545
USA
horie@lanl.gov

Lee Davison
39 Cañoncito Vista Road
Tijeras, NM 87059
USA
leedavison@aol.com

Naresh N. Thadhani
School of Materials Science
Georgia Institute of
  Technology
Atlanta, GA 30332
USA

Editors-in-Chief:
Lee Davison
39 Cañoncito Vista Road
Tijeras, NM 87059
USA
leedavison@aol.com

Yasuyuki Horie
MS D413
Los Alamos National Laboratory
Los Alamos, NM 87545
USA
horie@lanl.gov

Library of Congress Cataloging-in-Publication Data
High-pressure shock compression of solids VI : old paradigms and new challenges/
editors, Yasuyuki Horie, Lee Davison, Naresh Thadhani.
    p. cm. — (High pressure shock compression of condensed matter)
  Includes bibliographical references and index.
    ISBN 978-1-4612-6554-2   ISBN 978-1-4613-0013-7 (eBook)
    DOI 10.1007/978-1-4613-0013-7
    1. Materials—Compression testing. 2. Materials at high pressures. 3. Shock
(Mechanics) I. Horie, Y. (Yasuyuki) II. Davison, L.W. (Lee W.) III. Thadhani, Naresh.
IV. Series.
TA417.7.C65 H555   2002
620.1′1242—dc21   2002070730

ISBN 978-1-4612-6554-2   Printed on acid-free paper.

© Springer Science+Business Media New York 2003
Originally published by Springer-Verlag New York, Inc. in 2003
Softcover reprint of the hardcover 1st edition 2003

9 8 7 6 5 4 3 2 1   SPIN 10883955

Typesetting: Photocomposed copy prepared from the authors' Microsoft Word files.

www.springer-ny.com

Preface

It is increasingly recognized that some phenomena associated with high-pressure shock compression of solids are fundamentally different from those explained by conventional macroscopic descriptions based on the concept of laminar motions. The evidence for this arises from both experimental and theoretical investigations and the phenomenological differences are manifest at the mesoscale, a spatial and temporal scale between the atomic and the continuum levels. Important features that emerge at such mesoscopic (grain diameter) levels are, for example, fluctuating particle velocities and transient eddies, as well as the formation of a hierarchy of dissipative structures such as non-uniformly distributed lattice defects, residual stresses, and inelastic deformation fields. These phenomena are obviously controlled by the interactions of shock waves with local material properties, microstructure, and internal material interfaces.

The subject of inhomogeneity and stress fluctuation has long played an important role in the physics of fluid motion and is not new in the study of solid response under high pressure. Nevertheless, the dynamical issue of heterogeneous and non-equilibrium shock processes occurring on the mesoscale has largely been ignored despite the fact that they must strongly influence phenomena such as plastic flow, fracture, and phase transformation. Field quantities averaged over many grains are obviously inadequate for describing physical processes occurring at the grain level. A steady, plane shock wave at the macrolevel is neither steady nor plane at the mesolevel. Stochasticity raises its ugly head.

This book is an outgrowth of invited talks given during two meetings: the pre-conference workshop on "Shock Dynamics and Non-Equilibrium Mesoscopic Fluctuations in Solids," and the special, plenary session, "What is a Shock Wave?" held during the 12th American Physical Society Topical Conference on Shock Compression of Condensed Matter. Both meetings were held during June 23–29, 2001 in Atlanta. The topical conference was attended by about 450 scientists and engineers from 18 countries, and 25 specialists from the United States, Russia, and China participated in the workshop. The theme of both the workshop and the plenary session of the topical conference was re-evaluation of the paradigm(s) underlying our understanding of shock phenomena, examination of basic assumptions, and presentation of new calculations, measurements, and theories that challenge these assumptions. The key questions addressed during the meetings were

1. What experimental data are available and what are their implications?
2. Are there new mesoscale theories of shock dynamics?
3. How do the theories affect the existing fracture and phase transition paradigms?
4. What kind of new computational and material response models are needed?

The chapters of this book concern the themes of these meetings, but they go far beyond the level of lectures that could be presented during the technical meetings or covered in conference proceedings. These articles expose many underlying and unresolved questions, and illustrate that they can no longer be dismissed simply as "small or minor effects." Successful fitting of experimental data by itself does not guarantee that the model used is a physically correct model, particularly at the micro- and mesolevels. A case in point is the fact that many models, even those that we know work well for, say, looking at stress gauge records and/or free surface velocity data, are not adequate when viewed from the perspective of stress rate and velocity rate (acceleration) [R. A. Graham, private communication]. We hope that this book will inspire and challenge its readers and set them on a path to discover a deeper, more fundamental understanding of shock wave phenomena in solid materials.

The first chapter of the volume, by Lee Davison, presents the most traditional view of the subject of shock phenomena in solids. It is the shortcomings of this theory that motivate the research discussed in the remainder of the book. In the second chapter, by Jim Asay and Lalit Chhabildas, conventional experimental observations not satisfactorily explained by traditional theories of the response of elastoplastic continua are discussed. This is followed with a chapter by John Lee in which observations of structure and turbulence in shock and detonation waves are discussed. Brad Holian discusses molecular-dynamic calculations of shock phenomena and the understanding that they bring to micro- and mesoscale aspects of shock propagation in atomic lattices. This is followed by chapters by Yuri Mescheryakov and Tatyana Khantuleva, respectively, in which measurements of mesoscale fluctuations in shock-induced flow fields in solids are presented and a new theory to explain the observed phenomena is described. Yilong Bai et al. then present a discussion of the effects of random defects in material bodies on mechanical behavior, pointing out that the interaction of these defects with mesoscopic heterogenieties and stress fluctuations helps explain damage localization and fracture and the sometimes surprising sensitivity of mechanical responses to small differences in the initial defect state of materials. In the following chapter, Jack Gilman discusses the different mechanisms of elastic and plastic deformation and suggests that the kinematical aspects of these differences are not properly captured in conventional theories of elastoplasticity. This is followed by a chapter in which Ron Rabie comments on various mechanisms underlying the development of structure in plane laminar shocks and presents some new data on the structure of very strong shocks. Finally, Craig Tarver extends the previous discussions of non-reactive solids to include consideration of the effects of shock-induced chemical reactions.

We express our sincere thanks to all the authors for their effort to expand their talks and make timely materials available in its present form. We specially thank Prof. Y.L. Bai who could not attend the meetings due to a bureaucratic policy, but was still willing to contribute an article that summarizes his and his colleagues' important contributions to the subject of dynamic fracture.

<table>
<tr><td>Los Alamos, New Mexico</td><td>Yasuyuki Horie</td></tr>
<tr><td>Tijeras, New Mexico</td><td>Lee Davison</td></tr>
<tr><td>Atlanta, Georgia</td><td>Naresh N. Thadhani</td></tr>
</table>

Contents

Contributors

J.R. Asay and L.C. Chhabildas

> Sandia National Laboratories
> P.O. Box 5800
> Albuquerque, NM 87185-1181
> U.S.A.

Y.L. Bai, M.F. Xia, Y.J.Wei, and F.J. Ke

> State Key Laboratory for Non-linear Mechanics
> Institute of Mechanics, Chinese Academy of Sciences
> Beijing 100080
> People's Republic of China

Lee Davison

> 39 Cañoncito Vista Road
> Tijeras, New Mexico 87059
> U.S.A

John J. Gilman

> Materials Science and Engineering
> University of California at Los Angeles
> Los Angeles, California 90095
> U.S.A.

Brad Lee Holian

> Theoretical Division
> Los Alamos National Laboratory
> Los Alamos, New Mexico 87545
> U.S.A.

T.A. Khantuleva

> Department of Physics and Mechanics
> Institute of Mathematics and Mechanics
> St. Petersburg State University
> Bibliotechnaya Place
> St.Petersburg 198904
> Russia

John Lee

> Department of Mechanical Engineering
> McGill University
> 817 Sherbrooke St. W
> Montreal, Quebec H3A2K6
> Canada

Yuri I. Mescheryakov

> Institute of Mechanical Engineering Problems
> Russian Academy of Science
> V.O. Bolshoi 61
> Saint Petersburg, 199178
> Russia

Ronald L. Rabie

> HE Science and Technology—C920
> Los Alamos National Laboratory
> Los Alamos, NM 87545
> USA

Craig M. Tarver

> Lawrence Livermore National Laboratory
> P.O. Box 808, L-282
> Livermore, CA 94551
> USA

CHAPTER 1

Traditional Analysis of Nonlinear Wave Propagation in Solids

Lee Davison

1.1. Introduction

Before beginning a formal discussion of nonlinear wave propagation, it seems useful to describe some of the basic observations. Most of the work that has been done concerns the response of materials to compression because the experiments are convenient and permit access to states of larger deformation than can be attained in materials under tension. The simplest situation considered is that in which a wave is introduced into the material by applying a compressive force uniformly over the surface of a halfspace. When this force increases smoothly in time, the resulting wave is also smooth. However, it is observed that (with a few exceptions) the gradient in such a wave increases with increasing propagation distance. Eventually, the wavefront evolves into an almost discontinuous jump. One cannot expect formation of a true discontinuity, but this is often a useful mathematical approximation to reality.

The wavefront that does evolve attains its form as the result of a standoff in the competition between a tendency for it to become steeper and a tendency for it to disperse. We shall describe materials as *normal* if they become stiffer as compression increases. The effect of normal nonlinear compressibility is to cause wave steepening, whereas dissipative phenomena that introduce a characteristic time into the process cause the waveform to become less steep. When the compression wave is weak, the equilibrium waveform is often rather disperse, but, as stronger waves are considered, the effect of nonlinear response of the material becomes more significant and the waveform becomes much steeper, eventually appearing as a discontinuity. The opposite behavior is observed in the case of decompression waves. In normal materials, they spread as they propagate into the material.

A *shock* is a surface of discontinuity of material velocity (and, consequently, stress, density, etc.) propagating in a continuous medium. Shocks may arise through evolution of a smooth wave, or may be introduced directly by impact or contact of the material with a detonating explosive.

Early investigations of shock phenomena were mathematical rather than experimental, and a collection of these early papers has been prepared by Johnson and Chéret [1]. A somewhat more modern account is given in the classic book of Courant and Friedrichs [2]. Since this pioneering mathematical work, activity in the field has broadened to include additional theoretical contributions as well as thousands of experimental and computational investigations. All of this work comprises the body of knowledge that we now identify as shock physics.

In preparing this chapter, it has been my intention to discuss nonlinear wave propagation in the traditional fashion, providing explicit equations for calculating quantities of interest and solving problems illustrating fundamental aspects of shock waves. The basic principles are presented in a form appropriate to all unstructured continuua, but the examples are calculated for fluids. A justification for this is presented in Section 1.3.2.

1.2. Mechanical Principles

The traditional description of nonlinear wave propagation in solids is based on classical continuum mechanics. The theory employs long established *conservation laws* for mass, momentum, moment of momentum, and energy, and *constitutive equations* that describe the responses of specific materials. Most of the early work was directed toward fluids, but the field has since expanded to include elastic solids, plastic, viscoplastic, or viscoelastic solids, porous and granular materials, fiber-reinforced materials, etc., in addition to fluids. Consideration of transport processes such as viscosity and heat conduction lead to characteristic response times, as do kinetics of evolutionary processes such as accumulation of inelastic deformation, and progress of chemical reactions or phase transformations. Characteristic lengths often arise in theories intended to describe structured materials. In this case, responses include solitary waves and other phenomena quite different from any discussed in this chapter [3].

1.2.1. Kinematics

The kinematics of motion of unstructured materials are completely described by equations giving the positions of the material points as they vary in time. Continuous material bodies reside in a three-dimensional Euclidean space in which the places can be identified using a Cartesian coordinate frame. Two such frames are needed to describe motions of deformable continuua. The first is called a *reference, material,* or *Lagrangean* frame. Points of the body are identified by their coordinates, $\mathbf{X}$, in this frame when the body is in an arbitrary, but specific, configuration, which we designate its *reference configuration*. A second coordinate frame (called a *spatial, laboratory,* or *Eulerian* frame) is used to identify places in space. Motions of the body are described by a function giving the place, $\mathbf{x}$, in the spatial frame occupied by each material point, $\mathbf{X}$, in the course of time. The two frames and the configurations of the body are illustrated in Fig. 1.1.

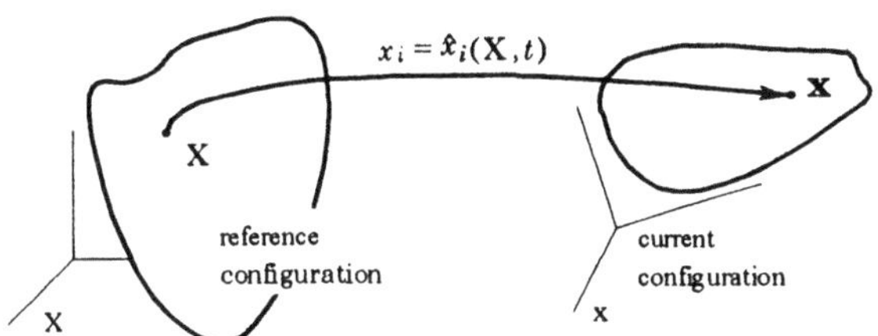

Figure 1.1. Reference and current configurations of a material body.

Although the material and spatial coordinate frames can be chosen arbitrarily, some special choices prove convenient. In this chapter, we shall adopt coincident Cartesian coordinate frames. In three dimensions, the motion of an unstructured body can be completely described by an equation of the form

$$x_i = \hat{x}_i(\mathbf{X}, t), \quad \text{for } i = 1,2,3 , \tag{1.1}$$

giving the location of each material point [4–6]. In this chapter we shall restrict attention to a class of motions called *uniaxial strains*. These are motions in which all material points move along parallel rectilinear paths. The motion on each of these paths is the same, so the motion as a whole can be described in terms of a single independent coordinate and time by equations of the form

$$x = \hat{x}(X, t) = X + U(X, t), \quad x_2 = X_2, \quad x_3 = X_3 . \tag{1.2}$$

It is a principle of continuum mechanics that no deformation can reduce an element of material to zero volume, so Eqs. 1.2 can be inverted to yield a representation of the motion in the form

$$X = \hat{X}(x, t) = x - u(x, t), \quad X_2 = x_2, \quad X_3 = x_3 . \tag{1.3}$$

All of the kinematical aspects of the uniaxial deformation of an unstructured continuum can be determined from Eqs. 1.2 or 1.3. This information includes particle velocity, strain, and rates and gradients of these quantities. Equation 1.2 is often referred to as the *Lagrangean* representation of the motion; it focuses attention on what happens to a specific material point. Equation 1.3 is the *Eulerian* representation of the motion; it focuses attention on what happens at a particular place in the spatial frame. Problems involving fluid flow are usually analyzed in the Eulerian frame, whereas problems of solid mechanics lend themselves to Lagrangean analysis. Lagrangean analysis is particularly appropriate when considering experimental data taken using gauges that are fixed to the material.

It is useful to reflect on what is *not* included in Eqs. 1.2 or 1.3. The only thing these equations say about a material point is its location. There is nothing to indicate a size or orientation, for example. These would be attributes of the points of a structured continuum, for which the description of the motion would have to include this information. Although there is a body of work concerning porous and granular materials, composite materials, etc., in which structure is specifically included in the kinematical description, the traditional paradigm does not allow for this. Many shock wave problems can be addressed in the one-dimensional context of uniaxial strain, but it is important to realize that one is really dealing with three-dimensional bodies. Proper treatment of the theory of material response requires consideration of invariance principles that are essentially three dimensional [5,7]. No problem arises in restricting properly set constitutive theories to a special case such as that of uniaxial strain, but a description of material response conceived in one dimension cannot easily be extended to a properly invariant three-dimensional theory.

The response of a material point depends on changes of its position relative to that of its neighbors, but not on the position itself, i.e., it is invariant to rigid translations and rotations. Changes in relative position are captured by the *deformation gradient* $F_{iJ} = \partial \hat{x}_i(\mathbf{X},t)/\partial X_J$. In the case of uniaxial strain, the only component required for the kinematical description is

$$F = \frac{\partial \hat{x}(X,t)}{\partial X} = 1 + \frac{\partial U(X,t)}{\partial X} . \tag{1.4}$$

From Eq. 1.4_1, we see that the relation between a line element dX in the reference configuration and its image, dx, in the current configuration is

$$dx = F\,dX , \tag{1.5}$$

i.e., an element of material lying along the X axis changes length by the factor F in the course of the deformation. The deformation gradient component adequately describes the uniaxial strain deformation but, in the three-dimensional context, the deformation gradient includes information about both elongation and rotation of material line elements. Invariance principles require that the rotation be eliminated from measures of deformation appearing in constitutive equations. Such measures of deformation are called, loosely, *strain*. Careful consideration of strain measures is essential to development of theories of the response of solid materials, but the response of compressible fluids can be expressed in terms of changes in their specific volume or density. Once the constitutive equations have been developed, the change in either of these quantities also provides an adequate measure of deformation for analyzing the response of solids to uniaxial strains.

The materials that we shall consider in this chapter are called, in the language of continuum mechanics, *simple materials*. This means that the response depends on the deformation gradient (and possibly on the rate at which it changes in time), but not on higher spatial derivatives than the first. No characteristic length arises in the description of simple materials so knowledge of the response of a homogeneously deformed body is sufficient to determine its response to all other deformations.

Equation 1.5 can be interpreted as an equation for the change in volume of an element of material of unit cross section, so we have

$$dv = F \, dV \, , \tag{1.6}$$

where dv is the current volume of a portion of the material that has volume dV in the reference configuration. Let us designate the specific volume (volume per unit mass) of material in the neighborhood of a material point by v when the volume is measured in the current configuration and by v_R when it is measured in the reference configuration. With this, Eq. 1.6 can be written

$$v = F v_R \, , \tag{1.7}$$

or, defining the current and reference densities $\rho = 1/v$ and $\rho_R = 1/v_R$, respectively,

$$\rho = \rho_R / F \, . \tag{1.8}$$

The details of strain measurement are lost in fluid models, but are very important in solid models. In fluid models, the only aspect of strain that plays a role is the change in density. In uniaxial strains, the density or specific volume ratio, $F = \rho_R / \rho = v / v_R$, conveys the same information as a strain measure. Note, however, that this quantity enters the theory differently depending on the strain measure used. It is often convenient to use the *Lagrangean compression*

$$\Delta = 1 - \frac{v}{v_R} \, , \tag{1.9}$$

in place of v or ρ. The Lagrangean compression is dimensionless and has the virtue of measuring deformation; v alone does not, since the deformation depends on both v and v_R. Superficial consideration may suggest that an uniaxially strained body is not sheared, but that is not so, as indicated by the sketches in Fig. 1.2.

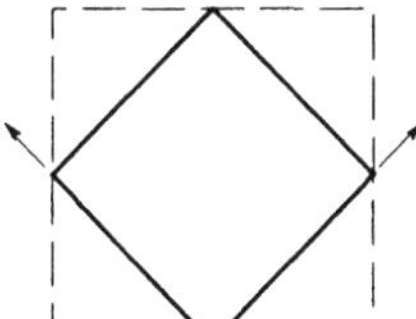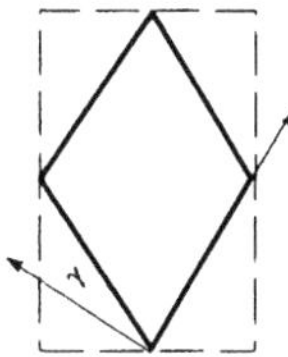

Figure 1.2. Illustration of uniaxial strain. The right-angle corners of the undeformed body are altered by the angle γ as a result of the deformation. This angle is a measure of the shear produced by the uniaxial strain. In contrast to simple shear, the uniaxially deformed body also experiences a volume change.

The velocity, $\dot{\mathbf{x}}$, of a material point $\mathbf{X}$ is simply the rate of change of position, $\mathbf{x}$, calculated holding $\mathbf{X}$ fixed. The particle velocity components for uniaxial strains are given by

$$\dot{x} = \frac{\partial \hat{x}(X,t)}{\partial t} = \frac{\partial U(X,t)}{\partial t}, \quad \dot{x}_2 = \dot{x}_3 = 0 \tag{1.10}$$

and the particle acceleration components are

$$\ddot{x} = \frac{\partial^2 \hat{x}(X,t)}{\partial t^2} = \frac{\partial^2 U(X,t)}{\partial t^2}, \quad \ddot{x}_2 = \ddot{x}_3 = 0 . \tag{1.11}$$

When the Eulerian description of the motion is used, we have

$$\dot{x}(x,t) = -F\frac{\partial \hat{X}(x,t)}{\partial t} = F\frac{\partial u(x,t)}{\partial t} \tag{1.12}$$

and

$$\ddot{x}(x,t) = \frac{\partial \dot{x}(x,t)}{\partial t} + \frac{1}{2}\frac{\partial \dot{x}^2(x,t)}{\partial x} \tag{1.13}$$

for the particle velocity and acceleration, respectively.

It is often necessary to consider the rate at which the deformation changes. The spatial gradient of the particle velocity, the *velocity gradient*, $l_{ij} = \dot{x}_{i,j} = \partial \dot{x}_i(x,t)/\partial x_j$, is a measure of the relative rate at which neighboring material points are moving. Usually we are interested in the symmetric and antisymmetric parts of this tensor, called *stretching*, $d_{ij} = (\dot{x}_{i,j} + \dot{x}_{j,i})/2$, and *spin*, $w_{ij} = (\dot{x}_{i,j} - \dot{x}_{j,i})/2$, respectively. For uniaxial strains, the stretching tensor has only the single nonzero component

$$d = \frac{\partial \dot{x}(x,t)}{\partial x} \tag{1.14}$$

and the spin tensor vanishes. A straightforward calculation shows that

$$\dot{F} \equiv \frac{\partial F(X,t)}{\partial t} = F\frac{\partial \dot{x}(x,t)}{\partial x} = Fd . \tag{1.15}$$

1.2.2. Stress

When forces are applied to a body, the effect is manifest throughout its volume, and is described using a *Cauchy stress tensor*, $\mathbf{t}$ (see Fig. 1.3). The principle of conservation of moment of momentum is satisfied if and only if this tensor is symmetric [4,5]. Constitutive descriptions are restricted to ensure that this symmetry condition is satisfied.

It is an essential point of solid mechanics that both normal and transverse (shear) stress components can be present on surfaces in the material. Either of

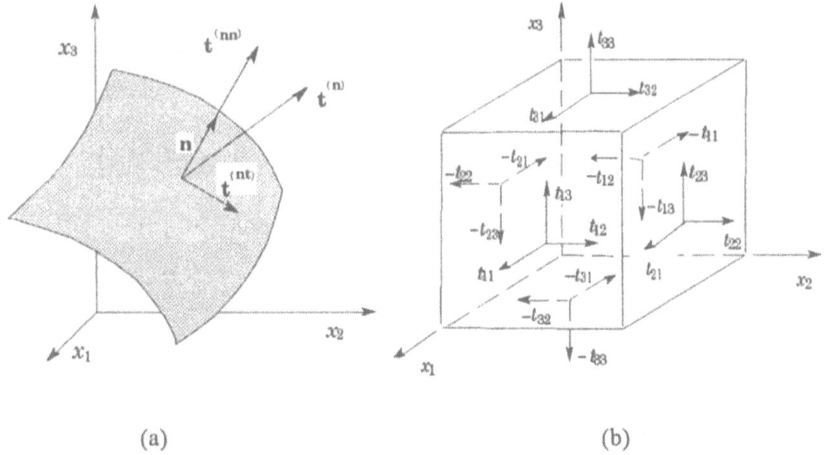

(a)   (b)

Figure 1.3. (a) The stress vector, $\mathbf{t}^{(n)}$, having components $t_i^{(n)} = t_{ij}n_j$, at a point on a surface and its normal, and tangential components., $t_i^{(nn)} = t_{kl}n_k n_l n_i$, and $t_i^{(nt)} = t_i^{(n)} - t_i^{(nn)}$, respectively. (b) Stress tensor components on planes normal to the spatial coordinate axes.

these components may be absent under certain loading conditions, but it is only in the special case of inviscid fluids that the shear components are absent in all cases. We shall define the mechanical pressure, p, by the equation

$$p = -\tfrac{1}{3}\,\mathrm{tr}\,\mathbf{t} = -\tfrac{1}{3}\left(t_{11} + t_{22} + t_{33}\right). \tag{1.16}$$

This quantity is obviously equal to the pressure in the case of a fluid in equilibrium, in which case all of the normal stress components are equal and no shear stress is present. The quantity $\mathrm{tr}\,\mathbf{t}$, hence p, is invariant to transformations of the spatial coordinate frame, so Eq. 1.16 provides a reasonable definition of the pressure for stress fields that involve shear. Using Eq. 1.16, one can write the stress components in the form

$$t_{ij} = -p\,\delta_{ij} + \bar{t}_{ij}, \tag{1.17}$$

where $\bar{t}$, called the *stress deviator tensor*, is a measure of the shear stress ($\delta_{ij} = 1$ when $i = j$ and 0 otherwise).

It is useful to note a special aspect of the representation of applied loads that is implicit in the description of Fig. 1.3. In some areas of mechanics it is supposed that loads imposed on the surface of a body include both forces and moments. Distributed moments are absent from the traditional continuum description illustrated in Fig. 1.3, although the distributed forces can give rise to a moment applied to a finite part of the body.

In states of uniaxial strain, it is usual to have nonzero transverse stress components. When the elastic properties are symmetric with respect to the strain axis, we have $t_{22} = t_{33}$. The shear stress can be shown to achieve its maximum value on planes lying at an angle of $45°$ with the strain axis and the magnitude of the shear stress on these planes is

$$\tau_{max} = -\tfrac{1}{2}(t_{11} - t_{22}).\tag{1.18}$$

Combining Eqs. 1.16 and 1.18 leads to the special form,

$$t_{11} = -p - \tfrac{4}{3}\tau_{max},\tag{1.19}$$

of Eq. 1.17 appropriate to this case. As with the general expression, it expresses the normal stress as the sum of contributions due to pressure and shear.

1.2.3. Conservation of Mass, Momentum, and Energy

The principles of conservation of mass, momentum, and energy form the basis of classical mechanics. (We have noted that symmetry of the Cauchy stress tensor is a necessary and sufficient condition for conservation of moment of momentum.)

Integral form of the conservation equations. The most basic expression of the conservation principles is in the form of integral equations. The mass of a layer of an uniaxially strained material body that occupies the interval (X_A, X_B) in the reference configuration is

$$\int_{x(X_A,t)}^{x(X_B,t)} \rho(x,t)\,dx \tag{1.20}$$

and the condition for conservation of the mass of this body can be written

$$\frac{d}{dt}\int_{x(X_A,t)}^{x(X_B,t)} \rho(x,t)\,dx = 0.\tag{1.21}$$

Similarly, we have the integral equation

$$\frac{d}{dt}\int_{x(X_A,t)}^{x(X_B,t)} \rho\dot{x}\,dx = t_{11}\big(x(X_B,t),t\big) - t_{11}\big(x(X_A,t),t\big)\tag{1.22}$$

for balance of momentum and the equation

$$\frac{d}{dt}\int_{x(X_{\mathrm{A}},t)}^{x(X_{\mathrm{B}},t)}\rho(\varepsilon+\tfrac{1}{2}\dot{x}^2)\,dx \tag{1.23}$$

$$= t_{11}\big(x(X_{\mathrm{B}},t),t\big)\,\dot{x}\big(x(X_{\mathrm{B}},t),t\big)-t_{11}\big(x(X_{\mathrm{A}},t),t\big)\,\dot{x}\big(x(X_{\mathrm{A}},t),t\big)$$

for balance of energy.

Smooth motions. By invoking suitable assumptions, one can reformulate the integral form of the constitutive principles as partial differential equations that are valid for smooth motions.

Let us begin by considering Eq. 1.21 for conservation of mass. Transforming this equation to the fixed domain of the reference configuration yields the new form

$$\frac{d}{dt}\int_{X_{\mathrm{A}}}^{X_{\mathrm{B}}}\rho(X,t)\,F(X,t)\,dX = 0 \tag{1.24}$$

for the expression of the principle. When the fields are smooth, the derivative can be calculated before integration, giving

$$\int_{X_{\mathrm{A}}}^{X_{\mathrm{B}}}\frac{\partial}{\partial t}\big[\rho(X,t)F(X,t)\big]\,dX = 0\,. \tag{1.25}$$

If we impose the requirement that the conservation principle apply to a body defined by *any* pair of boundary planes, the integrand must vanish, leaving us with the result

$$F(X,t)\frac{\partial\rho(X,t)}{\partial t}+\rho(X,t)\frac{\partial F(X,t)}{\partial t}=0\,. \tag{1.26}$$

This is the Lagrangean form of the equation representing the principle of balance of mass for smooth uniaxial deformations. Because $F = \rho_{\mathrm{R}}/\rho$ and $\partial F/\partial t = \partial\dot{x}/\partial X$, this result can be written in the more familiar and useful form

$$\frac{\partial\dot{x}}{\partial X}-\rho_{\mathrm{R}}\frac{\partial v}{\partial t}=0\,. \tag{1.27}$$

Similar manipulation of the equations for balance of momentum and energy leads to the Lagrangean form of the differential equations representing these conservation principles for smooth motions. The final result of this calculation is

$$\frac{\partial \dot{x}}{\partial X} - \rho_R \frac{\partial v}{\partial t} = 0$$

$$\frac{\partial t_{11}}{\partial X} - \rho_R \frac{\partial \dot{x}}{\partial t} = 0 \tag{1.28}$$

$$\rho_R \frac{\partial \varepsilon}{\partial t} - t_{11} \frac{\partial \dot{x}}{\partial X} = 0.$$

When these equations are transformed to Eulerian coordinates, we obtain the expressions

$$\frac{\partial \rho}{\partial t} + \frac{\partial}{\partial x}(\rho \dot{x}) = 0$$

$$\frac{\partial t_{11}}{\partial x} - \rho \frac{\partial \dot{x}}{\partial t} - \rho \dot{x} \frac{\partial \dot{x}}{\partial x} = 0 \tag{1.29}$$

$$\rho \frac{\partial \varepsilon}{\partial t} + \rho \dot{x} \frac{\partial \varepsilon}{\partial x} - t_{11} \frac{\partial \dot{x}}{\partial x} = 0.$$

In writing these equations, the change of mass density has been used to measure the deformation. This is done without loss of generality, because, in the case of uniaxial strain, any other measure of deformation can be expressed in terms of ρ and ρ_R. These equations constrain smooth plane flows, but are insufficient to determine the solution of specific problems. To complete the basis for determining specific solutions, these equations must be augmented by boundary and initial conditions and a constitutive equation (or equations) describing the response of a specific material or class of materials.

Smooth Steady Waves.   Plane acoustic and elastic waves analyzed using linear theories propagate unchanged in form, i.e., the solution depends only on the single variable

$$z = x - ct, \tag{1.30}$$

where c is a material constant called the wavespeed. It is interesting that the nonlinear equations may also have solutions of this form, but the waveforms are not arbitrary as in the case of the linear theory.

Let us consider the possibility of finding a solution of the form

$$\rho = \rho(z), \quad \dot{x} = \dot{x}(z), \quad p = p(z). \tag{1.31}$$

Substitution of these functions into Eqs. 1.29 leads to the result

$$\frac{d}{dz}\left[\rho(\dot{x}-c)\right]=0$$

$$\frac{d}{dz}\left[t_{11}-\rho\dot{x}^2+c\rho\dot{x}\right]=0 \tag{1.32}$$

$$\frac{d}{dz}\left[-\rho c(\varepsilon+\tfrac{1}{2}\dot{x}^2)+[\rho(\varepsilon+\tfrac{1}{2}\dot{x}^2)-t_{11}]\dot{x}\right]=0.$$

These equations are immediately integrable, leading to the results

$$[\![\rho]\!]c=[\![\rho\dot{x}]\!]$$

$$[\![\rho\dot{x}]\!]c=[\![\rho\dot{x}^2-t_{11}]\!] \tag{1.33}$$

$$[\![\rho(\varepsilon+\tfrac{1}{2}\dot{x}^2)]\!]c=[\![[\rho(\varepsilon+\tfrac{1}{2}\dot{x}^2)-t_{11}]\dot{x}]\!],$$

where we have adopted the notation $[\![\varphi]\!]=\varphi(z_a)-\varphi(z_b)$ for the change in any variable φ, between any two points z_a and z_b in the wave.

The Lagrangean equations, Eqs. 1.28, can be analyzed in the same way, seeking solutions that depend on the variable

$$Z=X-Ct. \tag{1.34}$$

The result is

$$\rho_R C[\![v]\!]+[\![\dot{x}]\!]=0$$

$$\rho_R C[\![\dot{x}]\!]+[\![t_{11}]\!]=0 \tag{1.35}$$

$$\rho_R C[\![\varepsilon+\tfrac{1}{2}\dot{x}^2]\!]+[\![t_{11}\dot{x}]\!]=0.$$

As with the partial differential equations, one needs a constitutive equation and boundary conditions to complete determination of the waveform satisfying the jump conditions of either Eqs. 1.33 or 1.35.

Shocks. At a shock discontinuity, the derivatives appearing in Eqs. 1.28 and 1.29 do not exist, so those equations are meaningless. Nevertheless, the integral forms of the governing equations are meaningful and the discontinuities that occur at a shock are constrained by the requirement that the integral equations be satisfied in the neighborhood of the shock.

Consider the wave shown schematically in Fig. 1.4. The planes $x_a=x(X_A,t)$ and $x_b=x(X_B,t)$ are chosen arbitrarily, but move with the material. Applying the equation for conservation of mass, Eq. 1.21, to the material between these two planes yields the result

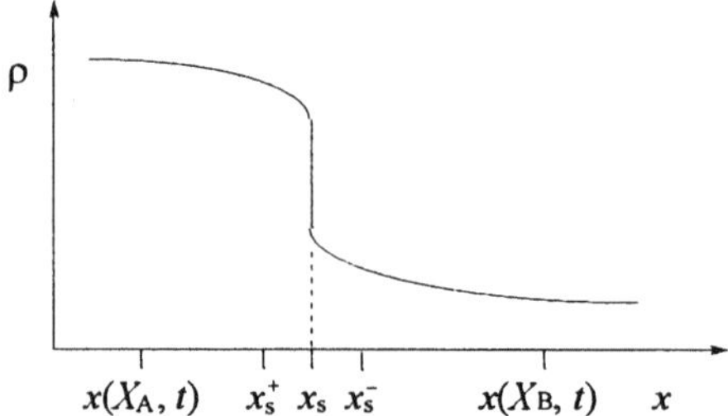

Figure 1.4. Shock embedded in a smooth flow.

$$\frac{d}{dt}\left\{ \int_{x_a(t)}^{x_s^+(t)} \rho(x,t)\,dx + \int_{x_s^+(t)}^{x_s^-(t)} \rho(x,t)\,dx + \int_{x_s^-(t)}^{x_b(t)} \rho(x,t)\,dx \right\} = 0 \qquad (1.36)$$

when the range of integration is subdivided as indicated. In the limit that $x_s^- - x_s^+ \to 0$, with the two planes remaining on their respective sides of the shock, the integral over the range (x_s^+, x_s^-) becomes arbitrarily small, vanishing in the limit. After differentiating the remaining two integrals with respect to the parameter t, we obtain

$$\int_{x_a(t)}^{x_s^+(t)} \frac{\partial \rho(x,t)}{\partial t}\,dx + \rho(x_s^+,t)\frac{dx_s^+(t)}{dt} - \rho(x_a,t)\frac{dx_a(t)}{dt}$$

$$+ \int_{x_s^-(t)}^{x_b(t)} \frac{\partial \rho(x,t)}{\partial t}\,dx + \rho(x_b,t)\frac{dx_b(t)}{dt} - \rho(x_s^-,t)\frac{dx_s^-(t)}{dt} = 0. \qquad (1.37)$$

Because the surfaces $x_a(t)$ and $x_b(t)$ move with the material, we have

$$\frac{dx_a(t)}{dt} = \dot{x}(X_A,t) \equiv \dot{x}_a \quad \text{and} \quad \frac{dx_b(t)}{dt} = \dot{x}(X_B,t) \equiv \dot{x}_b . \qquad (1.38)$$

The surfaces $x = x_s^\pm(t)$ move with the shock, so we have

$$\frac{dx_s^\pm}{dt} = u_S, \qquad (1.39)$$

where u_S is the *Eulerian shock velocity*. With this, Eq. 1.37 can be written

$$\int_{x_a(t)}^{x_s^+(t)} \frac{\partial \rho(x,t)}{\partial t}\,dx + \rho(x_s^+,t)u_S - \rho(x_a,t)\dot{x}_a$$

$$+ \int_{x_s^-(t)}^{x_b(t)} \frac{\partial \rho(x,t)}{\partial t}\,dx + \rho(x_b,t)\dot{x}_b - \rho(x_s^-,t)u_S = 0. \qquad (1.40)$$

Because the motion is smooth in each of the ranges of integration, the differential equation for conservation of mass is valid in each of these ranges, and we can make the substitution

$$\frac{\partial \rho}{\partial t} = -\frac{\partial (\rho \dot{x})}{\partial x} \tag{1.41}$$

so the integrals can be evaluated immediately. After some terms are cancelled, we are left with the equation

$$[\![\rho]\!]u_S - [\![\rho \dot{x}]\!] = 0 . \tag{1.42}$$

Similar manipulation of the integral equations for balance of momentum and of energy leads to the equations

$$[\![\rho \dot{x}]\!]u_S - [\![\rho \dot{x}^2 - t_{11})]\!] = 0 \tag{1.43}$$

and

$$[\![\rho(\varepsilon + \tfrac{1}{2}\dot{x}^2)]\!]u_S - [\![[\rho(\varepsilon + \tfrac{1}{2}\dot{x}^2) - t_{11}]\dot{x}]\!] = 0 . \tag{1.44}$$

Equations 1.42, 1.43, and 1.44 are called the *Eulerian jump conditions* for a shock. When the steady wavespeed, c, is identified with the shock velocity, u_S, we see that the jump conditions for a shock embedded in a steady flow are identical to the jump conditions of Eq. 1.33 for a steady wave. This justifies use of the term structured shock to describe a steady wave. One can also develop Lagrangean jump conditions for shocks (identical to Eqs. 1.35, with C and U_S identified), completing the parallel between steady structured waves and shocks, and providing results useful for analyzing shock propagation. Examination of the derivation given for the jump conditions shows that they are valid for unsteady shocks. When applied to a smooth wave, this wave must be steady.

Elimination of u_S and $\dot{x}$ from Eqs. 1.42 – 1.44 yields the relation

$$[\![\varepsilon]\!] = \frac{1}{2}(t_{11}^{+} + t_{11}^{-})[\![v]\!] \tag{1.45}$$

among the thermodynamic variables. This important result, called the *Rankine– Hugoniot equation*, is the vehicle for introducing the effect of a shock transition into a thermodynamic analysis. Manipulation of the jump conditions also yields the expressions

$$U_S = \left\{ \frac{1}{\rho_R} \frac{[\![p]\!]}{[\![\Delta]\!]} \right\}^{1/2} = \frac{1}{\rho_R} \frac{[\![p]\!]}{[\![\dot{x}]\!]} = \frac{[\![\dot{x}]\!]}{[\![\Delta]\!]} \tag{1.46}$$

for the *Lagrangean shock velocity*.

The Lagrangean and Eulerian shock velocities are related by the equation

$$U_S = \frac{\rho^-}{\rho_R}(u_S - \dot{x}^-),\tag{1.47}$$

and are seen to be the same when the shock is propagating into material at rest and at its reference density.

1.3. Analysis of Shocks

1.3.1. Hugoniot Curve

The three shock jump conditions apply to all materials governed by the underlying theory. Accordingly, they contain no information characterizing the differing responses of specific materials. The three jump conditions involve the variables ρ (or v), t_{11}, $\dot{x}$, ε, and the shock velocity, either U_S or u_S. If two of these variables are specified, the jump conditions suffice to determine the remaining three. Typically, one of the variables is given as a boundary condition related to the known stimulus producing the shock. If shocks of various strengths are introduced into the material and the response is recorded in terms of the resulting value of one of the other variables, one obtains points on a curve describing the locus of states achievable by shock transition from the given initial state. This locus is characteristic of the specific material studied, and is called an *Hugoniot curve* or simply an *Hugoniot*. It contains the minimum amount of information about the material that suffices to solve shock propagation problems in terms of the variables under consideration. An Hugoniot curve falls far short of a complete description of material response and additional thermodynamic information is required to determine temperature and entropy jumps at the shock. On the other hand, an Hugoniot curve can be determined from a comprehensive material model by thermodynamic and mechanical analysis.

There are 10 pairs of variables that provide equivalent representations of the Hugoniot curve. Any of these representations can be transformed to any of the other possible representations using the jump conditions. Three of these 10 Hugoniots are particularly important: the $t_{11} - v$ Hugoniot or, equivalently, the $t_{11} - \rho$ or $t_{11} - \Delta$ Hugoniot, is an important thermodynamic relation, the $t_{11} - \dot{x}$ Hugoniot is used for analysis of shock interactions, and the $U_S - \dot{x}$ Hugoniot is frequently the most direct representation of shock wave measurements.

It is important to remember that an Hugoniot curve is the locus of states achievable by shock transition from a given initial state. When the initial state is changed, the Hugoniot curve also changes. The Hugoniot is said to be centered on the initial state, and the one centered on the undeformed and unstressed state[*] is called the *principal Hugoniot curve*. Because classical mechanics is invariant to change of the spatial coordinate frame, an Hugoniot can be recentered to a different initial particle velocity by simple translation and/or reflection in any line $\dot{x} = \text{const}$. The translation is performed to center the Hugoniot on the desired value of $\dot{x}$ and the reflection is used to reverse the direction of shock propagation. When additional thermodynamic information is available, an Hugoniot can be transformed to a different initial thermodynamic state, and can be used as the basis for calculation of isotherms and isentropes. We shall address these issues in Section 1.4.

1.3.2. Shock Propagation

The basis for analysis of shock propagation and interaction is formed by the jump conditions in combination with an Hugoniot curve for the material in question. It is a trivial observation that the field equations, Eqs. 1.28 or 1.29, are satisfied by constant values of the variables. Accordingly, a global solution is obtained when two constant fields meet at a shock for which the jump conditions are satisfied. The simplest such case arises when a shock is introduced into the material by a sudden change in the force present on the boundary of a body in a uniform state. The work done by the applied force is manifest as an increase in the kinetic and internal energy of the material behind the shock.

An example of this sort occurs when a body (a target) that is undeformed, unstressed, and at rest ($\Delta^- = 0$, $p^- = 0$, and $\dot{x}^- = 0$) is impacted by another body (a projectile) moving at some known velocity, $\dot{x}_P$. If the projectile and the target are of the same material and are in the same thermodynamic state, a simple symmetry argument shows that the particle velocity imparted to the target will be $\dot{x}_P/2$. A measurement of any one of the other variables appearing in the jump conditions provides a point on the Hugoniot of the target material. A number of such measurements, made over a range of projectile velocities, defines the Hugoniot of the material over the range studied. One of the easiest quantities to measure, in addition to the projectile velocity, is the velocity of the shock produced in the target. In this case, it is the principal $U_S - \dot{x}$ Hugoniot that is determined. The result of tests of this sort conducted on many materials is that the $U_S - \dot{x}$ Hugoniot can often be represented by the line

$$U_S = C_B + S\dot{x}, \qquad (1.48)$$

where the constants C_B and S are characteristics of the material [8,9]. This Hugoniot is widely applicable in the midrange of shock strengths accessible to

---

[*] The pressures normally encountered in shock compression of solids or liquids are so great that an initial pressure near one atmosphere can be neglected. This simplification is not available for gases because they cannot exist at zero pressure.

experimental investigation and of interest in practical problems such as those of terminal ballistics and material processing.

No specific material response model is required to infer Eq. 1.48, or any other Hugoniot relation, from experimental observations. The usual experiments do not yield any information about stress components other than t_{11} that may be present in the material. Information about these quantities is inferred from postulated models of thermomechanical response.

In the range in which Eq. 1.48 is found to describe experimental observations, it is usual to use the compressible fluid approximation of solid material response. Although it seems inconsistent to discuss fluids in a chapter concerning wave propagation in solids, this is traditional, and fluid-like response of solids can be rationalized when the applied stress is sufficiently large. Let us begin by examining Eq. 1.17 or 1.19. These equations show that the normal stress that enters the governing field equations is the sum of a pressure term and a shear term. Solids respond elastically when the stress is small and, in this case, the pressure and shear are each maintained at a constant proportion of the total normal stress. Any material can support whatever pressure is imposed on it, but its ability to withstand shear is limited by the onset of inelastic flow or fracture. As the stress is increased, the contribution of pressure can increase indefinitely, but the shear contribution cannot. At some point, one may decide that the shear contribution can be neglected. When this is done, the only remaining stress is a pressure and the material is effectively a fluid.

Physically, one must think of the deformation somewhat as suggested in Fig. 1.5. Discontinuities in the displacement field arise that allow the body as a whole to deform in shear, as it must in uniaxial strain, while imposing only dilatation on any individual part of the body. The displacement discontinuities are present at very small scale and are not explicitly included in the kinematics.

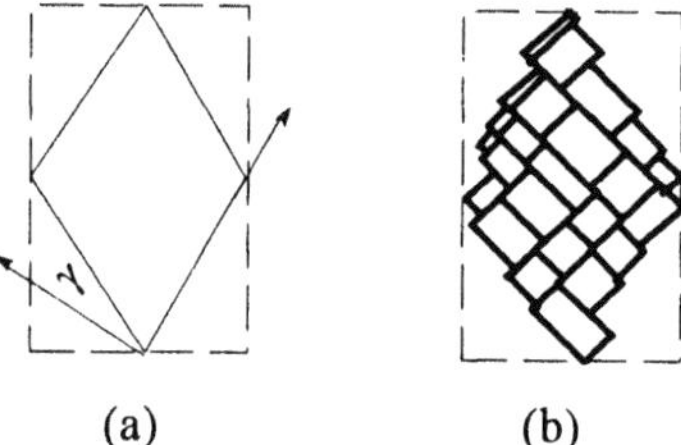

(a)   (b)

Figure 1.5. Figure 1.2 is an illustration of the shear that arises in a uniaxial deformation. The shear may be homogeneous as suggested by the configuration in part (a) of this figure (as one would expect in deformation of an elastic solid), or it may be discontinuous at the microscale, as suggested in part (b) of the figure. In the latter case, the discontinuities can arise as a result of plastic deformation or fracture. Note that the individual material blocks in the latter illustration are not sheared, so no shear stress is present.

They are recognized by imposing the requirement that pressure be the only contribution to the stress, but their occurrence means that the kinematics are valid only in some averaged sense.

In adopting the compressible fluid model for the response of solids at high stress, we assume that $t_{11} = -p$. When we transform the linear $U_S - \dot{x}$ Hugoniot of Eq. 1.48 to $p - v$ or $p - \Delta$ and $p - \dot{x}$ Hugoniots, we find that

$$p = \frac{(\rho_R C_B)^2 [\![-v]\!]}{\{1 - \rho_R S [\![-v]\!]\}^2} = \frac{\rho_R C_B^2 \Delta}{(1 - S\Delta)^2} \tag{1.49}$$

and

$$p = \rho_R (C_B + S\dot{x})\dot{x}. \tag{1.50}$$

These three representations of the Hugoniot curve are illustrated in Fig. 1.6. The secant lines shown in the figures (marked $\mathcal{R}$) connect the initial and final states associated with the shock transition, and are called *Rayleigh lines*.

Equation 1.46 shows that the shock speed is related to the slope of these Rayleigh lines, and consideration of the Rankine–Hugoniot equation, Eq. 1.45, shows that the increase in internal energy density that occurs upon passage of a shock is proportional to the area under the Rayleigh line in the $p - \Delta$ (or $p-v$) plane.

1.3.3. Shock Interactions and Hugoniot Measurement

Analysis of the propagation of shocks, of interactions between shocks, and of shocks with material boundaries, is performed using the jump conditions and the Hugoniot curves for the materials of interest. The analysis is accomplished much more easily than analysis of similar problems involving smooth waves because the jump conditions are algebraic rather partial differential equations.

The analysis of shock interactions is conducted using $p - \dot{x}$ Hugoniots. This is because the pressure and particle velocity must be continuous on the material

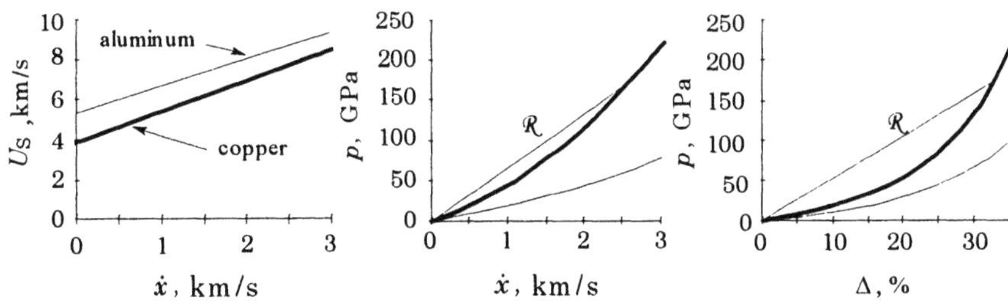

Figure 1.6. Hugoniot curves for copper and aluminum. Data are from McQueen et al. [9].

plane at which an interaction takes place. The pressure must be continuous because the surface at which the interaction occurs has no mass and the particle velocity must be continuous because the material bounded by the interaction surface neither separates nor interpenetrates.

We note that not all fields are continuous at interaction planes. In particular, the temperature is usually discontinuous in the absence of heat conduction. In this case the interaction plane persists as a *contact surface*. Shocks interact with contact surfaces in the same way as with other material interfaces.

Impact-generated shocks. Consider the impact problem illustrated in Fig. 1.7. At the moment of impact, shocks advancing into the aluminum and receding into the copper are generated. The state of the material between these two shocks lies at the intersection of the $p-\dot{x}$ Hugoniots for the two materials. The Hugoniot for the aluminum alloy is the principal Hugoniot, whereas the Hugoniot for the copper is the principal Hugoniot recentered to the initial velocity and reflected in a vertical line to represent a left-propagating shock. The event described in this example can be used to measure a point on the Hugoniot curve for the target material. Suppose that the velocity of the projectile is measured and the Hugoniot curve for the projectile material is known. Measurement of the shock velocity in the target material determines the Rayleigh line (see Eq. 1.46$_2$). The point at the intersection of this Rayleigh line and the projectile Hugoniot lies on the Hugoniot of the target material. A sequence of tests performed for a range of projectile velocities and/or projectile materials traces out the target Hugoniot.

Shock interaction at a material interface. The shock interaction problem illustrated in Fig. 1.8 represents a continuation of the impact problem discussed in the preceding paragraph. It is supposed that the shock in the aluminum alloy interacts with a copper plate that is placed to its right. The interaction at this second aluminum–copper interface produces both a transmitted and a reflected

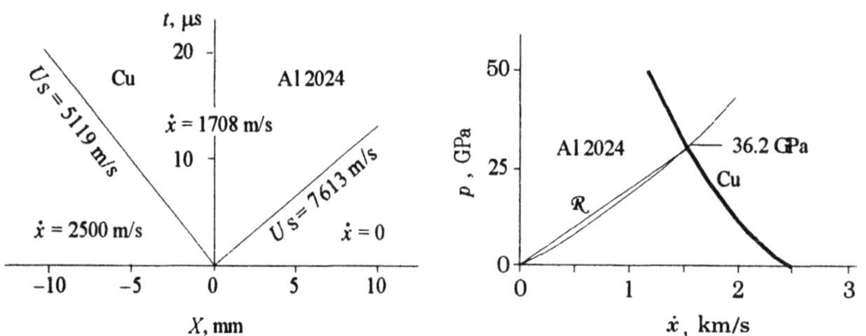

Figure 1.7. Impact of a copper plate having the velocity $\dot{x}_p = 2500$ m/s on a stationary aluminum alloy 2024 plate. (a) $X-t$ diagram and (b) $p-\dot{x}$ diagram.

shock, as shown on the X–t diagram of Fig. 1.8a. The state behind the shock propagating to the right in the second copper plate must lie on the principal Hugoniot for the copper, since it is propagating into material in its reference state. The shock reflected back into the aluminum alloy plate must lie on the Hugoniot for this material that is centered on state 1 and corresponds to a left-propagating shock. It can be shown that this Hugoniot curve is slightly less stiff than the one obtained by translating and reflecting the principal Hugoniot so that it passes through state 1. Nevertheless, the second-shock Hugoniot obtained by translation and reflection is a good approximation to the true second-shock Hugoniot and we shall use it. State 2, in the material between the reflected and transmitted shocks, lies at the intersection of this Hugoniot and the principal Hugoniot for the copper, as shown in Fig. 1.8b. It is important to note that the reflected disturbance is compressive. If the second copper plate were replaced by one of a material having an Hugoniot lying below that of the aluminum alloy, the reflected disturbance would be a decompression wave. In this case, it would be a smooth wave rather than a shock, as we shall see in Section 1.5.2.

Many other cases of shock interaction with interfaces or other shocks arise; all are analyzed in a manner similar to that illustrated in the preceding examples.

1.4. Thermodynamic Behavior of Compressible Fluids

Let us consider material response characterization beyond that provided by an Hugoniot curve, taking a model for fluids as our example. Analysis of fluid response is an exercise in thermodynamics [10]. One assumes that the material is always in equilibrium, i.e., the pressure changes instantaneously in response to a change in deformation. Nonequilibrium effects such as those of viscosity are not included in this thermodynamic analysis.

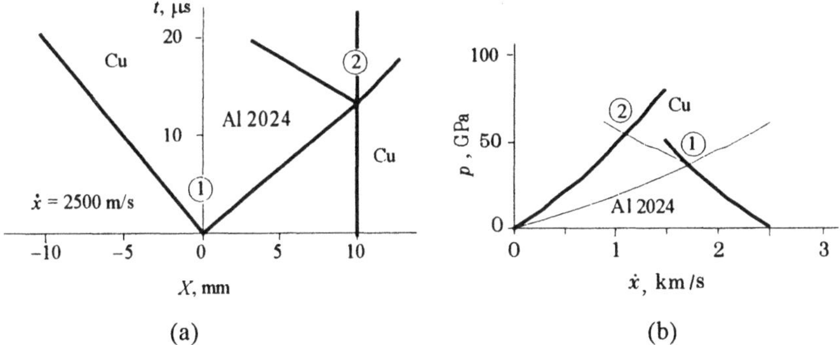

Figure 1.8. Interaction of a shock with a material interface. In state 1, produced by the initial impact, the particle velocity is 1708 m/s and the pressure is 36.2 GPa. In state 2, produced by the interaction at the second interface, the particle velocity is 1095 m/s and the pressure is 54.5 GPa.

The thermodynamic response of a fluid is captured by an *equation of state*. Many equivalent formulations are possible, but we shall base our discussion on the function

$$\varepsilon = \varepsilon(\Delta, \eta),\qquad(1.51)$$

where ε is the *internal energy density* (internal energy per unit mass), Δ is the *Lagrangean compression,* and η is the *entropy density.* Imposition of the requirement that the entropy production inequality be satisfied for all thermodynamic processes yields the (thermodynamic) pressure and temperature relations

$$p(\Delta, \eta) = \rho_R \frac{\partial \varepsilon(\Delta, \eta)}{\partial \Delta} \quad \text{and} \quad \theta(\Delta, \eta) = \frac{\partial \varepsilon(\Delta, \eta)}{\partial \eta},\qquad(1.52)$$

respectively. The thermodynamic pressure defined by Eq. 1.52_1 is not the same as the mechanical pressure defined by Eq. 1.16 in cases where part of the stress arises from viscous or other effects not captured by equilibrium thermodynamic analysis.

The differential expression,

$$d\varepsilon = \frac{1}{\rho_R} p\, d\Delta + \theta\, d\eta,\qquad(1.53)$$

a form of the *First Law of Thermodynamics*, is obtained from Eqs. 1.51 and 1.52.

By manipulation of Eqs. 1.51 and 1.52 one can, in principle, express the several thermodynamic quantities in terms of other variables. In particular, one can express the pressure in terms of compression and internal energy density. The derivative of this function appears in the equation

$$\gamma = \frac{1 - \Delta}{\rho_R} \frac{\partial p(\Delta, \varepsilon)}{\partial \varepsilon},\qquad(1.54)$$

defining the important thermodynamic coefficient called *Grüneisen's parameter,* γ. Lattice dynamical models of materials indicate that γ is a function of compression (or specific volume) alone, a result that accounts for much of its usefulness.

Mie–Grüneisen equation of state. Integration of Eq. 1.54, with $\gamma = \gamma(\Delta)$, leads to the result

$$p = p^{(H)}(\Delta) + \frac{\rho_R\, \gamma(\Delta)}{1 - \Delta} \left[\varepsilon - \varepsilon^{(H)}(\Delta) \right].\qquad(1.55)$$

In this equation, the pressure and internal energy density on the Hugoniot curve, $p^{(H)}(\Delta)$ and $\varepsilon^{(H)}(\Delta)$, appear because this curve was used to evaluate the undetermined function of Δ arising from the integration. The $p-\Delta$ Hugoniot curve can be measured, and the internal energy density on this curve can then be calculated using Eq. 1.45, the Rankine–Hugoniot equation. Equation 1.55 is called a *Mie–Grüneisen equation of state*, and plays an important role in the analysis of shock compression of solids.

Hugoniot obtained from an isotherm. An Hugoniot curve is easily calculated from a known isotherm if Grüneisen's parameter and the specific heat at constant volume for material in the reference state are known for the material of interest.

Substituting the known isotherm, $p = p^{(\theta)}(\Delta)$, for the temperature $\theta = \theta_R$ and passing through the point $p = p_R$, $\rho = \rho_R$, $\varepsilon = \varepsilon_R$ into the Mie–Grüneisen equation allows us to write

$$p^{(H)}(\Delta) = p^{(\theta)}(\Delta) + \frac{\rho_R\,\gamma(\Delta)}{1-\Delta}\left[\varepsilon^{(H)}(\Delta) - \varepsilon^{(\theta)}(\Delta)\right]. \qquad (1.56)$$

The internal energy density on the Hugoniot centered at the same reference state is given by the Rankine–Hugoniot equation, Eq. 1.45,

$$\varepsilon^{(H)}(\Delta) = \varepsilon_R + \frac{1}{2\rho_R}\left[p^{(H)}(\Delta) + p_R\right]\Delta. \qquad (1.57)$$

Evaluation of the thermodynamic derivative

$$\frac{\partial\varepsilon(\Delta,\theta)}{\partial\Delta} = \frac{1}{\rho_R}p - \frac{\gamma(\Delta)}{1-\Delta}C^{(v)}\theta \qquad (1.58)$$

on the $\theta = \theta_R$ isotherm, followed by integration with respect to Δ, gives

$$\varepsilon^{(\theta)}(\Delta) = \varepsilon_R + \frac{1}{\rho_R}\int_0^\Delta p^{(\theta)}(\Delta')\,d\Delta' - C^{(v)}\theta_R\int_0^\Delta \frac{\gamma(\Delta')}{1-\Delta'}\,d\Delta', \qquad (1.59)$$

where $C^{(v)}$ is the specific heat at constant volume (a function of temperature alone) evaluated for the temperature θ_R. When Eqs. 1.57 and 1.59 are substituted into Eq. 1.56 we obtain the equation

$$\left[1 - \frac{\gamma(\Delta)\Delta}{2(1-\Delta)}\right]p^{(H)}(\Delta) = p^{(\theta)}(\Delta)$$

$$+ \frac{\gamma(\Delta)}{(1-\Delta)}\left[\frac{1}{2}p_R\Delta - \int_0^\Delta p^{(\theta)}(\Delta')\,d\Delta' + \rho_R C^{(v)}\theta_R\int_0^\Delta \frac{\gamma(\Delta')}{1-\Delta'}\,d\Delta'\right] \qquad (1.60)$$

22   Lee Davison

for the Hugoniot. Note that calculation of an Hugoniot curve from an isotherm requires only knowledge of Grüneisen's parameter and the reference-state value of the specific heat at constant volume.

Temperature and entropy at states on an Hugoniot. Equations 1.45 and 1.53 can be combined to yield the differential equation

$$\frac{d\eta^{(\mathrm{H})}(\Delta)}{d\Delta} = \frac{\Delta - \Delta^-}{2\rho_R\,\theta^{(\mathrm{H})}(\Delta)}\left[\frac{dp^{(\mathrm{H})}(\Delta)}{d\Delta} - \frac{p^{(\mathrm{H})}(\Delta) - p^-}{\Delta - \Delta^-}\right] \tag{1.61}$$

for the entropy density at points on the Hugoniot curve $p = p^{(\mathrm{H})}(\Delta)$ centered on the state $p = p^-$, $\Delta = \Delta^-$.

Before Eq. 1.61 can be solved, it is necessary to determine the temperature $\theta = \theta^{(\mathrm{H})}(\Delta)$. Evaluation of the equation of state $\theta = \theta(\Delta, \eta)$ on the Hugoniot curve, differentiation with respect to Δ, and substitution from Eq. 1.61 gives

$$\frac{d\theta^{(\mathrm{H})}(\Delta)}{d\Delta} + \frac{\gamma(\Delta)}{1-\Delta}\theta^{(\mathrm{H})}(\Delta) = \frac{\Delta - \Delta^-}{2\rho_R C^{(v)}}\left[\frac{dp^{(\mathrm{H})}(\Delta)}{d\Delta} - \frac{p^{(\mathrm{H})}(\Delta) - p^-}{\Delta - \Delta^-}\right], \tag{1.62}$$

where we have identified the thermodynamic derivatives $\partial\theta(\Delta, \eta)/\partial\eta = C^{(v)}/\theta$ and $\partial\theta(\Delta, \eta)/\partial\Delta = -\gamma\theta/(1-\Delta)$. Equation 1.62 is a first-order ordinary differential equation. In the case that $C^{(v)}$ is constant, the equation is linear and can be reduced to quadrature:

$$\theta^{(\mathrm{H})}(\Delta) = \exp\left[-\int_{\Delta^-}^{\Delta}\varphi(\Delta')d\Delta'\right]\left\{\theta^{(\mathrm{H})}(\Delta^-) + \int_{\Delta^-}^{\Delta}\left[\psi(\Delta')\exp\left(\int_{\Delta^-}^{\Delta'}\varphi(\Delta'')d\Delta''\right)\right]d\Delta'\right\}.$$

$$\tag{1.63}$$

In this equation

$$\varphi(\Delta) = \exp\left[\frac{\gamma(\Delta)}{1-\Delta}\right] \quad \text{and} \quad \psi(\Delta) = \frac{\Delta - \Delta^-}{2\rho_R C^{(v)}}\left[\frac{dp^{(\mathrm{H})}(\Delta)}{d\Delta} - \frac{p^{(\mathrm{H})}(\Delta) - p^-}{\Delta - \Delta^-}\right].$$

When $C^{(v)}$ depends on temperature, according to the Debye theory, for example, Eq. 1.62 is no longer linear, but can still be integrated numerically.

Isentrope determined from an Hugoniot curve. One can determine the isentrope through a point characterized by the pressure p^+ and the compression Δ^+ on the Hugoniot curve centered on the state (p^-, Δ^-) by combining the Mie–Grüneisen and Rankine–Hugoniot equations. The isentrope sought is designated $p = p^{(\eta)}(\Delta)$. Integration of Eq. 1.52$_1$ shows that the internal energy density on the isentrope is given by the equation

$$\varepsilon^{(\eta)}(\Delta) = \varepsilon^{(H)}(\Delta^+) + \frac{1}{\rho_R} \int_{\Delta^+}^{\Delta} p^{(\eta)}(\Delta') d\Delta' . \qquad (1.64)$$

Evaluation of Eq. 1.55 on the isentrope, and use of Eq. 1.45, gives

$$\left[1 - \frac{\gamma(\Delta)(\Delta - \Delta^-)}{2(1 - \Delta)}\right] p^{(H)}(\Delta)$$

$$= p^{(\eta)} - \frac{\gamma(\Delta)}{1 - \Delta}\left[\int_{\Delta^+}^{\Delta} p^{(\eta)}(\Delta') d\Delta' + \frac{1}{2}\left[p^+(\Delta^+ - \Delta^-) + p^-(\Delta^+ - \Delta)\right]\right]. \qquad (1.65)$$

When $p^{(H)}(\Delta)$ and $\gamma(\Delta)$ are known, this equation can be solved for the $p-\Delta$ isentrope [usually it is easiest to differentiate to obtain an ordinary differential equation for $p^{(\eta)}(\Delta)$]. Note that the isentrope is obtained from the Hugoniot using only the additional information embodied in an expression for Grüneisen's parameter. It is often assumed that $\gamma(v)/v = \gamma_R/v_R$, where γ_R and v_R are Grüneisen's parameter and the specific volume for the material in a reference state. When this expression for $\gamma(v)$ is adopted, differentiation of Eq. 1.65 yields the result

$$\frac{dp^{(\eta)}(\Delta)}{d\Delta} - \gamma_R p^{(\eta)}(\Delta) = \left[1 - \frac{\gamma_R}{2}(\Delta - \Delta^-)\right]\frac{dp^{(H)}(\Delta)}{d\Delta} - \frac{\gamma_R}{2}[p^{(H)}(\Delta) + p^-], \qquad (1.66)$$

a linear equation that is easily reduced to quadrature as was done for the temperature calculation. When the principal Hugoniot curve takes the often-used form of Eq. 1.49 (for which $p^- = 0$ and , $\Delta^- = 0$), the isentrope through the point (p^+, Δ^+) on the Hugoniot curve is given by

$$p^{(\eta)}(\Delta)$$

$$= \rho_R C_B^2 \exp[\gamma_R(\Delta - \Delta^+)]\left\{\frac{p^+}{\rho_R C_B^2} + \int_{\Delta^+}^{\Delta}\left[\frac{1 + (S - \gamma_R)\Delta'}{(1 - S\Delta')^3}\exp[-\gamma_R(\Delta' - \Delta^+)]\right] d\Delta'\right\},$$

$$(1.67)$$

an equation easily integrated numerically.

Analysis of Eq. 1.65 shows that the branch of this isentrope for states of compression exceeding Δ^+ lies below the Hugoniot and the branch for states of compression less that Δ^+ lies above the Hugoniot for normal materials.

Analyses similar to those presented in this and the preceding example permit calculation of isotherms from Hugoniots, transformation of a given Hugoniot to one centered on a different state, and other thermodynamic curves.

Shock stability. In the discussions of Section 1.3, it was assumed that stable shocks actually exist. This is not true for all possible solutions of the jump conditions. Although this issue might usefully have been addressed earlier, it was necessary to defer the discussion until the isentrope developed in this section became available.

One answer to the question of existence is obtained by studying the jump in entropy that occurs as a shock passes a material point. Useful insight is also gained from comparison of the magnitude of the shock velocity with that of the sound velocity in the material adjacent to the shock. The entropy jump and the wavespeed criteria are consistent in the condition that they impose for existence (or stability) of shocks.

Entropy jump. The entropy production inequality holds that a solution of the jump equations that results in a decrease in entropy is inadmissible. Examination of Eq. 1.61 (a sketch is helpful) shows that the entropy density for states on the Hugoniot increases with increasing compression when the Hugoniot is concave upward and decreases with increasing compression when it is concave downward. Because a process that results in a decrease in entropy is inadmissible, we conclude that compression shocks cannot exist when the Hugoniot is concave downward. The same analysis of decompression shocks shows that they cannot exist for materials having an Hugoniot that is concave upward. As we shall see, the cases in which shocks cannot exist are exactly those that admit smooth waves as solutions.

Comparison of shock speed and soundspeeds. The Lagrangean isentropic soundspeed is given by the equation

$$C^2(\Delta) = \frac{1}{\rho_R} \frac{dp^{(\eta)}(\Delta)}{d\Delta} . \tag{1.68}$$

The derivative in this equation can be related to Hugoniot properties using Eq. 1.65. Analysis shows that the soundspeed in the material ahead of the shock is less than the shock velocity, and that in the material behind the shock exceeds the shock velocity. The situation is depicted in the X–t plot of Fig. 1.9. We see that the shock will overtake any wavelet propagating in the material ahead of the shock, so the shock cannot spread in the forward direction. The situation behind the shock is more complicated. A wavelet propagating toward the shock will catch up to it, but wavelets can also propagate backward from the shock. It is these wavelets that communicate the boundary condition to the shock. If the shock is stronger than is consistent with the boundary condition, a decompression wavelet can propagate backward from the shock, weakening it until its strength becomes consistent with the boundary condition. If the shock is weaker than is consistent with the boundary condition, compression wavelets will propagate forward, catch up to the shock and strengthen it. In either case, the shock will evolve to the strength that is consistent with the boundary condition.

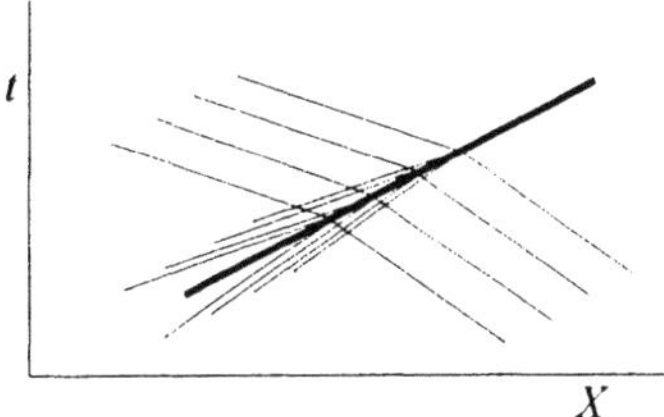

Figure 1.9. Schematic $X\!-\!t$ diagram showing a segment of a shock trajectory (heavy line) and line segments having slopes corresponding the wavespeeds in the material ahead of the shock and behind it. Wavelets moving along the trajectories shown (in the direction of increasing time) are overtaken by the shock if they are ahead of the shock. If they are behind the shock, they can either overtake the shock or be overtaken by it. The flow behind the shock is controlled by the boundary condition.

1.5. Analysis of Smooth Waves

Let us now turn to smooth solutions of the field equations. In particular, we consider Eqs. 1.29, which describe the fields in Eulerian coordinates. In the absence of dissipation, smooth flows are isentropic, so the pressure is related to the density by a constitutive equation of the form

$$p = p^{(\eta)}(\rho) . \tag{1.69}$$

When this equation is combined with Eqs. $1.29_{1,2}$, the latter become

$$\frac{\partial \rho}{\partial t} + \dot{x}\frac{\partial \rho}{\partial x} + \rho\frac{\partial \dot{x}}{\partial x} = 0$$

$$c^2 \frac{\partial \rho}{\partial x} + \rho\frac{\partial \dot{x}}{\partial t} + \rho\dot{x}\frac{\partial \dot{x}}{\partial x} = 0, \tag{1.70}$$

where we have written

$$c^2(\rho) = \frac{dp^{(\eta)}(\rho)}{d\rho} . \tag{1.71}$$

For thermodynamically stable materials $dp^{(\eta)}/d\rho$ is positive, so $c(\rho)$ is a real number (taken to be positive) that we shall be able to identify as the *Eulerian isentropic soundspeed*. The third of the field equations, Eq. 1.29_3, is automatically satisfied by the solution of the first two equations and, thus, is redundant.

When the first of Eqs. 1.70 is multiplied by $c = c(\rho)$ and successively added to, and subtracted from, the second of these equations, we obtain the equivalent set of equations

$$c\left[\frac{\partial\rho}{\partial t}+(c+\dot{x})\frac{\partial\rho}{\partial x}\right]+\rho\left[\frac{\partial\dot{x}}{\partial t}+(c+\dot{x})\frac{\partial\dot{x}}{\partial x}\right]=0$$

$$c\left[\frac{\partial\rho}{\partial t}+(-c+\dot{x})\frac{\partial\rho}{\partial x}\right]-\rho\left[\frac{\partial\dot{x}}{\partial t}+(-c+\dot{x})\frac{\partial\dot{x}}{\partial x}\right]=0.$$

(1.72)

To proceed with the solution, it is useful to introduce *characteristic coordinates*, x^* and t^*, in the $x-t$ plane. These coordinates are chosen so that

$$\frac{dx}{dt}=c(\rho)+\dot{x}\quad\text{on}\quad t^*=\text{const.}$$

$$\frac{dx}{dt}=-c(\rho)+\dot{x}\quad\text{on}\quad x^*=\text{const.}$$

(1.73)

We can write these equations in the equivalent form

$$\frac{\partial x}{\partial x^*}-(c+\dot{x})\frac{\partial t}{\partial x^*}=0$$

$$\frac{\partial x}{\partial t^*}+(c-\dot{x})\frac{\partial t}{\partial t^*}=0.$$

(1.74)

When Eqs. 1.72 are transformed to the x^*-t^* plane, they become

$$\frac{\partial}{\partial x^*}\left(\dot{x}+\int_{\rho_R}^{\rho}\frac{c(\rho')}{\rho'}d\rho'\right)=0$$

$$\frac{\partial}{\partial t^*}\left(\dot{x}-\int_{\rho_R}^{\rho}\frac{c(\rho')}{\rho'}d\rho'\right)=0,$$

(1.75)

and can be integrated immediately to give

$$\dot{x}+\int_{\rho_R}^{\rho}\frac{c(\rho')}{\rho'}d\rho'=\Phi^+(t^*)$$

$$\dot{x}-\int_{\rho_R}^{\rho}\frac{c(\rho')}{\rho'}d\rho'=\Phi^-(x^*),$$

(1.76)

where the $\Phi^\pm$ are undetermined functions. The left members of these equations, which are constant on lines of constant t^* and x^*, respectively, and are called *Riemann invariants*. If the value of the Riemann invariant is known at any point on one of its associated characteristic curves, it is known at all points on this curve. When both Riemann invariants have been determined, the particle velocity at a point x^*, t^* is given by the equation

$$\dot{x} = \frac{1}{2}\left[\Phi^-(x^*) + \Phi^+(t^*)\right], \tag{1.77}$$

and $\rho(x^*, t^*)$ is obtained by solving the equation

$$\int_{\rho_R}^{\rho} \frac{c(\rho')}{\rho'}\, d\rho' = \frac{1}{2}\left[\Phi^+(t^*) - \Phi^-(x^*)\right]. \tag{1.78}$$

1.5.1. Simple Compression Wave

Waves propagating into material in a uniform state form a class called *simple waves*. This name is justified by the fact that the fields in the wave depend on only a single coordinate. Simple waves arise, for example, when a slab in a uniform state is subjected to a continuously increasing load and when a steady shock is reflected from an unrestrained boundary. In some cases of particular interest, simple waves evolve into shocks.

Consider the problem illustrated in the x–t diagram of Fig. 1.10a. Let the material be at rest and unstressed and occupy the halfspace $x \geq 0$ at $t = 0$, and let the velocity history

$$\dot{x}(t) = \begin{cases} \dot{x}_b(t) = 0, & t \leq 0 \\[2mm] \dot{x}_b(t) > 0, & 0 < t \leq t^+ \\[2mm] \dot{x}_b(t) = \dot{x}_b(t^+), & t \geq t^+ \end{cases} \tag{1.79}$$

be imposed on the boundary. We see that the position of the boundary is given by

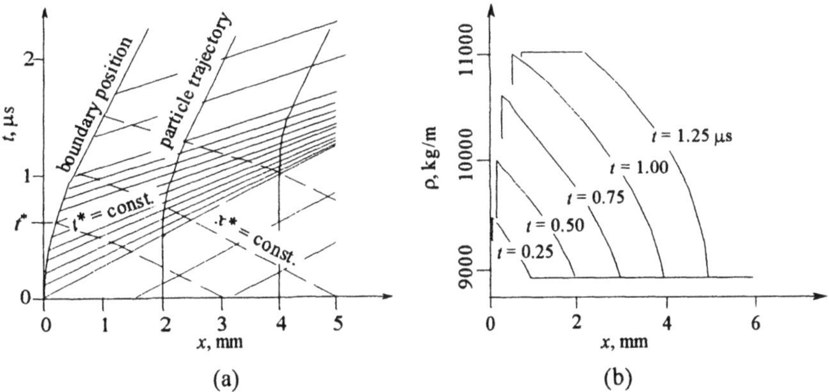

Figure 1.10. (a) The x–t diagram for a wave introduced into copper in a uniform state by subjecting the boundary to uniform acceleration for 1 µs and maintaining its velocity at the value achieved (1000 m/s). (b) Density waveforms in copper that are associated with the x–t diagram.

28   Lee Davison

$$x_b(t) = \begin{cases} 0, & t \le 0 \\ \displaystyle\int_0^t \dot{x}_b(t')dt', & 0 < t \le t^+ \\ \displaystyle\int_0^{t^+} \dot{x}_b(t')dt' + \dot{x}_b(t^+)(t - t^+), & t \ge t^+ . \end{cases} \tag{1.80}$$

The leading characteristic (one of the $t^* = \text{const.}$ family) is the solution of Eq. 1.73_1, which, in view of the initial conditions $\rho = \rho_R$ and $\dot{x} = 0$, is

$$x = c(\rho_R)t . \tag{1.81}$$

The $x^* = \text{const.}$ characteristics are solutions of Eq. 1.73_2. In the region ahead of the wave, these characteristics are the straight lines

$$x = x^* - c(\rho_R)t , \tag{1.82}$$

where we have identified each x^* characteristic with its point of intersection with the line $t = 0$. Once they cross the $t^* = 0$ characteristic and enter the wave, the density and particle velocity begin to vary and these characteristics curve in some, as yet undetermined, way.

The Riemann invariant on the x^* characteristic is given by Eq. 1.79_2:

$$\Phi^-(x^*) = \dot{x} - \int_{\rho_R}^{\rho} \frac{c(\rho')}{\rho'} d\rho' . \tag{1.83}$$

Evaluating this invariant at a point on the characteristic that lies in the region ahead of the wave gives

$$\Phi^-(x^*) = 0 = \dot{x} - \int_{\rho_R}^{\rho} \frac{c(\rho')}{\rho'} d\rho' . \tag{1.84}$$

This invariant does not actually depend on x^* because of the uniformity of the initial field values. At the point where this characteristic intersects the boundary, we have

$$\dot{x}_b(t^*) - \int_{\rho_R}^{\rho_b(t^*)} \frac{c(\rho')}{\rho'} d\rho' = 0 . \tag{1.85}$$

Because $\dot{x}_b(t^*)$ is known, this equation can be solved for $\rho_b(t^*)$, the mass density of the material on the boundary at $t = t^*$.

Let us identify the $t^* = \text{const.}$ characteristics with the time at which they intersect the boundary. The Riemann invariant on these t^* characteristics, given by Eq. 1.79$_1$, is

$$\Phi^+(t^*) = \dot{x} + \int_{\rho_R}^{\rho} \frac{c(\rho')}{\rho'} \, d\rho' . \tag{1.86}$$

At the time this characteristic intersects the boundary, we know that $\dot{x} = \dot{x}_b(t^*)$ and $\rho = \rho_b(t^*)$ so

$$\Phi^+(t^*) = \dot{x}_b(t^*) + \int_{\rho_R}^{\rho_b(t^*)} \frac{c(\rho')}{\rho'} \, d\rho' , \tag{1.87}$$

and substitution into Eq. 1.86 gives

$$\dot{x}_b(t^*) + \int_{\rho_R}^{\rho_b(t^*)} \frac{c(\rho')}{\rho'} \, d\rho' = \dot{x} + \int_{\rho_R}^{\rho} \frac{c(\rho')}{\rho'} \, d\rho' . \tag{1.88}$$

Using Eq. 1.84, which holds throughout the wave, we obtain

$$2\dot{x} = \dot{x}_b(t^*) + \int_{\rho_R}^{\rho_b(t^*)} \frac{c(\rho')}{\rho'} \, d\rho' ,$$

which shows that $\dot{x}$ is a function of t^* alone, i.e., it has the constant value

$$\dot{x}(t^*) = \dot{x}_b(t^*) \tag{1.89}$$

along the entire t^* characteristic. That being the case, Eq. 1.88 becomes

$$\int_{\rho_R}^{\rho_b(t^*)} \frac{c(\rho')}{\rho'} \, d\rho' = \int_{\rho_R}^{\rho} \frac{c(\rho')}{\rho'} \, d\rho' , \tag{1.90}$$

and we see that ρ is also has a constant value,

$$\rho(t^*) = \rho_b(t^*) , \tag{1.91}$$

on the t^* characteristic.

Substitution of these results into Eq. 1.73$_1$ shows that the t^* characteristics are the straight lines

$$x = \left[c(\rho(t^*)) + \dot{x}(t^*) \right] (t - t^*) + x_b(t^*) , \tag{1.92}$$

as plotted in the x–t diagram.

The foregoing results are sufficient to plot temporal or spatial waveforms produced by application of the boundary condition of Eq. 1.79. The first step is to solve Eq. 1.85 for $\rho_b(t^*)$ by numerical means. With this and $\dot{x}(t^*)$, as given by Eq. 1.89, we can plot several t^* characteristics. Suppose, for example, that we want to plot the waveform $\rho(x_a, t)$. We can determine the value of t at which each of the t^* characteristics intersects the line $x = x_a$ by solving Eq. 1.92. Then, the density at this point is given by Eq. 1.91. Repetition of this process for several values of t^* yields enough density values to plot the waveform, as has been done in Fig. 1.10b.

The solution of Eq. 1.85 depends on the function $c(\rho)$. To calculate an example, let us use the isentrope of Eq. 1.67. The squared Eulerian soundspeed at points on this isentrope is given by

$$c^2(\rho) = \frac{\rho_R\,\gamma_R}{\rho^2}\,p^{(\eta)}(\rho) + \rho_R^2\,C_B^2\left\{\frac{(1+S-\gamma_R)\rho - (S-\gamma_R)\rho_R}{[(1-S)\rho + S\rho_R]^3}\right\}. \tag{1.93}$$

With this result, we can evaluate the integral in Eq. 1.85 and determine $\rho_b(t^*)$ once the boundary condition $x_b(t^*)$ is given.

As a specific example, the response to the input velocity ramp

$$\dot{x}_b(t) = \begin{cases} 0, & t \le 0 \\[2mm] \dot{x}^+\dfrac{t}{t^+}, & 0 \le t \le t^+ \\[2mm] \dot{x}^+, & t \ge t^+ \end{cases} \tag{1.94}$$

is illustrated in Fig. 1.10b. An important thing to note from this figure is that the waveform becomes steeper as the wave propagates into the material. This is characteristic of compression waves propagating in normal materials, and leads eventually to formation of a shock, as will be discussed in Section 1.5.3. As a preliminary to this, we calculate the time and place at which the waveform develops a vertical tangent.

The particle-velocity waveform at a given time, t, (regarded as a fixed parameter) can be represented by an equation of the form

$$\dot{x} = \dot{x}(t^*(x; t)),$$

so its slope can be written

$$\frac{\partial \dot{x}(x, t)}{\partial x} = \frac{\dfrac{d\dot{x}(t^*)}{dt^*}}{\dfrac{dx(t, t^*)}{dt^*}}. \tag{1.95}$$

The denominator in this expression is obtained by differentiating Eq. 1.92, and is

$$\frac{\partial x(t, t^*)}{\partial t^*} = \left[\frac{dc(t^*)}{dt^*} + \frac{d\dot{x}(t^*)}{dt^*}\right](t - t^*) - c(t^*) - \dot{x}(t^*) + \frac{dx_b(t^*)}{dt^*} \,. \qquad (1.96)$$

The last two terms cancel, in accord with the solution in the simple wave region, and substituting this result into Eq. 1.95 gives the equation

$$\frac{\partial \dot{x}(x, t)}{\partial x} = \frac{\dfrac{d\dot{x}(t^*)}{dt^*}}{\left[\dfrac{dc(t^*)}{dt^*} + \dfrac{d\dot{x}(t^*)}{dt^*}\right](t - t^*) - c(t^*)} \qquad (1.97)$$

for the slope of the waveform. The vertical tangent occurs at the time when the denominator vanishes,

$$t = t^* + \frac{c(t^*)}{\dfrac{dc(t^*)}{dt^*} + \dfrac{d\dot{x}(t^*)}{dt^*}} \,. \qquad (1.98)$$

The corresponding value of x, as obtained from Eq. 1.92, is

$$x = x_b(t^*) + \frac{c(t^*)[c(t^*) + \dot{x}(t^*)]}{\dfrac{dc(t^*)}{dt^*} + \dfrac{d\dot{x}(t^*)}{dt^*}} \,. \qquad (1.99)$$

For the example problem that we have been discussing, this point is on the characteristic $t^* = 0$, and falls at $t = 1.315\,\mu s$ and $x = 5.181\,mm$.

Although calculation of the trajectories of the x^* characteristics is not required to obtain the simple wave solution, they will prove useful in discussing shock formation. The x^* characteristics, called cross characteristics because their trajectory crosses the simple wave, are determined by solving Eq. 1.73_2. Because x^* and c are most naturally expressed as functions of t^*, it is convenient to adopt the representation

$$x = x(t^*) \qquad (1.100)$$

for these characteristics. The derivative dx/dt can be written

$$\frac{dx}{dt} = \frac{\dfrac{dx}{dt^*}}{\dfrac{dt^*}{dt}} \,, \qquad (1.101)$$

so we have

$$\frac{dx}{dt^*} = \frac{dx}{dt}\frac{dt}{dt^*} \,, \qquad (1.102)$$

and substitution of this result into Eq. 1.73_2 gives

$$\frac{dx}{dt^*} = [\,\dot{x}(t^*) - c(t^*)\,]\frac{dt}{dt^*}.$$

(1.103)

Substituting the derivative of Eq. 1.92 with respect to t^* into Eq. 1.103 gives

$$\frac{dx}{dt^*} + \Phi(t^*)t = \Psi(t^*)$$

(1.104)

with

$$\Phi(t^*) = \frac{1}{2c(t^*)}\left[\frac{d\dot{x}(t^*)}{dt^*} + \frac{dc(t^*)}{dt^*}\right]$$

$$\Psi(t^*) = \frac{1}{2} + \Phi(t^*)t^*.$$

(1.105)

Equation 1.104 is a linear, first-order, ordinary differential equation that has the solution

$$t(t^*) = \exp\left[-\int_0^{t^*}\Phi(\xi)d\xi\right]\left\{\frac{x^*}{2C_B} + \int_0^{t^*}\Psi(\vartheta)\exp\left[\int_0^{\vartheta}\Phi(\xi)d\xi\right]d\vartheta\right\},$$

(1.106)

where we have imposed the initial condition (see Fig. 1.10)

$$t(0) = \frac{x^*}{2C_B},$$

(1.107)

with C_B being the soundspeed of the material ahead of the wave. All of the functions required to evaluate $\Phi(t^*)$ and $\Psi(t^*)$ can be obtained from the isentrope of Eq. 1.67, the associated isentropic soundspeed of Eq. 1.93, and the simple wave solution. Numerical evaluation of Eq. 1.106 then leads to the cross characteristics plotted in Fig. 1.10a.

A similar analysis leads to trajectories of the material particle positions. In this case, the differential equation defining the trajectory is

$$\frac{dx}{dt} = \dot{x},$$

(1.108)

and the analysis leads to the trajectory

$$t(t^*) = \exp\left[-2\int_0^{t^*}\Phi(\xi)d\xi\right]\left\{\frac{x(0)}{C_B} + \int_0^{t^*}[2\vartheta\,\Phi(\vartheta)+1]\exp\left[2\int_0^{\vartheta}\Phi(\xi)d\xi\right]d\vartheta\right\},$$

(1.109)

through the simple wave, where $x(0)$ is the particle position at $t = 0$.

1.5.2. Shock Reflection

Let us consider the reflection produced when a steady shock encounters an unrestrained boundary. We know that a decompression wave propagating into a normal material cannot be a stable shock, so we begin with the premise that it is a smooth wave. The material behind the incident shock is in a uniform state, so the reflected wave propagating into this material will be a simple wave. In the previous example, the simple wave was produced gradually by the ramp loading imposed. The present example differs in that the decompression wave is produced instantaneously when the shock encounters the boundary. This means that the characteristics defining the wave all emerge from the same point on the x–t diagram, as illustrated in Fig. 1.11a. This special form of a simple wave is called a *centered simple wave*. The fields in the wave satisfy the equations of balance of mass and momentum, which we shall write in the Lagrangean frame. Using Δ in place of ρ as the measure of deformation, we have,

$$\frac{\partial \dot{x}}{\partial X} + \frac{\partial \Delta}{\partial t} = 0$$

$$\frac{\partial p}{\partial X} + \rho_R \frac{\partial \dot{x}}{\partial t} = 0.$$

(1.110)

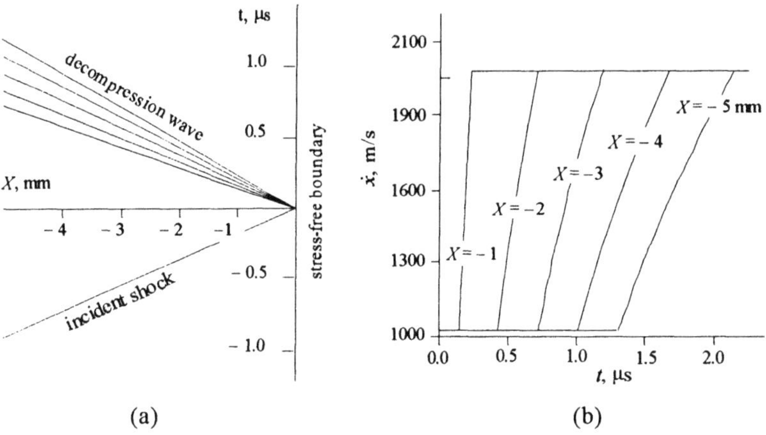

Figure 1.11. (a) The X–t diagram for reflection of a shock from an unrestrained boundary. The example is drawn for a 50 GPa shock in copper. The state ahead of the shock is defined by $p = 0$, $\Delta = 0$, and $\dot{x} = 0$, and the state behind the shock is $p^+ = 50\,\text{GPa}$, $\Delta^+ = 0.187$, and $\dot{x}^+ = 1024\,\text{m/s}$. (b) Decompression waveforms. The state behind the decompression wave is defined by $p = 0$, $\Delta = -0.00922$, and $\dot{x} = 2084\,\text{m/s}$. Note that the velocity of the material is slightly (1.8%) in excess of twice that of the velocity of the material behind the incident shock. This observation justifies the experimental practice of inferring that the velocity of the material behind the shock is one-half of the measured velocity of the unrestrained surface.

Because the deformation is smooth, p is related to Δ by the decompression isentrope through the state behind the shock. In this case, we have

$$\frac{\partial p}{\partial X} = \frac{\partial p^{(\eta)}(\Delta)}{\partial X} = \frac{dp^{(\eta)}(\Delta)}{d\Delta}\frac{\partial \Delta}{\partial X} = \rho_R C^2(\Delta)\frac{\partial \Delta}{\partial X}, \qquad (1.111)$$

where $C^2(\Delta) = v_R\, dp^{(\eta)}(\Delta)/d\Delta$ is the squared *Lagrangean isentropic sound-speed*. With this, the field equations to be solved are

$$\frac{\partial \dot{x}}{\partial X} + \frac{\partial \Delta}{\partial t} = 0$$

$$C^2(\Delta)\frac{\partial \Delta}{\partial X} + \frac{\partial \dot{x}}{\partial t} = 0. \qquad (1.112)$$

Following the lead of our previous work, we seek a solution of Eqs. 1.112 that has a constant value along each of the rays through the point $X = 0$, $t = 0$, which are the $x^* = \text{const.}$ characteristics. A solution of this form will depend on the single variable

$$Z = -\frac{X}{t}. \qquad (1.113)$$

In this case, Eqs. 1.112 become

$$\frac{d\dot{x}}{dZ} - Z\frac{d\Delta}{dZ} = 0$$

$$C^2(\Delta)\frac{d\Delta}{dZ} + Z\frac{d\dot{x}}{dZ} = 0. \qquad (1.114)$$

Elimination of $d\dot{x}/dZ$ from the first equation of this pair gives

$$[C^2(\Delta) - Z^2]\frac{d\Delta}{dZ} = 0, \qquad (1.115)$$

which can have a nontrivial solution $\Delta(Z)$ only if the coefficient vanishes: $C^2(\Delta) - Z^2 = 0$. Because we are seeking a solution in the form of a left-propagating wave, we choose the positive square root, thus obtaining the implicit solution

$$C(\Delta) = +Z \qquad (1.116)$$

for $\Delta(Z)$. Substitution of Eq. 1.116 into Eq. 1.114$_2$ gives

$$Z\frac{d\Delta}{dZ} + \frac{d\dot{x}}{dZ} = 0, \qquad (1.117)$$

or

$$\frac{d\dot{x}}{d\Delta} = Z = C(\Delta), \tag{1.118}$$

from which we obtain

$$\dot{x} = \dot{x}^+ + \int_{\Delta}^{\Delta^+} C(\Delta')d\Delta'. \tag{1.119}$$

This completes the solution of the problem, since Eq. 1.116 can be inverted to give Δ as a function of $-X/t$. Then, for any particle and time, X and t, in the wave, the integral in Eq. 1.119 can be calculated, thus yielding $\dot{x}$. Some example waveforms are shown in Fig. 1.11b.

1.5.3. Formation and Attenuation of Shocks

Many important nonlinear wave propagation phenomena involve flows that are not isentropic. Particular cases of interest involve nonsteady shocks—those that are changing strength as they propagate. Simple analytical methods fail in these cases, but it is possible to develop some understanding of the issue if attention is restricted to disturbances of only moderate strength. In this case, the isentrope can be approximated by a function that is quadratic in Δ. To this degree of approximation, the Hugoniot is the same as the isentrope, so the entropy jump at the shock is neglected. This leaves us with an isentropic flow. Two cases involving shocks of varying strength are considered in the following subsections.

1.5.3.1. Shock Formation

Examination of the analysis of the simple compression wave discussed in Section 1.5.1 shows that the characteristics are converging and must eventually intersect. The extension of the diagram to later times that is given in Fig. 1.12a shows these intersections. The waveforms shown in Fig. 1.10b have been calculated for times before the characteristics intersect, but the same procedure can be applied to calculation of waveforms for times within, and beyond, the region of intersection. Several such waveforms have been plotted in Fig. 1.12b. One sees that the solution is triple-valued for points beyond that at which the first intersection occurs (given by Eqs. 1.98 and 1.99). The low-density branch corresponds to the state ahead of the wave, and the high-density branch corresponds to the final state achieved behind the wave. At intermediate densities, the waveform bulges forward, giving rise to a multivalued solution.

It is clear that this result is inadmissible on physical grounds; one cannot accept a solution that yields three different values for a field variable at the same time and place. The resolution of this problem is to embed a shock into the waveform so the solution is single valued except for the usual indeterminacy at the shock itself. The shock is inserted into the flow in such a way that mass is conserved. The equation for mass conservation is

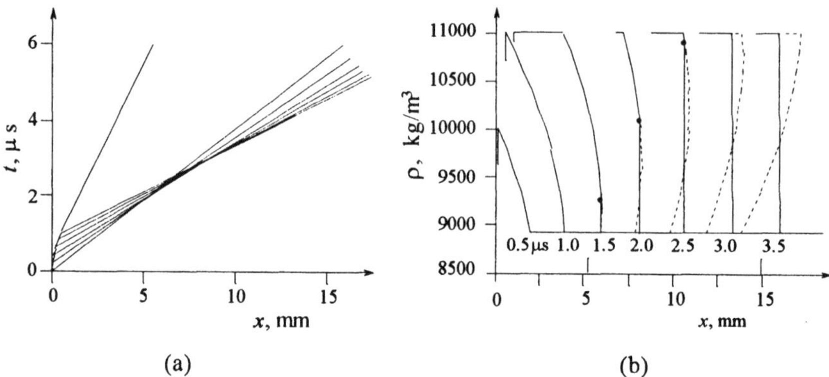

Figure 1.12. (a) The x–t diagram of Fig. 1.10a, extended to later times at which the characteristics intersect. Not shown are the characteristics in the region ahead of and behind the wave. In the region beyond the point where the characteristics begin to intersect, these characteristics overlay those shown, leading to a triple-valued solution. (b) The waveforms of Fig. 10b, extended to later times at which the solution to the wave equation becomes multivalued. The dotted lines designate the inadmissible triple-valued section of the waveforms. The shocks shown have been embedded in the field in such a way that the multivalued portion of the solution is eliminated and mass is conserved. The vertical marks at the left indicate the boundary position for each waveform.

$$\int_{x_b(t)}^{x_S(t)} \rho(x, t)\, dx = \rho_R\, x_S(t), \qquad (1.120)$$

where $x_b(t)$ is the position of the boundary and $x_S(t)$ is the position of the shock. Once the density field has been determined at a several times from the characteristic analysis, this equation is easily solved for shock position at each of these times. The shock amplitude is obtained by connecting the upper and lower branches of the triple-valued solution. It is essential to this process that the entropy jump at the shock is neglected, thus restricting the validity of the result to shocks of only moderate strength.

1.5.3.2. Shock Attenuation

Shock attenuation (often called *hydrodynamic attenuation*) occurs when a shock is overtaken by a smooth decompression wave. Consider the case in which an unstressed projectile plate of thickness L and having velocity $\dot{x}_P$ impacts an unstressed halfspace of the same normal material that is at rest. The disturbance produced is illustrated in the X–t diagram of Fig. 1.13.

At impact, shock waves form at the interface and propagate forward into the target and backward into the projectile plate. The state behind these shocks is characterized by the parameters

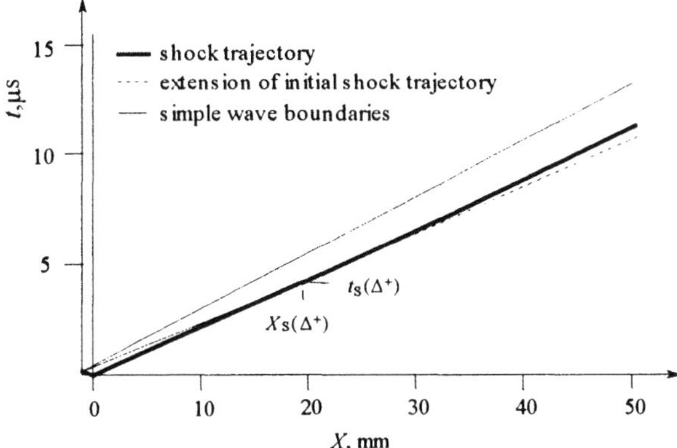

Figure 1.13. An X–t diagram illustrating the overtaking of a shock by a centered simple decompression wave. The diagram is drawn for the case in which a copper projectile plate 1-mm thick and moving at 1000 m/s impacts a stationary copper target.

$$\dot{x}^+ = \dot{x}_P/2, \quad p^+ = \rho_R U_S \dot{x}^+, \quad \text{and} \quad \Delta^+ = \dot{x}^+/U_S. \tag{1.121}$$

When the receding shock encounters the stress-free back surface of the projectile plate, a centered simple decompression wave forms and propagates forward. Because the soundspeed in the compressed material behind the advancing shock exceeds the shock velocity, the smooth wave will eventually overtake the shock. In this wave the advancing characteristics are given by

$$X = C(\Delta)(t - t^*) - L \tag{1.122}$$

and the particle velocity is related to the compression by the equation

$$\dot{x} = \dot{x}^+ - \int_{\Delta}^{\Delta^+} C(\Delta')d\Delta'. \tag{1.123}$$

Because the overtaking wave is one of decompression, we expect that the interaction will cause a decrease in shock strength and velocity. Because of the varying shock strength, the flow behind the shock will not be isentropic. This complication is avoided in the weak shock approximation that we shall adopt. To determine the shock trajectory and the waveform, we need to calculate the soundspeed from the second-order expression for the isentrope. If we base this on the Hugoniot $U_S = C_B + S\dot{x}$, we obtain

$$p^{(\eta)}(\Delta) = \rho_R C_B^2 \Delta(1 + 2S\Delta), \tag{1.124}$$

so the Lagrangean soundspeed is given by

$$C(\Delta) = C_\mathrm{B}(1+2S\Delta) \tag{1.125}$$

and

$$\int_\Delta^{\Delta^+} C(\Delta')\,d\Delta' = C_\mathrm{B}(\Delta^+ - \Delta)\left[1 + S(\Delta^+ - \Delta)\right]. \tag{1.126}$$

With this, Eq. 1.123 becomes

$$\dot{x} = \dot{x}^+ - C_\mathrm{B}\left\{\Delta^+ - \Delta^- + S[(\Delta^+)^2 - \Delta^2]\right\}. \tag{1.127}$$

The shock trajectory (following the point at which the interaction begins) can be expressed parametrically by giving t and X as a functions of Δ in the form $t = t_\mathrm{S}(\Delta)$ and $X = X_\mathrm{S}(\Delta)$. Substitution of these equations into Eq. 1.122 gives

$$X_\mathrm{S}(\Delta) = C(\Delta)[t_\mathrm{S}(\Delta) - t^*] - L. \tag{1.128}$$

Through differentiation, we obtain

$$\frac{dX_\mathrm{S}(\Delta)}{d\Delta} = \frac{dC(\Delta)}{d\Delta}[t_\mathrm{S}(\Delta) - t^*] + C(\Delta)\frac{dt_\mathrm{S}(\Delta)}{d\Delta}. \tag{1.129}$$

Along the shock trajectory

$$dX = U_\mathrm{S}(\Delta)dt, \tag{1.130}$$

so Eq. 1.129 becomes the linear, first-order ordinary differential equation,

$$\frac{dt_\mathrm{S}(\Delta)}{d\Delta} + \varphi(\Delta)t_\mathrm{S}(\Delta) = \varphi(\Delta)t^*, \tag{1.131}$$

where we have written

$$\varphi(\Delta) \equiv \frac{dC(\Delta)}{d\Delta}\frac{1}{C(\Delta) - U_\mathrm{S}(\Delta)}. \tag{1.132}$$

Combining expansions of the various quantities appearing in Eq. 1.132 shows that

$$\varphi(\Delta) = \frac{2}{\Delta}\left(1 - \frac{1}{2}S\Delta\right) \tag{1.133}$$

to within the accuracy of the analysis.

The solution of Eq. 1.131 is

$$t_\mathrm{S}(\Delta) = t^* + \left(\frac{\Delta^+}{\Delta}\right)^2[t_\mathrm{S}(\Delta^+) - t^*]\exp[-2S(\Delta^+ - \Delta)], \tag{1.134}$$

where

$$t_S(\Delta^+) = \frac{C(\Delta^+)\,t^* + L}{C(\Delta^+) - U_S^+}$$

(1.135)

is the time at which the leading characteristic of the decompression wave intersects the shock. The associated value of X on the shock trajectory is given by the Eq. 1.128.

From the known speed of the initial shock and the soundspeed behind this shock, the X–t diagram can be plotted up to the time at which the decompression wave begins to interact with the shock. The diagram is completed when Eqs. 1.134 and 1.128 are used to plot the trajectory of the attenuating shock.

The amplitude of the attenuating shock at a point on its trajectory can be characterized by the value of Δ at the point. If one selects several values of Δ in the range $0 \leq \Delta \leq \Delta^+$, the associated points on the trajectory can be plotted using Eqs. 1.134 and 1.128. The value of Δ associated with each of these points can be used to calculate the pressure (using Eq. 1.124) and the particle velocity (using Eq. 1.127). The waveform behind the shock is given by the simple wave solution. Several pressure waveforms associated with the problem described in Fig. 1.13 are plotted in Fig. 1.14.

1.5.4. Analysis of Structured Shocks

In the introduction to this chapter, it was noted that dissipative effects can cause a shock to spread into a smooth steady wave. In this section, we consider a simple example of such a wave. The steady waveform is allowed to extend indefinitely in space, as suggested in Fig. 1.15a.

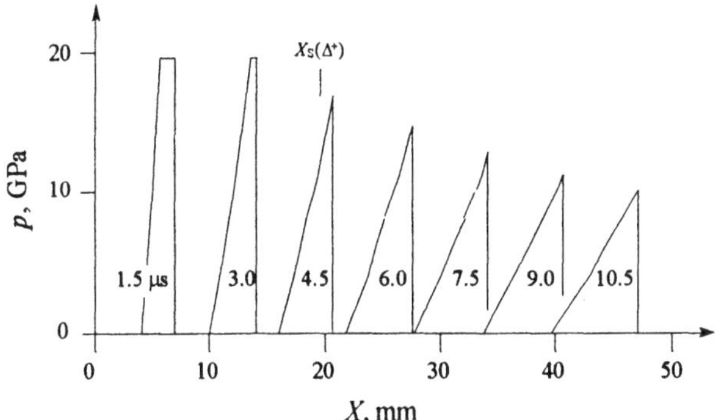

Figure 1.14. A sequence of pressure–distance waveforms illustrating shock attenuation. The distance at which the leading characteristic of the decompression wave catches up to the shock and the attenuation process begins is marked $X_S(\Delta^+)$.

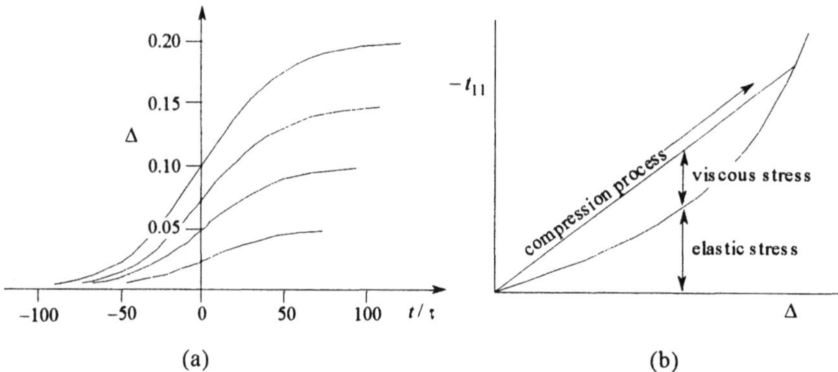

Figure 1.15. (a) Steady waveforms in a linearly viscous material. (b) Illustration of the compression process occurring in a smooth steady wave.

Restricting attention to a mechanical theory, we shall assume that t_{11}, Δ, and $\dot{x}$ all vanish in the limit $X \to \infty$ and that $t_{11} \to t_{11}^+$, $\Delta \to \Delta^+$, and $\dot{x} \to \dot{x}^+$ as $X \to -\infty$. We further assume that the material is in equilibrium at these limits, so the deformation rates vanish as $X \to \pm\infty$. We have seen that the relevant jump conditions, Eqs. 1.35, are satisfied for any two points in the waveform. Taking one of the points infinitely distant in the $+X$ direction, we have

$$\dot{x}(Z) = U_S\, \Delta(Z)$$

$$t_{11}(Z) = -\rho_R\, U_S\, \dot{x}(Z),$$

(1.136)

where $Z = X - U_S t$.

Elimination of $\dot{x}$ from this pair of equations gives

$$t_{11}(Z) = -\rho_R\, U_S^2\, \Delta(Z).$$

(1.137)

In the limit $X \to -\infty$ this becomes

$$t_{11}^+ = -\rho_R\, U_S^2\, \Delta^+,$$

(1.138)

giving the equation

$$U_S^2 = -\frac{1}{\rho_R}\frac{t_{11}^+}{\Delta^+}$$

(1.139)

for the steady wavespeed. Equation 1.137 is a parametric equation for the Rayleigh line connecting the initial and final states in the steady wave. It shows that the process that occurs at a material point as the wave passes follows this line in the $t_{11} - \Delta$ plane. This is in contrast with the situation for a shock, in which the concept of process is meaningless, being replaced one of discontinuous transition.

To proceed with the steady wave analysis, we must specify a constitutive equation describing the material in which the wave is propagating. We shall assume that, in the case of uniaxial compression, this equation takes the form of a relation expressing t_{11} as a function of Δ and $\dot{\Delta}$. We shall further assume that it takes the form of a sum of an equilibrium term and a rate-dependent term, i.e.,

$$t_{11} = t_{11}^{(E)}(\Delta) + t_{11}^{(V)}(\Delta, \dot{\Delta}),\tag{1.140}$$

where $t_{11}^{(V)}(\Delta, 0) = 0$. Because the effect of the viscous term is to resist compression, we have $t_{11}^{(V)}(\Delta, \dot{\Delta}) < 0$ when $\dot{\Delta} > 0$.

In a steady wave, we have

$$\dot{\Delta} = -U_S \frac{d\Delta}{dZ},\tag{1.141}$$

so Eq. 1.140 can be written in the form

$$t_{11}(Z) = t_{11}^{(E)}\big(\Delta(Z)\big) + t_{11}^{(V)}\left(\Delta(Z), \frac{d\Delta(Z)}{dZ}\right)\tag{1.142}$$

in this case.

When this relation is substituted into Eq. 1.137 we obtain the ordinary differential equation

$$t_{11}^{(E)}\big(\Delta(Z)\big) + t_{11}^{(V)}\left(\Delta(Z), \frac{d\Delta(Z)}{dZ}\right) = -\rho_R U_S \, \Delta(Z)\tag{1.143}$$

for the waveform $\Delta(Z)$. Once this equation is solved, the particle velocity waveform can be obtained from Eq. 1.136 and the stress waveform can be obtained from Eq. 1.142.

As a specific example, let us consider a material described by a third-order elastic relation to which a linear viscous stress is added. For compression waves, this relation is

$$t_{11} = \left[-C_{11}\,\Delta + \frac{1}{2}(3C_{11} + C_{111})\Delta^2\right] + v\,\frac{\dot{\Delta}}{1 - \Delta}.\tag{1.144}$$

When this equation is substituted into Eq. 1.143, we obtain an ordinary differential equation that can be integrated immediately to give

$$Z = \int_{\Delta^+/2}^{\Delta} \frac{\tau U_S \, d\Delta'}{(\Delta^+ - \Delta')(1 - \Delta')\Delta'},$$

where the characteristic time, τ, is given by $\tau = 2v/(3C_{11} + C_{111})$.

A graph of this result is shown in Fig. 1.15. Also shown is a schematic stress–compression plot illustrating the elastic part of the stress and the Rayleigh line, which is the path followed during passage of the wave. The viscous part of the stress is the difference between the two. The steady waveform arises as a standoff in the competition between the tendency of the nonlinear elastic behavior to produce wave steepening and that of the viscous behavior to produce dispersion. The compression rate in the wave adjusts itself so the viscous stress is just the required amount. When the wave amplitude is large, a large viscous stress, and correspondingly high deformation rate, is required. As the viscosity coefficient decreases, the waveform becomes increasingly steep, tending to a shock in the limit. This behavior, along with the fact that the jumps across the wave satisfy the same jump conditions as for shocks, justifies the description of these waves as *structured shocks*.

1.6. Steady Detonation Waves

A *detonation wave* in an explosive is a shock followed by a *chemical reaction zone* and a region of unsteady flow, the latter often in the form of a centered simple decompression wave. When the chemical reactions proceed rapidly, the reaction zone is very thin.

A *steady detonation* is steady in the sense that neither the jumps nor the shock velocity vary with time and the reaction zone is translated forward at the shock velocity. Even in the case of a steady detonation, the decompression wave following the reaction zone is unsteady. In this section, we shall discuss two simple models of the steady detonation process. The first and simplest, the Chapman–Jouguet (CJ) theory, is based on the premise that the chemical reaction occurs instantaneously. The waveform associated with this simplest case is illustrated in Fig. 1.16. The second model, represented by the Zel'dovich–von Neumann–Döring (ZND) theory, is an extension of the CJ theory that takes finite reaction rates into account.

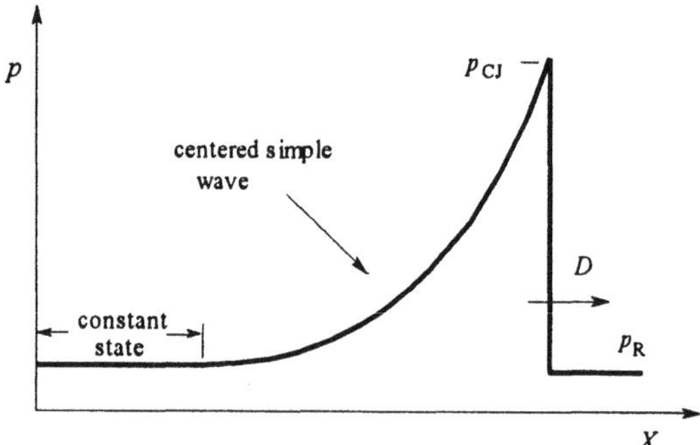

Figure 1.16. Illustration of the pressure waveform predicted by the Chapman–Jouguet theory combined with the Taylor analysis of the following decompression wave.

1.6.1. The Chapman–Jouguet Detonation

In the most idealized view of a detonation, each particle of the explosive undergoes an instantaneous transition from its initial unreacted form to reaction products as the shock passes. The equation of state of the reaction products is entirely different from that of the unreacted explosive and the internal energy of the products includes the chemical energy liberated by the reactions. The shock at the front of a detonation wave obeys the same jump conditions as those that describe nonreactive shocks. As with the nonreactive case, these conditions provide three constraints relating the five variables p^+, $\dot{x}^+$, v^+ (or $\rho^+ = 1/v^+$), ε^+, and D, where we write $U_S \equiv D$ in conformity with the practice of detonation physics. The material is described by an Hugoniot curve relating any pair of the above variables in the states that can be reached by a shock transition from the unreacted state. The three jump conditions and the Hugoniot curve suffice to determine the state behind the shock to within one variable, which we have regarded as a measure of the shock strength. In the case of nonreactive materials, steady shock propagation is sustained by energy supplied by forces applied to material at the boundary behind the shock and the shock strength is determined by the imposed boundary condition. The situation differs in the present case because a detonation wave is sustained by chemical energy rather than by forces imposed on the boundary. The new concept introduced in this section concerns establishment of a relation between chemical energy release and shock strength.

Let us restrict attention to the case that the unreacted explosive is at rest in its reference state: $\dot{x}^- = 0$, $p^- = p_R$, $v^- = v_R$, and $\varepsilon^- = \varepsilon_R = 0$. The jump conditions have been given in Lagrangean form as Eqs. 1.35. The usual manipulation of these equations gives the equation

$$p^+ - p_R = \rho_R\, D^2 \left(1 - \frac{v^+}{v_R}\right) \tag{1.145}$$

describing the Rayleigh line and the Rankine–Hugoniot equation

$$\varepsilon^+ = \frac{1}{2}(p^+ + p_R)v_R\left(1 - \frac{v^+}{v_R}\right) \tag{1.146}$$

relating the thermodynamic variables.

Because the detonation products form a gas, the polytropic gas theory provides a convenient and widely used equation of state that, in this case, takes the form

$$\varepsilon(p, v) = \frac{pv}{\Gamma - 1} - q - \frac{p_R v_R}{\Gamma - 1}, \tag{1.147}$$

where the polytropic exponent, Γ, is a material constant. The positive parameter q is the internal energy per unit mass that is liberated by the chemical reaction of a unit mass of explosive when the pressure and specific volume are held at the reference values p_R and v_R.

When Eq. 1.147 is substituted into the Rankine–Hugoniot equation, the result is an Hugoniot curve that can be written in the form

$$\left(\frac{p^+}{p_R}+\mu^2\right)\left(\frac{v^+}{v_R}-\mu^2\right)=1-\mu^4+\frac{2\mu^2 q}{p_R\, v_R}, \tag{1.148}$$

where $\mu^2 \equiv (\Gamma-1)/(\Gamma+1)$. This reaction product Hugoniot is centered on the state of the unreacted explosive but does not pass through this state. At the specific volume v_R the pressure on the Hugoniot exceeds the center-point value, p_R, by the amount $\rho_R(\Gamma-1)q$.

As noted previously, the foregoing equations are not sufficient to determine all five of the variables that describe the detonation shock. Some means must be found to complete the solution. To begin, let us consider Hugoniot and Rayleigh lines shown in the p–v plot of Fig. 1.17. We know that the state behind the detonation shock must lie on both the Hugoniot and the Rayleigh line corresponding to the detonation velocity. The lowest-velocity Rayleigh line shown does not intersect the Hugoniot, so it cannot correspond to a solution. The intermediate-velocity Rayleigh line intersects the Hugoniot at exactly one point, designated CJ and called the *Chapman–Jouguet*, or *CJ, point;* it corresponds to an unique solution to the problem. The highest-velocity Rayleigh line intersects the Hugoniot at two points, designated S and W, respectively, for two possible solutions called the *strong detonation* and the *weak detonation*.

Because the CJ state lies on both the Rayleigh line and the Hugoniot curve for the detonation products, we can eliminate the pressure from this pair of equations (Eqs. 1.148 and Eq. 1.145), producing the result

$$1=\frac{1}{2\mu^2 q}\left(1-\bar{v}_{CJ}\right)\left[D_{CJ}^2\left(\bar{v}_{CJ}-\mu^2\right)-(1+\mu^2)p_R\, v_R\right], \tag{1.149}$$

where $\bar{v}_{CJ}=v_{CJ}/v_R$ This is a quadratic equation for v_{CJ} that depends on D_{CJ}^2 as a parameter. When D_{CJ}^2 takes the value

$$D_{CJ}^2=(\Gamma^2-1)q+\Gamma p_R v_R+\left[(\Gamma^2-1)^2 q^2+2\Gamma(\Gamma^2-1)\, p_R v_R\right]^{1/2}, \tag{1.150}$$

the unique solution

$$\bar{v}_{CJ}=\frac{\Gamma}{\Gamma+1}\left(1+\frac{p_R v_R}{D_{CJ}^2}\right) \tag{1.151}$$

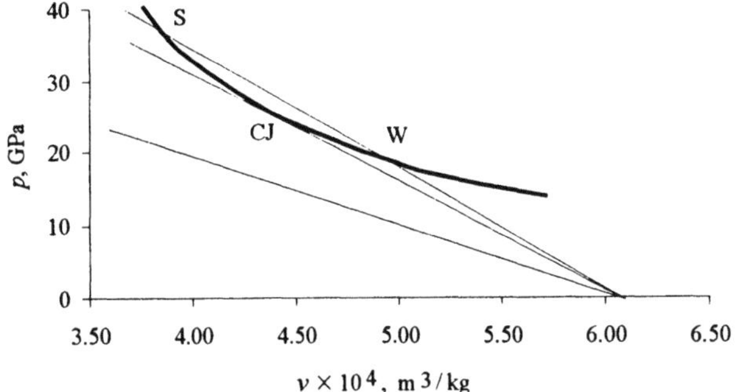

$$v \times 10^4, \text{ m}^3/\text{kg}$$

Figure 1.17. Detonation product Hugoniot and three Rayleigh lines. The lower Rayleigh line does not intersect the Hugoniot, the middle line is tangent to the Hugoniot at the point designated CJ, and the upper line intersects it twice at points designated S and W. The Hugoniot is drawn for the explosive TNT prepared at the density $\rho_R = 1600 \text{ kg/m}^3$. The equation of state is characterized by $\Gamma = 2.6$ and $q = 4.9 \times 10^6 \text{ J/kg}$.

is obtained. The CJ pressure lies at the point on the detonation product Hugoniot at which $v = v_{CJ}$:

$$\frac{p_{CJ}}{p_R} = \frac{1 - \mu^2 \bar{v}_{CJ} + (2\mu^2 q / p_R v_R)}{\bar{v}_{CJ} - \mu^2}. \tag{1.152}$$

Because gaseous explosives cannot exist at zero pressure, the effect of initial pressure cannot be ignored when developing a theory for detonation of these materials. Nevertheless, the limiting case for vanishing initial pressure agrees with the exact results to within a few percent for gas detonations when the initial pressure is 1 atm. The results for the limiting case $p_R \to 0$,

$$D_{CJ}^2 = 2(\Gamma^2 - 1)q, \quad v_{CJ} = \Gamma v_R /(\Gamma + 1), \quad p_{CJ} = 2\rho_R (\Gamma - 1)q, \tag{1.153}$$

are those usually seen in the literature on solid explosives because the equation of state for the unreacted material is meaningful for $p = 0$ and initial pressures of ~1 atm are entirely negligible in comparison with the CJ pressure.

It is important to note that one need not restrict attention to detonations propagating at the CJ speed. We have seen that no solution exists for $D < D_{CJ}$. When $D > D_{CJ}$, there are solutions corresponding to strong and weak detonations, respectively. The strong detonation is produced when a boundary condition involving a pressure exceeding p_{CJ} is imposed. Weak detonations also exist, and may indeed be the most commonly occurring case. Unfortunately, discussion of this issue requires consideration of details of multiple chemical reactions and/or other effects that lie beyond the scope of this chapter.

The simplest examination of the various detonation points involves comparing the detonation velocity with the isentropic soundspeed in the material behind the detonation shock. The Lagrangean isentropic soundspeed for a polytropic gas is given by

$$C^{(\eta)} = \sqrt{(\Gamma p v)}. \tag{1.154}$$

When CJ parameters are substituted into this equation, we find that the Lagrangean soundspeed and the detonation shock speed are the same. A disturbance existing in the flow immediately behind the shock neither overtakes it nor falls further behind it, suggesting marginal longitudinal stability of the solution.

1.6.2. Zel'dovich–von Neumann–Döring Detonation

In developing the Chapman–Jouguet model it was assumed that the chemical reaction occurred instantaneously in a shock transition from the initial state of the unreacted explosive to a point on the Hugoniot of the detonation products. In this section we develop the Zel'dovich–von Neumann–Döring (ZND) theory of detonation, in which the reaction proceeds to completion at a finite rate after being initiated by passage of the shock.

This model immediately presents problems that do not arise in the Chapman–Jouguet analysis. In particular, we shall need the Hugoniot for the unreacted explosive, the equation of state of the partially reacted material, and a kinetic equation describing the rate at which the reaction proceeds. In each case, determining these equations is difficult, involving both chemical and physical considerations. Practical analyses of detonations are usually carried out in the Chapman–Jouguet context, but understanding of detonation physics requires consideration of reaction processes.

We begin by seeking a solution to this problem in which the shock and the following reaction zone form a steady structured wave. As we have seen, the transition from an initial state to any state occurring in a steady wave satisfies the same conditions as a shock transition between the two states. The steady-wave transition from the unreacted material to the fully reacted detonation products can be analyzed in exactly the same way as in the Chapman–Jouguet theory, and leads to the same results. Strong, weak, and CJ states can be identified. The propagation speed and the pressure, specific volume, and particle velocity at the point in the steady waveform at which the reaction is complete attain the same values as in the Chapman–Jouguet theory. The point in pursuing the ZND theory is not simply to calculate this final state. Rather, interest centers on determination of the structure of the reaction zone, and this is done by means of a steady wave analysis.

The state immediately behind the ZND detonation shock, called the *von Neumann spike point* (see Fig. 1.18), is one in which the explosive has been

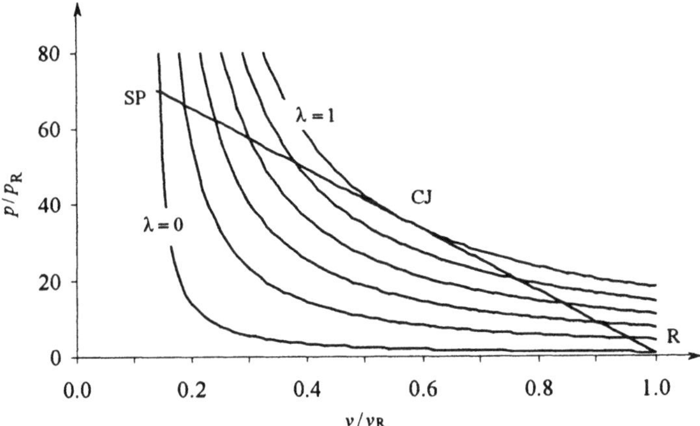

Figure 1.18. Partial reaction Hugoniots for a stoichiometric methane–oxygen detonation. The parameters used are: $p_R = 101325\,\text{Pa}$, $v_R = 0.9169\,\text{kg/m}^3$, $\Gamma = 1.3$, and $q = 5.236\,\text{MJ/kg}$. The von Neumann spike point is designated SP.

compressed but has not yet reacted. It lies on the Hugoniot of the unreacted explosive.

We shall consider a single reaction converting the explosive into detonation products. The degree to which this reaction has progressed is characterized by a variable λ, ranging from 0 to 1, called the *extent of reaction*. This variable is introduced into the equation of state for the detonation products to characterize both the varying characteristics of the reactant–product mixture and the degree to which the heat of detonation has been liberated. Its value increases continuously over its range as the reaction zone propagates past a given material particle. The chemical energy liberated by reaction varies from none at $\lambda = 0$ to the heat of detonation, q, at $\lambda = 1$.

We shall conduct our discussion of the ZND theory in the context of a gaseous explosive such as a methane–air mixture because this case lends itself to the simplest presentation. The internal energy function for the mixture comprising the partially reacted explosive gas and the detonation products is taken in the form of Eq. 1.147, except that the heat of detonation is liberated in proportion to the value of λ at the material point under consideration:

$$\bar{\varepsilon}(v, p, \lambda) = \frac{pv}{\Gamma - 1} - q\lambda - \frac{p_R v_R}{\Gamma - 1} . \tag{1.155}$$

As with all steady waves, the process follows the Rayleigh line of Eq. 1.145. In the present case, the detonation shock produces a transition from the initial state to the spike point and the reaction occurs in a smooth, steady decompression wave connecting the spike point to the CJ point. The internal energy density change satisfies the Rankine–Hugoniot equation, Eq. 1.146, throughout this process.

Substituting Eq. 1.155 into Eq. 1.146 yields a result like Eq. 1.148, but with q replaced by λq. We write this equation in the form

$$p^{(\mathrm{H})}(\bar{v},\lambda) = p_\mathrm{R}\,\frac{1-\mu^2\bar{v}+(2\mu^2 q\lambda/p_\mathrm{R}v_\mathrm{R})}{\bar{v}-\mu^2}\,, \qquad (1.156)$$

where $\bar{v}=v^+/v_\mathrm{R}$. The Hugoniot curves obtained for specific values of λ, called *partial-reaction Hugoniots*, are centered on the state $(p_\mathrm{R},v_\mathrm{R})$, but are offset from this point by an amount that depends on λq. A plot of several partial-reaction Hugoniot curves is given in Fig. 1.18.

As the reaction zone passes a material point, the state attained on each of these Hugoniot curves falls at the intersection of the Hugoniot and the Rayleigh line for the steady wave. Equations 1.156 and 1.145 can be solved for the pressure and specific volume at any point in the wave.

Corresponding values of p and v in the waveform lie at the intersection of the partial-reaction Hugoniot and the Rayleigh line. Equating the pressure given by Eq. 1.145 with that given by Eq. 1.156 yields the expression

$$\lambda = \frac{1}{2\mu^2 q}(1-\bar{v})\big[D_{\mathrm{CJ}}^2(\bar{v}-\mu^2)-p_\mathrm{R}v_\mathrm{R}(1+\mu^2)\big] \qquad (1.157)$$

for the extent of reaction as a function of $\bar{v}$.

Conventional chemical-kinetic equations express $\dot{\lambda}$ in terms of the thermodynamic state variables, so we will need the material derivative of Eq. 1.157:

$$\dot{\lambda} = \frac{1}{(\Gamma-1)q}\big[\Gamma(D_{\mathrm{CJ}}^2+p_\mathrm{R}v_\mathrm{R})-(\Gamma+1)D_{\mathrm{CJ}}^2\,\bar{v}\big]\dot{\bar{v}}\,. \qquad (1.158)$$

In the steady wave $\dot{\bar{v}}=-D_{\mathrm{CJ}}\,d\bar{v}/dZ$, so the foregoing equation can be written

$$dZ = \frac{D_{\mathrm{CJ}}}{(\Gamma-1)q\dot{\lambda}}\big[(\Gamma+1)D_{\mathrm{CJ}}^2\,\bar{v}-\Gamma(D_{\mathrm{CJ}}^2+p_\mathrm{R}v_\mathrm{R})\big]d\bar{v}\,. \qquad (1.159)$$

This equation can be integrated to give the steady waveform $\bar{v}=\bar{v}(Z)$ if $\dot{\lambda}$ is expressed as a function of $\bar{v}$.

A typical kinetic relation for a simple chemical reaction is the first-order *Arrhenius equation*

$$\dot{\lambda} = k\,(1-\lambda)\,\exp\!\left(-\frac{\varepsilon^\dagger}{\mathcal{R}\,\theta}\right), \tag{1.160}$$

where k and $\varepsilon^\dagger$ are material constants called the *frequency factor* and *activation energy*, respectively, and $\mathcal{R}$ is the ratio of specific heats. To use this relation, we need expressions for λ and the temperature, θ, in terms of the value of $\bar{v}$ at points in the wave. We begin this task by calculating $\theta^{(H)}(\bar{v},\lambda)$ on each of the partial-reaction Hugoniots. Differentiating the result of applying the polytropic ideal gas equation of state, $\varepsilon = C^{(v)}\theta$, to states on the Hugoniot yields the result

$$\frac{\partial\theta^{(H)}(\bar{v},\lambda)}{\partial\bar{v}} = \frac{1}{C^{(v)}}\frac{\partial\varepsilon^{(H)}(\bar{v},\lambda)}{\partial\bar{v}}.$$

When $\varepsilon^{(H)}(\bar{v},\lambda)$ is obtained from the Rankine–Hugoniot equation, this result can be expressed in the form

$$\frac{\partial\theta^{(H)}(\bar{v},\lambda)}{\partial\bar{v}} = \frac{v_R}{2C^{(v)}}\left\{(1-\bar{v})\frac{\partial p^{(H)}(\bar{v},\lambda)}{\partial\bar{v}} - p^{(H)}(\bar{v},\lambda) - p_R\right\}$$

and, using Eq. 1.156, this becomes

$$\frac{\partial\theta^{(H)}(\bar{v},\lambda)}{\partial\bar{v}} = -\frac{(1-\mu^2)v_R}{2C^{(v)}}\left\{p_R v_R + \left[(1-\mu^4)p_R v_R + 2\mu^2\lambda q\right]\frac{1}{(\bar{v}-\mu^2)^2}\right\}.$$

This equation is immediately integrable, yielding the result

$$\theta^{(H)}(\bar{v},\lambda) - \theta^{(H)}(1,\lambda) =$$
$$-\frac{(1-\bar{v})}{(\Gamma+1)C^{(v)}}\left\{p_R v_R + \left[\frac{2\Gamma}{\Gamma+1}p_R v_R + (\Gamma-1)q\lambda\right]\frac{1}{(\bar{v}-\mu^2)}\right\}. \tag{1.161}$$

When the initial condition

$$\theta^{(H)}(1,\lambda) = \frac{q\lambda}{C^{(v)}} + \theta_R$$

is substituted into Eq. 1.161, calculation of the temperature on the Hugoniots is complete.

The steady reaction zone can now be obtained by substituting Eq. 1.161 into Eq. 1.160 and that result into Eq. 1.159. Integration (by numerical means) then gives the waveform. Some results of this analysis are shown in Fig. 1.19.

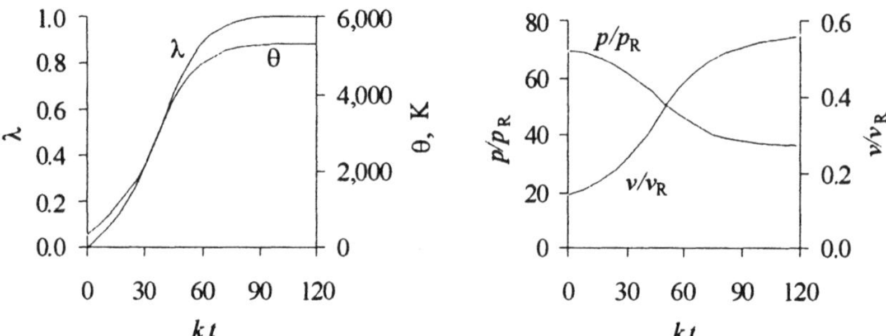

Figure 1.19. Temporal waveforms showing reaction zone structure calculated using the parameters $p_R = 100325\,\mathrm{Pa}$, $v_R = 0.9169$, $\Gamma = 1.3$, $\varepsilon^\dagger/\mathcal{R} = 15000\,\mathrm{K}$, $C^{(v)} = 1039\,\mathrm{J/kg}$, and $q = 5.236 \times 10^6\,\mathrm{J/kg}$ The plots are shown for the case that the wave arrives at the point in question at $t = 0$. The parameter $1/k$ arising in the kinetic equation has the units of time and provides the scale used. Since, for the kinetic equation used, the solution approaches the CJ state asymptotically, no precise reaction zone length is defined by the theory.

1.6.3. Taylor Decompression Wave

To complete the analysis of the detonation waveform we must consider the flow behind the detonation shock and reaction zone. The decompression process will proceed along the isentrope passing through the CJ state and expanding to the background pressure. This decompression wave is called a *Taylor wave*. The equation for the decompression isentrope is $p v^\Gamma = p_{CJ}(v_{CJ})^\Gamma$.

The X–t diagram for this problem is shown in Fig. 1.20. Since the CJ state prevailing behind the detonation shock is constant, the following wave is a centered simple wave. The Lagrangean isentropic soundspeed is given by $C^{(\eta)} = \sqrt{(\Gamma p v)}$, so we see that the leading characteristic of this wave lies along the shock trajectory and the trailing characteristic remains fixed in the reference configuration. Applying the results of Section 1.5.1 to the present problem gives the simple wave fields

$$v = v_{CJ}\left(D_{CJ}\frac{t}{X}\right)^{2/(\Gamma+1)}$$

$$p = p_{CJ}\left(\frac{1}{D_{CJ}}\frac{X}{t}\right)^{2\Gamma/(\Gamma+1)} \tag{1.162}$$

$$\dot{x} = \dot{x}_{CJ}\frac{\Gamma+1}{\Gamma-1}\left[\frac{2\Gamma}{\Gamma+1}\left(\frac{1}{D_{CJ}}\frac{X}{t}\right)^{(\Gamma-1)/(\Gamma+1)} - 1\right].$$

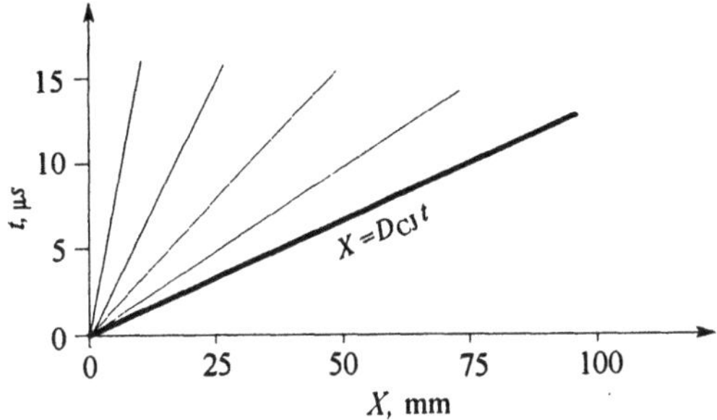

Figure. 1.20. Lagrangean space–time diagram for a Chapman–Jouguet detonation wave.

The region in which this solution applies extends from the CJ point (the detonation shock or the end of the ZND reaction zone) to the trailing characteristic $X = 0$, so the characteristic lies in the range

$$0 \le \frac{X}{t} \le D_{CJ} . \tag{1.163}$$

Results obtained from Eqs. 1.162 are plotted in Fig. 1.21.

1.7. Concluding Remarks

In writing this chapter, I have taken the most traditional view of the subject of shock compression of solids. Various basic aspects of the physics, formation, propagation, and decay of shocks have been illustrated.

Scientific interest in shock compression of solids was originally motivated by military applications involving stresses and rates of deformation that are much higher than those encountered in most engineering applications of mechanics. It is fortunate that the early interest in shock compression of solids involved high stresses, because the response of solids under these conditions is

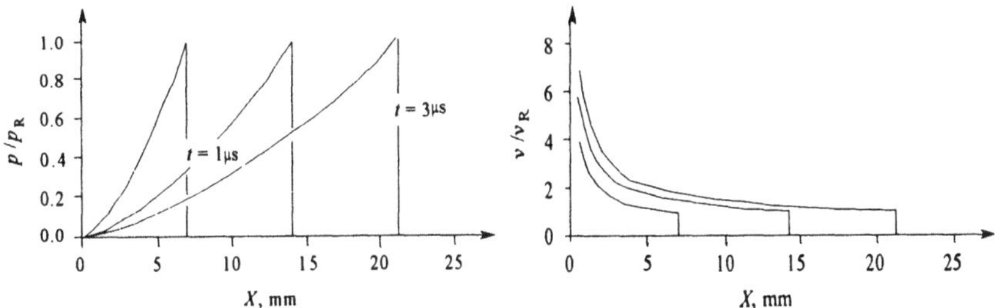

Figure. 1.21. Lagrangean waveforms at 1, 2, and 3 µs after introduction of a CJ detonation wave into an explosive characterized by the parameters $D_{CJ} = 7000\,\text{m/s}$ and detonation product polytropic exponent $\Gamma = 2.8$.

similar to that of fluids. Since the theory of compressible fluid behavior was well developed, much of the early research activity was directed toward experimental characterization of the response of solids to strong shock compression [8]. The discovery of new phenomena in the course of this early experimental work motivated development of both more refined experimental observations and more comprehensive theoretical descriptions of material response. This work was very successful and continues to this day.

Discussions of current work can be found in this volume, in other volumes in the present series (see p. ii of this volume), in proceedings of conferences sponsored by the Topical Group on Shock Compression of Condensed Matter of the American Physical Society (the most recent being [12]), proceedings of the symposia on detonation (the most recent being [13]), and elsewhere.

Current experimental work following in the tradition of the earlier studies addresses response of a broad range of materials to shock compression in the range accessible using explosives and projectiles driven by compressed gas or propellants [14]. Extensions of this work involve effects at very high pressure and development of equations of state based on quantum theory of matter. These equations of state address such issues as chemical reactions, phase transitions, ionization, and changes in atomic structure [16].

Since the earliest days of research in shock compression of solids, there has been considerable interest in the lower stresses at which the peculiarities of solid responses are evident and their effects pronounced. Elastoplastic response in the form of precursor waves to moderately strong shocks was observed [17] and explained in terms of both continuum-mechanical and microscopic theories [18,19]. Experimental investigation of the response of solids in the range where elastic and plastic effects are of comparable importance continues, with attention being accorded to both dynamic observations and microscopic examination of materials recovered after being subjected to shock loading [20]. Considerable effort is being devoted to development of continuum theories based on microscopic models of material behavior. Many observed phenomena have been explained theoretically and high fidelity numerical simulation of thermomechanical processes is routine. Nevertheless, there is no universally accepted theory of elastoplastic or elastoviscoplastic response. Some old and extensively studied problems, such as that of the mechanism of decay of elastic precursors, remain to be resolved.

The deformation processes that occur in brittle materials such as ceramics differ in important ways from those of ductile metals. These materials are subject to intensive study both because they present interesting scientific questions and because they have a variety of important technological applications [21,22].

Some solids exhibit nonlinear viscoelastic responses over a moderate range of stress, and a body of theoretical and experimental work addressing this problem has accumulated [23].

Nonlinear elastic responses of solids to shock compression are limited to the range of compression below that at which flow and/or fracture are observed. The theory of plane nonlinear elastic wave propagation is essentially the same as that presented for fluids. A number of strong, crystalline solids have been investigated and characterized. Since these materials exhibit various electrical effects, investigations of these phenomena have also been conducted [24].

The tensile stress field arising when two plane decompression waves collide provides one of the simplest environments for the study of fracture. The applied stress and the duration of its application can be varied over a range that permits the fracture process to be initiated and then arrested at any stage of its progress. Informative dynamic measurements can be made routinely, and samples can be recovered for post-test examination. Despite the complexity of dynamic fracture and the elastoplastic deformation that produces it, considerable success has been realized in understanding the underlying physics, modeling it theoretically, and simulating it numerically [25–27].

Many classes of materials have been accorded special treatment addressed to their peculiarities of response or applications. These include porous and granular materials [3,28,29], explosives [30–32], composites [33], and materials of interest in the geological and planetary sciences [34,35]. In recent years, considerable attention has been accorded to solid-state chemical reactions induced by shock compression [36].

The need to solve problems of shock compression of solids motivated the early development of supercomputers, and these problems continue to stimulate this development. Much of the activity in the area of shock physics has been directed toward theories and numerical methods for solving these problems. This work has been spectacularly successful, and application of the most complex and comprehensive theories of material response to analysis of problems of considerable geometrical complexity is routinely accomplished [37,38].

Development of methods and apparatus for experimental investigation of shock phenomena has been essential to continued progress in this field. Many people have contributed to this work and its success is attributable to both their ingenuity and dedication. Unrelated development of electronic and optical instrumentation and of lasers has played an essential role.

There remain many issues to resolve. Observations of materials recovered after shock compression, evidence of chemical reactions in the solid state, observations of particle velocity dispersion in laser-interferometric measurements, results of molecular-dynamic analyses, and other evidence all point to significant mesoscale irregularities, instability, and turbulence in shock-induced motions. Nevertheless, the theory of the subject is based almost entirely on the premise of smooth, regular motion. The remainder of this volume concerns research directed toward remedying these shortcomings.

References

[1]   J.N. Johnson and R. Chéret (eds.), *Classic Papers in Shock Compression Science*, Springer, New York (1998).

[2]   R. Courant and K.O. Friedrichs, *Supersonic Flow and Shock Waves*, Interscience, New York (1948).

[3]   V.F. Nesterenko, *Dynamics of Heterogeneous Materials*, Springer-Verlag, New York (2001).

[4]   C. Truesdell and R.A. Toupin, The Classical Field Theories, in *Handbuch der Physik* **III/1** (ed. S. Flügge), Springer-Verlag, Berlin (1960).

[5]   L.E. Malvern, *Introduction to the Mechanics of Continuous Media*, Prentice-Hall, Englewood Cliffs, NJ (1969).

[6]   D.S. Drumheller, *Introduction to Wave Propagation in Nonlinear Fluids and Solids*, Cambridge University Press, Cambridge (1998).

[7]   C. Truesdell and W. Noll, in *Handbuch der Physik* **III/3** (ed. S. Flügge), Springer-Verlag, Berlin (1965).

[8]   M.H. Rice, R.G. McQueen, and J.M. Walsh, in *Solid State Physics* **6** (eds. F. Seitz and W. Turnbull), Academic Press, New York, pp. 1–63 (1958).

[9]   R.G. McQueen, S.P. Marsh, J.W. Taylor, J.N. Fritz, and W.J. Carter, in *High Velocity Impact Phenomena* (ed. R. Kinslow), Academic Press, New York, pp. 293–417 with appendices on pp. 515–568 (1970).

[10]   R.N. Thurston, in *Handbuch der Physik* **IVa/4** (ed. S. Flügge), Springer-Verlag, Berlin, pp. 109–308 (1974).

[11]   J.R. Asay and M. Shahinpoor, eds., *High-Pressure Shock Compression of Solids*, Springer-Verlag, New York (1993).

[12]   *Shock compression of Condensed Matter—2001* (eds. M.D. Furnish, N.N. Thadhani, and Y. Horie), American Institute of Physics, Melville, NY (2002).

[13]   *Eleventh International Detonation Symposium*, U.S. Office of Naval Research report ONR33300-5 (2000).

[14]   L.M. Barker, M. Shahinpoor, and L.C. Chhabildas, in [11], pp. 43–73.

[15]   L. Davison and M. Shahinpoor, eds., *High-Pressure Shock Compression of Solids—III*, Springer-Verlag, New York (1997).

[16]   S.K. Sikka, B.K. Godwal, and R. Chidambaram, in [15], pp. 1–35.

[17]   S. Minshall, *J. Appl. Phys.* **26**, pp. 463–469 (1955).

[18]   J.W. Taylor, *J. Appl. Phys.* **34**, pp. 2727–2731 (1963)

[19]   J.N. Johnson, in [11], pp. 217–264.

[20]   G.T. Gray III, in [11], pp. 187–215.

[21]   T. Mashimo, in [15], pp. 101–146.

[22]   J. Cagnoux and J.-Y. Tranchet, in [15], pp. 147–169.

[23] J.W. Nunziato, E.K. Walsh, K.W. Schuler, and L.M. Barker, in *Handbuch der Physik* **IVa/4** (ed. S. Flügge), Springer-Verlag, Berlin, pp. 1–108 (1974).

[24] R.A. Graham, *Solids Under High-Pressure Shock Compression*, Springer-Verlag, New York (1993).

[25] T.H. Antoun, L. Seaman, D.R. Curran, G.I. Kanel, S.V. Razorenov, and A.V. Utkin, *Dynamic Fracture of Materials*, Springer-Verlag, New York, in press.

[26] L. Davison, D.E. Grady, and M. Shahinpoor, eds., *High-Pressure Shock Compression of Solids—II: Dynamic Fracture and Fragmentation*, Springer-Verlag, New York (1996).

[27] A.K. Zurek and M.A. Meyers, in [26], pp. 25–70.

[28] L. Davison, Y. Horie, and M Shahinpoor, eds., *High-Pressure Shock Compression of Solids—IV: Response of Highly Porous Solids to Shock Loading*, Springer-Verlag, New York (1997).

[29] L.S. Bennett, K. Tanaka, and Y. Horie, in [28], pp. 105–175.

[30] R. Engelke and S.A. Sheffield, in [15], pp. 171–239.

[31] S.A. Sheffield, R.L. Gustavsen, and M.U. Anderson, in [28], pp. 23–61, (1997).

[32] M.R. Baer in [28], pp. 63–82.

[33] F.L. Addessio and J.B. Aidun, in [15], pp. 241–275.

[34] T.J. Ahrens, in [11], pp. 75–113.

[35] L. Davison, Y. Horie, and T. Sekine, eds., *High-Pressure Shock Compression of Solids—V: Shock Chemistry with Application to Meteorite Impacts*, Springer-Verlag, New York (2003).

[36] N.N. Thadhani and T. Aizawa, in [28], pp. 257–287.

[37] J.M. McGlaun and P. Yarrington, in [11], pp. 323–353.

[38] J.R.Asay and G.I. Kerley, *Int. J. Impact Engng.* **5**, pp. 69–99 (1987).

CHAPTER 2

Paradigms and Challenges in Shock Wave Research*

James R. Asay and Lalit C. Chhabildas

Shock compression science is a relatively new discipline that enables studies of materials in thermodynamic regimes inaccessible by other methods. The underlying motivation for these studies is to promote a physical understanding of dynamic material response. A fundamental understanding of material compression allows development of material models used to predict the response of materials subjected to dynamic loading. The past fifty years have produced a challenging research environment, resulting in the discovery and implementation of a broad spectrum of experimental, theoretical, and computational methods for this purpose. As a result, a large variety of phenomenological models have been developed for use in numerical simulations of material response under complex loading conditions. Even after a half-century of intensive research, however, there are still a large number of shock processes that cannot be resolved *a priori* with our existing level of understanding. A few examples include the prediction of shock transition times, compressive or tensile strengths under shock loading, and the kinetics of shock-induced phase transitions. This has led us to examine the issue of "what is a shock?" in this chapter by looking at the fundamental assumptions. These include assumptions of uniaxial displacement for applied planar loading, steady shock wave propagation, homogeneous shocked states, and thermodynamic equilibrium in the shocked state. We examine the traditional approaches used for describing shock processes in solids, including the apparent anomalous observations resulting from these interpretations. Both experimental and computational results are used to show that shocks produce highly heterogeneous states of compression, which strongly influence observable material properties in the shocked state. Understanding these effects will require improved experimental diagnostics to probe deformation mechanisms *in situ* and in real time. We conclude by discussing what kinds of measurements will be needed to resolve these issues.

---

* Sandia is a multi-program laboratory operated by Sandia Corporation, a Lockheed Martin Company, for the United States Department of Energy under Contract DE-AC04-94AL85000.

2.1. Introduction

The field of shock compression science has its roots in World War II driven by nuclear weapons applications to describe the high-pressure response of materials in regimes inaccessible by other methods. These requirements led to the development of a variety of experimental and theoretical tools [15,30,44,48,55,131, 142] for probing the transitory nature of shock compression under the simple conditions of planar shock loading. For these loading conditions, the conservation equations of mass, momentum, and energy provide a simple relationship to determine the pressure, energy, and density states produced by shock compression [11,39,87]. To obtain the equation-of-state (EOS) and other material property information from planar shock loading, several assumptions are implicitly made. The assumptions include (1) planar uniaxial motion, (2) steady one-dimensional shock propagation, and (3) a homogeneous, uniform shocked state in thermodynamic equilibrium. Based on these assumptions a large number of EOS and other material properties have been addressed experimentally and the shock wave method is now well established as an extremely useful and mature tool for high-pressure research.

Many of the original questions posed by the early investigations of shock wave research have been addressed to a large degree, with the result that a basic understanding now exists for the thermophysical, mechanical, and compressive properties of shocked materials. These developments include highly refined continuum models of dynamic material response that accurately predict most of the aggregate features of shock compression. The resulting models have been incorporated into a variety of computational methods to routinely solve complex engineering problems. In general, however, material models being used in these applications do not provide unique descriptions of dynamic material response even though they may accurately represent physical phenomena over a small range of loading conditions. Despite considerable progress in modeling efforts, there are still a large number of problems that cannot be solved using existing continuum models of shock wave response, particularly from a first-principles approach. To mention a few examples, it is still not possible to predict (1) shock wave risetimes in either fluids or solids over a range of pressures, (2) dynamic compressive or tensile strengths of shocked materials, or (3) kinetics of shock-induced phase transitions. Presently, these properties must be evaluated experimentally for use in semiempirical continuum models.

The physical processes associated with high-rate deformation are not well understood, particularly at the different length and time scales that affect macroscopic or continuum properties. In addition, the present state of experimental diagnostics is not to the point that we can routinely and reliably probe deformation processes on a microscopic scale, especially in the picosecond to nanosecond regime. Recent advances in computer capacity and capability [12,13,79, 165] offer 3-D computer simulations as a practical, albeit expensive, tool for probing details of the deformation process [79], but these solutions must be "validated" with experimental data at different length scales. Since we do not presently have the ability to directly probe these scales in real time during shock compression, it will be necessary to infer the actual deformation processes for some time to come from a combination of time-resolved continuum measurements, shock recovery methods, 3-D computational techniques, and newly evolving shock diagnostics.

To truly realize the goal of predictive continuum models, it will be necessary to accurately represent the physical processes that occur in real time during shock deformation. It will also be necessary to critically examine the fundamental postulates used to interpret shock compression phenomena in order to ensure that the correct mathematical framework is used to describe physical processes. Fundamental questions that need to be addressed include:

- Are so-called "steady waves" really steady? What is the mechanism that produces steady waves?

- What are the actual physical and kinematic effects influencing the shock transition?

- Are shock-induced motions purely translational for uniaxial strain loading?

- Are homogeneous, equilibrium states produced immediately after shock transition, and if so, how do we know?

- Can thermodynamic principles be used to accurately interpret material response induced by shock compression?

Answering these questions is a formidable task that will require a dedicated effort on the part of experimentalists, theorists, and numerical analysts. Many researchers have given considerable thought to these issues over the past several decades, which are worth capturing before proceeding with the main thrust of the chapter.

In discussing the significant differences between the fast transformation rate for the fast α–ε transition in iron under shock loading versus the sluggish rate for quasi-static loading Bridgman [25] stated: "It seems to me to be a widely held opinion that transitions involving changes of lattice would be unlikely to occur in times as short as microseconds ... the question of what causes such discontinuities seems to be somewhat obscure...the precise mechanism by which reaching the plastic flow may induce the discontinuity seems not to have been worked out."

Teller [151] provided some of the earliest commentary on the role of localized shear states being responsible for accelerating reaction rates. His comment was prompted by published data of P.W. Bridgman who observed that sugar exploded when compressed hydrostatically to pressures of a few tens of kbar.

Teller postulated that the shear strain produced during compression might be much larger than the displacement of atoms and commented: "It is quite possible that the work performed by the shear stresses make the greatest contribution to the energy store of the substance" and "it would be of interest to find methods by which to measure the energy that has been made available to the substance by the shear stresses."

Kormer [107] was aware of non-homogeneous states being produced during shock compression and stated: "The shock wave is a powerful generator of defects, formed during the strong plastic deformation taking place at the wave front. These disturbances of the ideal crystal lattices, as under normal conditions, determine to a large extent the electrical, optical, and other physical characteristics of the material. The generation of imperfections brings about an acceleration of phase transformations and is the reason for the relatively high conductivity, the absorbing power, and possibly the polarization of shock-compressed dielectrics reported by a number of investigators."

In discussing the effects of shock-induced phase transitions, Al'tshuler [2] made the comment: "The formation of the high pressure phase is preceded by the passage of the first plastic wave. Its shock front is a surface on which point, linear and two-dimensional defects, which become crystallization centers at super-critical pressures, are produced in abundance. Apparently, the phase transitions in shock waves are always similar in type to martensite transitions. The rapid transition of one type of lattice into another is facilitated by non-diffusional martensite rearrangements; they are based on the cooperative motion of many atoms to small distances."

While considering the effect of shock compression on the thermodynamic properties of materials, Graham [57] provided a thoughtful perception of shock processes as follows: "Descriptions of the state of shock-compressed matter have traditionally been based on concepts of either a 'benign shock' or a 'catastrophic shock'. In the benign shock concept the shock transition is considered to carry the material from one equilibrium thermodynamic state to another in a manner analogous to quasi-static compression. In the catastrophic shock concept the shock transition is considered to be a source of considerable lattice disorder, which introduces a dis-equilibrium defect state. In this case, the shock transition carries matter into a non-equilibrium thermodynamics state in which defects have a substantial if not over-riding influence on materials response. Such non-equilibrium thermodynamic processes may not have a counterpart in other environments". Graham's visionary thoughts and comments provide a theme for this chapter.

Swegle and Grady [148] explained the "anomalous" aluminum reshock data in the following way: "The underlying mechanisms governing shock viscosity or the risetime of the shock waves, and their behavior in the shock process are not yet well understood. Viscous-like flow within the shock is thought to be associated with the microscopic processes of dislocation multiplication and motion, twinning, vacancy production, precipitate alteration, etc. Most probably it is a complex event involving the collective participation of several of these mechanisms ... and models, when coupled with numerical wave propagation codes, and with the parameters properly adjusted, have been successful in reproducing wave profile data such as those presented in this study. Such success unfortunately, does not guarantee that either the responsible mechanisms are being modeled, or that the correct physical laws governing the mechanisms have been identified."

In further commenting about shock compression of solids, Graham [60] states: "The overall picture that emerges is one in which shock-compressed mater and shock-compression processes are a great deal more complex than the mechanical measurements indicate. A picture of a homogeneous compression of solid lattices with some modification for the stress tensor to account for the solid-state aspects is simply insufficient. It is excessively naive. From the physical and chemical property viewpoint, the influence of large concentrations of defects from atomic to mesoscopic to macroscopic dominate the description."

In a more recent discussion of dynamic material models for describing shock deformation processes in polycrystalline metals, Panin [136] says: "Numerous attempts to link organically the dislocation theory to the continuum mechanics have been futile. It has long been believed that this is associated with purely mathematical problems of the macroscopic description of the collective behavior of strain-induced defects. However, recently it has become evident that our understanding of the basic act of plastic deformation is not true and is described by an inappropriate method. This has resulted in an inadequate description of the behavior of ensembles of strain-induced defects and inaccurate microscopic interpretation of the phenomenological relationships governing the continuum mechanics."

We believe that these thought-provoking comments convey the necessary perspective and identify some of the principal issues to be resolved. They also consistently support the theme that "accepted ideas of homogeneous, uniform states produced by shock compression are not strictly correct and therefore models based on these simple concepts are not adequate" [136]. Unfortunately, these ideas have not yet been translated into practical solutions. However, several emerging ideas [12,13,79,113,132,136,165] are important steps in examining our thinking about these fundamental procesess.

In this chapter, we present experimental and numerical evidence that shows major departures from accepted ideas of shock compression. The data we discuss are "anomalous" if the normal assumptions are rigorously used to either interpret shock data or to develop macroscopic material models of shock deformation. For example, mathematical treatments of planar shock loading of solids are usually based on variants of the elastic–plastic model with the assumptions that the shocked material undergoes uniaxial motion, is homogeneous, and that the strain tensor can be separated into elastic and plastic components. Furthermore, it is postulated that the total volumetric strain induced by the shock can be separated into elastic and plastic components. We will show that these assumptions are only approximately true in most cases and are far from adequate in others. Specifically, the presented results indicate that localized deformations occurring on a sub-continuum scale dominate mechanical and physical states of the shocked material. This observation partially explains several anomalous effects, including loss and recovery of material strength during shock compression. Another question concerns the fundamental material processes causing the shock transition itself. Are properties, such as plastic viscosity, the controlling factors [27] or are kinematic effects, such as localized particle rotations and dispersion, responsible for the observed risetimes [165]? These questions must be answered by thorough investigations of the deformation processes.

To systematically answer these questions, future objectives of shock wave research should include development and application of shock physics capabilities to the explicit problem of multiscale effects of dynamic deformation [24]. This will require a reformulation of the basic conservation equations to include the non-homogeneous and non-equilibrium effects occurring during shock compression. To emphasize this point, we have taken the approach of using reported shock compression experiments and results from numerical simulations to show that the existing assumptions and interpretations of material response are only approximate in several applications. In particular, we provide experimental evidence that highly non-equilibrium states are produced during and after shock loading, leading to time-dependent effects in the shocked state. This observation is also apparent in recent 3-D simulations of shock propagation in polycrystalline materials, which further elucidate the non-uniform and non-equilibrium processes occurring during shock compression. These results support the assertion that many of the assumptions mentioned above are incomplete and suggest that a paradigm shift is necessary to change the way we think about the shock process [136].

The basic question concerns the dynamic processes occurring in the shock transition. In this regard, we found it useful to take a specific material and examine the predictions of various models for a variety of loading conditions, including initial shock loading, unloading, and reloading from the shocked state. This approach provides insight into the mechanisms that may be responsible for the observed effects produced after shock loading and by inference information on the processes produced during shock loading. We use the term "may" because without specific measurements, we can only infer what is happening during the shock transition. Through this approach we raise basic questions about the assumptions used to model shock response at the continuum level. We also discuss evolving capabilities that may help understand these issues in the future. These include advanced experimental methods for examining material response from the atomic to continuum scales. With the ever-advancing computational capacities and capabilities, three-dimensional computer simulations of shock processes become viable for complementing experimental studies of these processes. The technical problems to be solved delve into the basic physical and chemical processes of shock compression and should constitute a significant thrust of modern-day research. These challenges provide exciting opportunities for the next-generation researcher who is willing to address difficult physical, chemical, and kinematic effects occurring during shock compression.

The chapter is organized as follows. We first provide a very brief discussion of the basic assumptions of shock propagation in fluids and solids. We present experimental results that challenge these assumptions, followed by numerical simulations that elucidate possible deformation mechanisms. Finally, a discussion of new experimental approaches that may be useful for providing additional information about the deformation processes is given.

2.2. Basic Assumptions

Our intent is not to give an in-depth review of shock compression formalism, since there are many good references in this regard [for example, 11,15,21,39, 44,81–83,85–87,131,172]. Instead, the goal is to show several principal results of shock wave research that cannot be easily interpreted with the conventional assumptions. These examples therefore illuminate the principal issues that must be resolved for future progress in model development. We first provide a brief review of the traditional approaches for describing shock waves in fluids and solids in order to establish a basis for discussing shock wave data that appear to be "anomalous" from the perspective of existing assumptions. This helps provide the reader with a context for evaluating the experimental and computational results that we present later.

It is first useful to define the different scales of shock deformation that will be discussed. Complex, real dynamic events are for the most part described at the macroscopic or continuum level for multidimensional numerical simulations. Existing continuum models used in computer codes are essentially phenomenological and generally employ simple descriptions of material response to represent physical processes such as dynamic strength effects or dynamic phase transformations. However, it is often speculated that dynamic failure and phase changes nucleate at atomic and/or microscopic scales and grow to larger dimensions that cannot be approximated as an ensemble at an atomic scale during the passage of a shock wave [165]. This requires that we understand material response over the full range from the finest, i.e., atomic level, the microscopic (e.g., dislocation) level, the sub-granular, or mesoscopic level, and finally leading to the continuum level. The main inference is that deformation in a shock wave is complex, involves different length scales, and therefore needs to be understood at all levels.

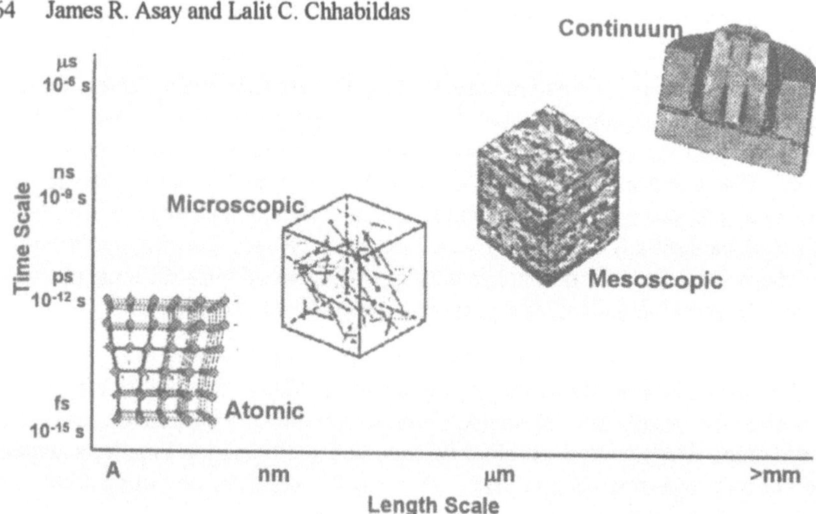

Figure 2.1. Scale lengths representing dynamic material deformation. The scales shown for the different processes are not rigidly fixed, but are meant to convey general regions of spatial and temporal response.

The relative length and time scales for the various scales are illustrated in Fig. 2.1. Although the definition of these scales is somewhat arbitrary, they provide a context for later discussions. More detailed discussion of the physical significance of these length scales can be found in the literature [69,112,113, 120,134–136]. There is significant research that probes material response at the atomic, microscopic, and continuum levels. In these studies, dislocation models have been developed that incorporate physics at an atomic level, which usually assumes that the dislocation distributions are homogeneous on a continuum level. Material models developed to describe macroscopic or continuum response also assume that any local deformations are much smaller than the continuum length scale, as illustrated in Fig. 2.1, so that dissipative mechanisms can be represented using continuum concepts such as viscous damping. The mesoscopic scale is intermediate between the microscopic and continuum scales and is probably the most challenging to study. This scale is not well defined, but includes deformation features that must be explicitly recognized to develop phenomenology at macroscopic scales. Mesoscopic scale descriptions usually cover dimensions of tens to hundreds of microns. Examples include individual grains in granular or polycrystalline materials and relatively large inclusions in metals or composite materials. This scale could even include deformation features induced by the shock itself, such as micro-slip bands or twins in single crystals [113]. Although these features could be explicitly treated in 3-D computer codes, it is desirable to represent the effects of mesoscopic scale motions into continuum models through the proper mathematical formalism.

The extent of systematic shock wave research at the mesoscopic scale is extremely limited, even though there is substantial evidence that this scale has a dominating influence on shock response. Besides, it is this scale that will naturally bridge the physical phenomenology occurring at smaller scales to the observed response at the full-scale continuum level. It should be clear from the results presented that the potential development of a full predictive capability for shock loading hinges on a realistic description of material response at the mesoscopic level for a large class of materials.

2.2.1 Shocks in a Fluid

The simplest description of a shock wave involves propagation in a fluid. In most shock wave investigations, particularly for equation-of-state (EOS) applications, the shock is assumed to be a mathematical discontinuity; however, this is known to be approximate since real material effects such as visco-plasticity result in finite shock wave risetimes [15,27,34,81–83,148]. A large number of papers have been written on the topic of shock propagation in fluids [15,21,39,80,85,131,146], so only a brief description will be given here. For 1-D planar motion, the principal assumption is that the shock is a steady transition from an initial equilibrium thermodynamic state to a final equilibrium state, as depicted in Fig. 2.2. For planar shock motion, it is assumed that a constant pressure P is applied over a large area of a specimen initially at an initial pressure P_0, specific volume (reciprocal density) V_0, internal energy E_0, and particle velocity u_{p0}, which produces uniaxial compression in the x direction. The pressure loading induces a planar wave or shock front propagating in the x direction. That compresses the fluid to pressure P, volume V, internal energy E and particle velocity u_{p0}. It is further assumed that the final compressed state is homogeneous

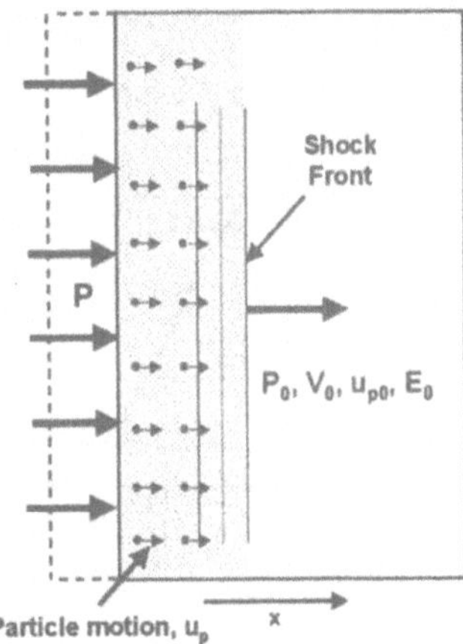

Figure 2.2. Planar shock propagation in a fluid. Both the initial and final states are assumed to be in thermodynamic equilibrium. The dashed lines indicate the initial size of the compressed region.

and that the fluid motions induced by the shock front are strictly translational in the direction of motion. If the wave motion is steady in time, the final pressure state produced by shock compression lies on a curve referred to as the principal Hugoniot [39], which is the locus of end states produced by steady shock compression from the same initial state.

If a strict discontinuity in pressure is imposed on one surface of a planar sample, it is assumed that a steady planar shock wave structure will evolve because the dissipative mechanisms controlling shock transition are balanced by nonlinear fluid response. The evolution distance for a steady shock structure to form is determined by the magnitude of the dissipative and non-linear effects, but is usually not detectable for step loading of fluids at pressures of a few GPa (1 GPa equals 10 kbar). Once a steady shock is formed, it will propagate unchanged in structure, as illustrated in Fig. 2.3. The isentrope shown in Fig. 2.3a represents the pressure–volume states obtainable by a reversible, adiabatic process achieved by continuous pressure loading in an inviscid fluid. If a continuous increase in pressure is applied to the surface rather than a step loading, a steady shock will still evolve to the same structure induced by step loading. This process is illustrated in Fig. 2.3b. Although we have idealized the shock formation process for the purpose of this discussion, Courant and Friedrichs [34] have shown that mathematical formulation of the steady wave solutions in fluids involves idealizing assumptions about the existence and uniqueness of the solution and its stability. The reader is referred to this article for an in-depth discussion.

When a steady shock structure is formed, the resulting pressure, density, and energy states will lie on the shock Hugoniot curve [11,39]. These states are determined through the conservation laws for mass, momentum, and energy that connect the initial and final states through the following equations:

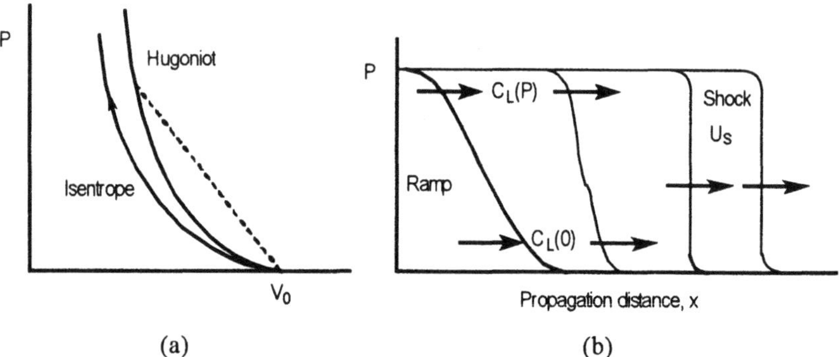

Figure 2.3. Conditions for formation of a steady shock wave from a planar ramp wave input. (a) Isentropic loading curve and Hugoniot curve for a fluid. (b) Evolution of a ramp wave into a steady shock wave. The Lagrangian soundspeed c_L increases with increasing pressure, which causes a steepening of the initial ramp into a shock.

$$P - P_0 = \rho_0 U_S (u_p - u_{p0}),\tag{2.1}$$

$$\frac{\rho}{\rho_0} = 1 - \frac{u_p - u_{p0}}{U_S - u_{p0}},\tag{2.2}$$

$$E - E_0 = \frac{1}{2}(P + P_0)(V_0 - V),\tag{2.3}$$

where P is the hydrostatic pressure, U_S the shock velocity, u_p the mass or particle velocity, ρ the density, V the specific volume $(1/\rho)$, and E the specific internal energy; the subscript 0 refers to the initial state. These equations are based on the assumptions that the fluid motion is one-dimensional and that the transition is from one thermodynamic equilibrium state on the EOS surface to another equilibrium state. Sometimes, this process is referred to as a "benign shock" transition [60,61], which has been a long-standing basis for using shock compression to determine the high-pressure equations of state [11,39]. These relations have been used to produce equations of state for a large variety of materials, including explosives [11,38,60,142]. In many applications, a linear relationship is observed between the shock velocity and particle velocity that allows solution of Eqs. 2.1–2.3.

The internal energy increase in a fluid due to an irreversible shock compression process consists of recoverable energy that results from reversible adiabatic compression and the thermal energy generated by dissipative processes occurring in the shock deformation [39,44,131,147]. The energy dissipated by the shock is usually assumed to be the shaded area shown in Fig. 2.4—the area between the straight line connecting the initial and final states and the Hugoniot curve. For steady shock compression, material compression follows the straight-line path referred to as the Rayleigh line. As indicated in Fig. 2.4, a point $(P,'V')$ during steady shock compression lies on the Rayleigh line and is given by

$$P' = P_0 + \rho_0 U_S^2 \left(1 - \frac{V'}{V_0}\right),\tag{2.4}$$

provided that the particle velocity ahead of the shock is zero.

It is important to note that the graphical concept of steady waves and the corresponding loading paths result from the continuum concept of one-dimensional steady-shock compression in fluids. As we will see in the next section, shock propagation in solids results in considerable deviation from this idealization.

2.2.2. Shocks in a Solid

The relatively simple ideas of shock propagation in fluids become considerably more complex in solids. For uniaxial loading, the state of compression is represented by a stress tensor with principal components of σ_x in the longitudinal

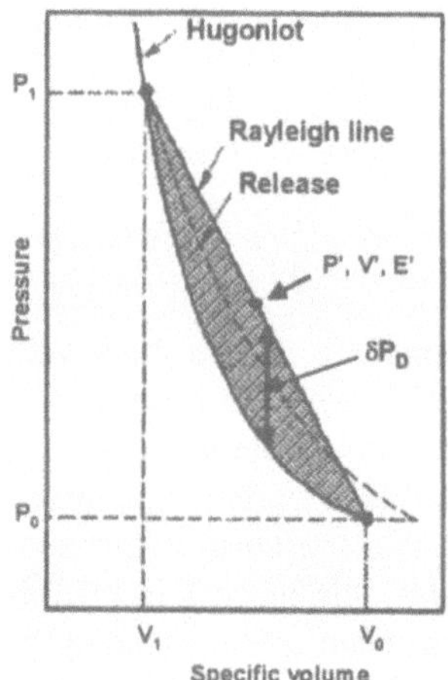

Figure 2.4. Energy dissipated in a steady wave. The Raleigh line connects the initial and final states of the shock in the pressure-specific volume plane. The shaded area is approximately equal to the energy dissipated during shock compression.

direction, and σ_y and σ_z in the lateral directions. For isotropic materials, the lateral stress components σ_y and σ_z are equal. For stresses below the onset of dynamic yielding, the dynamic uniaxial strain response is entirely elastic with the result that a single elastic shock is produced. The jump conditions Eqs. 2.1–2.3 apply to solids as well, with P replaced by σ_x [39]. At stress levels greater than the dynamic yield stress, also defined as the Hugoniot elastic limit (HEL), the deformation is inelastic and exhibits a rich variety of features that are dominated by rate-dependent deformation. For stresses up to a few times the HEL, a two-wave structure that evolves with time is observed in most metals, as illustrated in Fig. 2.5a. This stress regime is usually referred to as the weak shock case [166,167]. Ideally, the first shock is purely elastic and is referred to as the *elastic precursor* for historical reasons. The HEL is generally not constant with propagation distance and usually decays with propagation distance due to rate-dependent deformation mechanisms. A large number of papers dealing with elastic precursor decay have been written and the reader is referred to several key papers on the subject [6,7,65,66,90,94,148]. Reasonably good rate-dependent yield models based on dislocation dynamics have been developed to describe elastic precursor decay. For solids, a major fraction of the dissipated energy illustrated in Fig. 2.4 is converted to plastic work and stored as an increase in heat, with the remainder as cold work stored in crystal defects [52,73,140]. The exact details of the partition in energy are not always known and depend on the details of the deformation process [52,73,140].

The second wave representing inelastic or dissipative deformation is generally referred to as the "plastic" wave. However, several investigators have shown

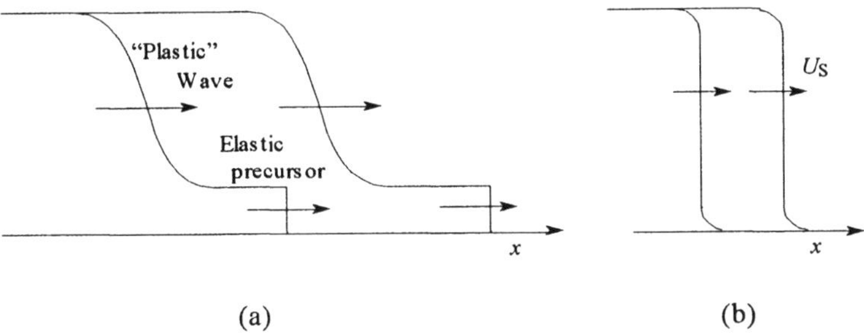

Figure 2.5. Elastic–plastic wave propagation in solids (a) Weak shock propagation. (b) Strong shock propagation.

that the notion of a plastic "wave" is not representative of the actual processes occurring in the wave [52,82,166,167]. The mathematical equations describing this deformation are parabolic solutions, which are representative of diffusion processes rather than hyperbolic solutions describing wave propagation. Gilman [52] argues that energy is not conserved during plastic deformation, so that an energy conservation function cannot describe this process. Thus, the notion of a plastic wave conveys an erroneous impression. Nevertheless, it is now common terminology to identify the second disturbance as a "plastic wave". Because the focus of this chapter is on other issues related to whether the basic assumptions apply to these processes, we will use this terminology.

There are basic issues involved in how to best describe the actual processes of deformation above the elastic limit. As we will discuss later, there is considerable evidence to show that the traditional notions of "plastic" response for purely translational motions should be replaced by more realistic descriptions of the deformation [113]. A final point is that even though the two-wave structure shown in Fig. 2.5a is not steady in a coordinate system moving at the particle velocity behind the first wave, the second or plastic wave can be steady in that system, so that the jump conditions given earlier can be used for this wave [89]. For sufficiently high shock stresses, the elastic precursor is over-driven and a single plastic wave is formed as shown in Fig. 2.5b. Wallace [166–168] treats the flow processes occurring for these two cases in great detail.

For later discussion, we summarize the basic assumptions commonly used to describe elastic-plastic response for isotropic solids at the continuum level:

1. shock deformation is homogeneous at a spatial scale large compared to the source of the plastic dissipation,

2. the total strain can be separated into elastic and plastic components with no plastic dilatation, and

3. hydrostatic pressure behind the shock can be approximated as the mean value of the longitudinal and lateral stresses,

4. the resolved shear stress in the shocked state is one half the difference between the longitudinal and lateral stresses.

Using these assumptions, the longitudinal stress in a shock experiment is equal to the pressure plus four-thirds of the resolved shear stress. For longitudinal wave propagation in the x direction, the above assumptions can be expressed mathematically as:

$$\varepsilon_x = \varepsilon_x^e + \varepsilon_x^p, \tag{2.5}$$

$$0 = \varepsilon_x^p + \varepsilon_y^p + \varepsilon_z^p, \tag{2.6}$$

$$\sigma_x = P + \frac{4}{3}\tau, \tag{2.7}$$

$$\tau = \frac{1}{2}\left(\sigma_x - \sigma_y\right), \tag{2.8}$$

where ε_x is the longitudinal strain defined as $\varepsilon_x = 1 - (\rho_0/\rho)$, σ_x is the stress in the x-direction, σ_y and σ_z are the corresponding lateral stress components and are equal for isotropic solids, and τ is the resolved shear stress. Equations 2.5–2.8 are the basis for describing the elastic–plastic response of solids [14,166–168].

Many studies have shown that these assumptions yield the approximate continuum response. However, there are also many counterexamples, which show they can confuse and perhaps even limit the understanding of dynamic material response. In particular, Gilman [52] and others have questioned these assumptions and in particular the additive assumption of elastic and plastic strains. Gilman argues that elastic and plastic strains are not additive because elastic strain results from a strain energy function, whereas the plastic strain does not. Graham [60,61] questions the validity of describing the anisotropic stress state as a perturbation to the pressure state. It is important to keep in mind that although these assumptions represent the overall features of shock wave evolution and propagation, they may be misleading, divert our thinking away from the real deformation processes that occur, and lead to some of the problems we will discuss [113,136]. In the following sections we will address the questions:

1. Is the assumption of planar motion an adequate description of 1-D shocks?

2. Are homogeneous, equilibrium states produced by steady shocks?

3. Is the traditional emphasis on mechanical deformation sufficient to describe real shock processes?

4. Do risetimes of "plastic" waves represent physical deformation mechanisms, such as viscoplastic mechanisms, and

5. Are existing models of solid response, e.g., the elastic-plastic model, adequate descriptions of real material response?

The later discussions regarding these assumptions will show that for polycrystalline metals:

- The stress state behind shock waves in solids is not generally homogeneous or uniform;

- Shock-wave structures are non-planar in simple planar experiments usually assumed to be planar motion;

- Viscoplastic or other dissipative mechanisms may not be the dominant factor controlling the shock wave structure;

- The shock process may more appropriately be thought of as an ensemble of stationary waves that collectively form a stationary structure;

- Deformation features at the mesoscopic scale may lead to material properties, such as compressive yield behavior, that are time dependent, in contrast to representations of the elastic-plastic model; and

- Different material descriptions can be used to describe similar shock wave features both at the continuum level and at the mesoscopic scale.

2.3. Specific Examples

The advent of time-resolved techniques for measurements of shock wave profiles in the 1960s [30,58] has been a significant development because for the first time this technique allowed investigators to probe the rate-dependent response of shock deformation at the continuum level. A variety of time-resolved gauges have been developed, including stress gauges, interferometers, spectroscopy, and capacitive gauges [30], to investigate a large variety of materials. In particular, Barker [16] was the first to use time-resolved interferometric techniques to investigate steady-wave evolution in a 6061-T6 aluminum alloy. His work was a comprehensive study of shock wave structure for different peak stresses and different thickness that determined the formation and evolution of the two-wave structure discussed earlier. These profile measurements provided critical data to test various continuum models proposed for the dynamic viscoplastic response of aluminum.

A summary of Barker's results is illustrated in Fig. 2.6. Barker showed that, to within the experimental resolution, the "plastic waves" were essentially steady over the stress range of 9–90 kbar (in the moving coordinate system established by the elastic precursor). Several investigators [71,89] subsequently used these data to develop many of the continuum concepts of rate-dependent plasticity in the low-pressure stress range. Grady [54] observed from these and other experiments that the risetime of the "plastic" wave varied inversely as the fourth power of the Hugoniot stress and that this relation was apparently applicable to a broad spectrum of metals and non-metals. Other researchers [43,156] have used similar concepts to demonstrate the generality of the fourth power relation. There have been numerous attempts to explain this relationship within the context of dynamic viscosity or other dissipative mechanisms. However, there is as yet no fundamental understanding for this relation.

2.3.1. Dynamic Material Models

The extensive time-resolved wave profile measurements made over the past forty years have also allowed development of numerous dynamic material models of shock compression. Generally, these models are based on a Maxwellian formalism, in which the dissipative term is described by the difference in stress from an equilibrium stress value [15,21,71,72,131,146]. Generally, these models do a relatively good job of reproducing wave profile data, as illustrated in Fig. 2.7. Also shown in the figure are calculated profiles reported by Herrmann and Lawrence, [72] using a constitutive equation based on a Maxwell construction, expressed as

$$\dot{\sigma}' = \frac{4}{3}G\frac{\dot{\rho}}{\rho}\left[a\left(\frac{\sigma - \sigma_e}{\sigma^*}\right)^n\right],\qquad(2.9)$$

where σ' is the stress deviator, G is the shear modulus, and σ is the longitudinal stress. This constitutive model, when implemented in a 1-D wave code [71,72], gives very good agreement with the experimental data, again illustrating the ability of continuum models to describe the plastic deformation profiles.

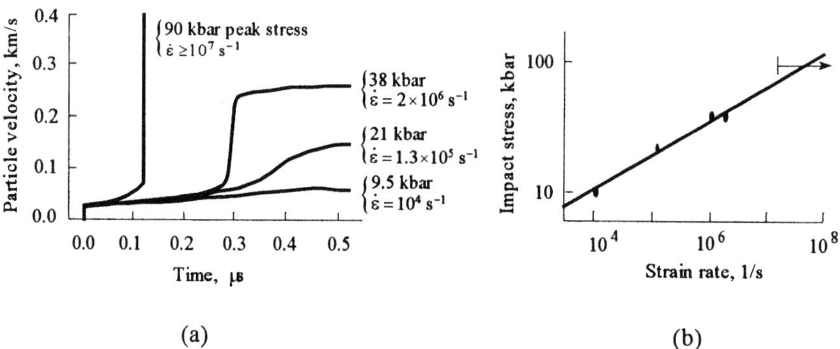

Figure 2.6. Plastic wave profiles in aluminum [16]. (a) Dependence of the plastic profile on shock pressure. (b) Dependence of the shock stress versus inverse time (related to strain rate).

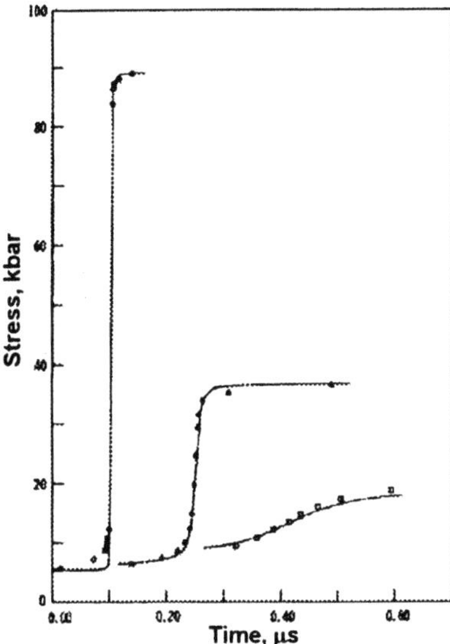

Figure 2.7. Comparison of a continuum model of dynamic deformation with steady wave profiles in 6061-T6 aluminum [71,72]. The particle velocity records reported by Barker [16] are transformed into stress wave profiles using an impedance-matching technique. The initial stress for each record corresponds to the amplitude of the elastic precursor.

Herrmann's model falls within a general class of viscoplastic models for describing "plastic" wave propagation in metals. We will not discuss these in detail and the reader is referred to the literature, including a recent investigation by Kanel' and coworkers [99]. Essentially all of these models can reproduce the wave profile data such as Barker's experiments rather well. Grady [56] has recently shown that this class of models can be described in a general form as

$$\dot{\sigma} = \sigma'_e \dot{\varepsilon} + k(\varepsilon)\dot{\varepsilon}^m - \frac{1}{\tau}[\sigma - \sigma_e(\varepsilon)], \qquad (2.10)$$

where the prime represents a spatial derivative and $k(\varepsilon)$, τ and m are adjustable parameters [56]. The limitation of these models is that they are usually not predictive for applications outside the range they were developed in. Furthermore, many of these models generally will not provide an accurate description of material responses other than steady shock loading, including shockless loading or off-Hugoniot loading, as we will show.

More physically based material models have been developed to describe the time-resolved wave profile measurements. Specifically, dislocation theory provides a foundation for developing microscopic-level deformation models. Johnson and Barker [89] applied a functional form of dislocation dynamics based on the Orowan equation first used by Taylor [149] to describe elastic precursor decay in metals and applied it to Barker's data. Their dislocation model can be expressed as

$$\dot{\sigma} - \frac{1}{\rho_0}(\lambda + 2\mu)\dot{\rho} = \frac{8}{3}\mu b v(\tau) N_{\mathrm{m}} , \qquad (2.11)$$

where λ and μ are the Lamé constants, b is the Burgers vector, $v(\tau)$ is the dislocation velocity, and N_{m} is the mobile dislocation density. They used reasonable values of dislocation parameters to describe the "plastic" wave profiles and obtained good agreement with the profiles measured by Barker, as illustrated in Fig. 2.8. The use of dislocation models to model continuum wave profiles demonstrates that microscopic descriptions of plastic flow can accurately describe steady shock wave evolution. However, as with the phenomenological models discussed above, these microscopic descriptions are also not predictive when used outside the stress regime over which the parameters are adjusted or if the initial material properties are changed.

The two modeling examples above highlight the main problem that exists in trying to understand shock deformation processes. Representation of the data by different models with a single set of experiments illustrates the non-uniqueness of these solutions. Although time-resolved shock wave profile techniques provide useful information about deformations at the macroscopic scale, they are not sufficient to uniquely determine actual deformation mechanisms. It is also important to re-emphasize that both the continuum and dislocation models discussed above assumed homogeneous shock response on a scale large compared to the deformation features (submicron scale). The analyses also implicitly assumed that the deformation was one-dimensional at all deformation scales, that elastic and plastic strains were separable, and that the stress tensor could be separated into hydrostatic and shear stress terms.

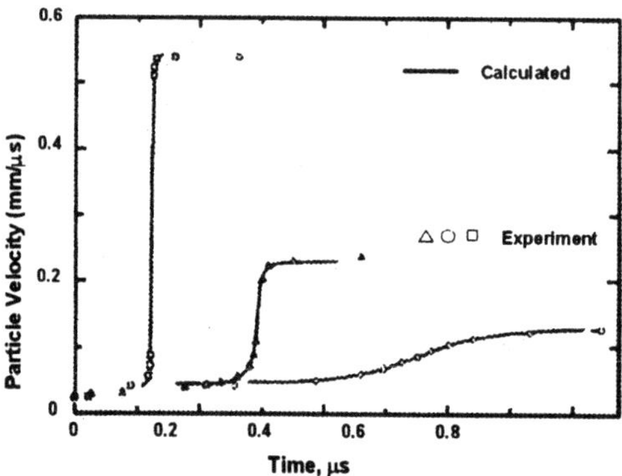

Figure 2.8. Use of a dislocation model to describe steady waves in as a function of driving stress. The wave profiles have been normalized to the initial amplitude of the elastic precursor, which is about 6 kbar. After Johnson and Barker [89].

2.3.2. Testing the Basic Assumptions

It is becoming obvious that additional information, such as real-time, *in situ* measurements of the deformation is needed to discriminate between proposed models. Because this capability is not available, another approach was used in the 1970s to investigate the state after shock compression and to thus provide a more rigorous test of the continuum models. This was accomplished by first passing a shock wave through a material and then subjecting the shocked material to further unloading or reloading from this state. These tests can then be used to determine if the existing models realistically represent processes occurring during shock compression. As a specific example, we will illustrate the approach in the context of the elastic–plastic model discussed earlier and show that a comparison of measured unloading and reloading wave profiles with predictions of this model indicates significant differences that can be attributed to physical processes occurring in the shock.

The simple elastic–plastic model predicts that the initial unloading from a shocked state should be elastic, followed by purely plastic deformation. For reloading from the shocked state, the expected response should be entirely plastic because the initial shock compression beyond the HEL produces a material state on the yield surface. These experiments therefore make it possible to infer the deformation process occurring during shock deformation. More complex elastic–plastic response, such as isotropic work hardening, produces an increase in the yield strength, but the compression state is still assumed to remain on the yield surface [10,47,111].

The basic experimental configuration for performing unloading and reloading from the shocked state is shown in Fig. 2.9. In these experiments, a planar impactor is backed with either a low-impedance or a high-impedance flyer plate, as illustrated in Fig. 2.9a. Upon impact, a shock is formed in both the sample and the impactor. Reflection of the shock from the backing flyer-plate material produces either a rarefaction wave to propagate into the shocked state or a reshock wave into the shocked state, depending on its shock impedance. The wave structures expected for these cases are illustrated in Fig. 2.9b. For unloading, an elastic–plastic wave structure is predicted from the E–P model so the initial unloading is elastic until reverse yielding occurs. For reloading from the shocked state, a "plastic" reshock is predicted from the model (shown in the figure with a finite risetime because of plastic viscosity).

Lipkin and Asay [111], and later Asay and Chhabildas [10], used this technique to evaluate the accuracy of an elastic–plastic model for aluminum. One set of measured unloading and reshock profiles [10] is reproduced in Fig. 2.10. The unloading wave profile illustrates that initial release from the shocked state deviates from the simple elastic–plastic model discussed above, resulting in a quasi elastic–plastic behavior usually referred to as a "Bauchinger effect". Re-

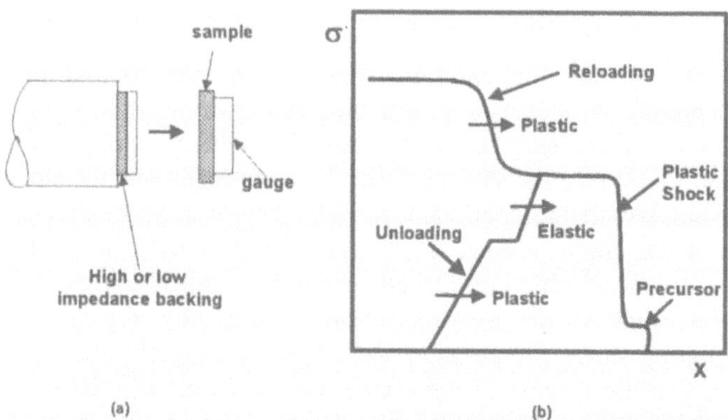

Figure 2.9. Experimental technique for producing shock loading followed by unloading or reloading. (a) Experimental configuration. Subsequent unloading or reloading from the initial shocked state is obtained through choice of the flyer plate backing material shock impedances (b) Expected wave structures for unloading and reloading from the initial shocked state.

compression from the shocked state illustrates an even larger deviation from the expected response, since an elastic precursor is observed prior to the main "plastic" shock. The observation of significant elastic stresses implies that the assumption of a pure plastic state produced during initial impact is not entirely correct. Although rate-dependent elastic–plastic response can be a contributing factor in explaining the elastic recompression response, Lipkin and Asay [111] found that these effects were not sufficient to explain the measured elastic re-compression.

The stress–volume response corresponding to the wave profiles shown in Fig. 2.10 is illustrated in Fig. 2.11. The dashed line is a yield surface consistent with the simple elastic–plastic model [47]. The solid line represents an assumed experimental loading path corresponding to the data in Fig. 2.10 that deviates from this simple elastic–plastic description [10,111]. One interpretation of the data is that the final stress state, σ_0, produced upon first shock loading lies within the yield envelope due to as yet unknown mechanism occurring during initial loading. Upon recompression from this state, deformation will therefore be quasi-elastic until the yield surface is reached at point "A". The experimental data also show that unloading deviates from the ideal elastic–plastic model described by the dashed line. It should be noted that unloading experiments by themselves would not have highlighted a discrepancy with the elastic–plastic model. We will discuss two possibilities for the observed reloading deviation from the perspective of the basic question of: "what is a shock?", which suggests that the von Mises yield criterion commonly used in the elastic–plastic model may be too simple to describe the observed recompression effect.

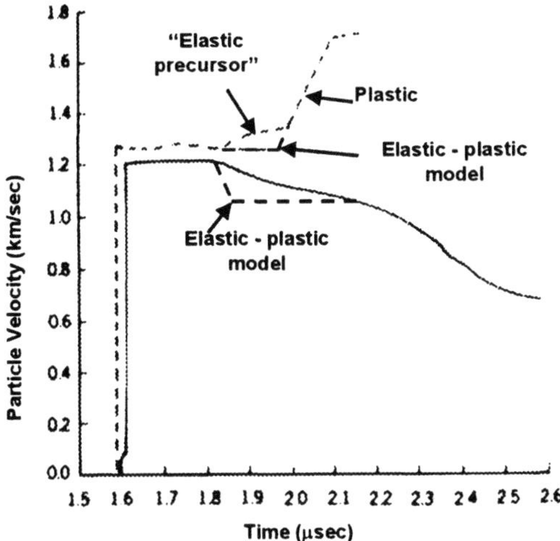

Figure 2.10. Experimental particle velocity profiles for unloading and reloading in aluminum. The initial stress state is about 230 kbar [10]. Two different experiments are shown.

The experimental results for shock loading/unloading/reloading can be used to determine the initial shear stress, τ_0, in the shocked state and the maximum resolved shear strength, τ_c for the material. τ_0 can be thought of as the amount of shear stress the material can support immediately after shock compression, whereas τ_c is the equilibrium value of the shear strength long after shock compression. If the shock produces localized shear states that vary throughout the crystal, or if there are time-dependent processes to the evolution of the critical shear strength in the shocked state, the shear stress and the maximum shear strength can differ at the stress level σ_0. We will discuss the possibilities of these two cases. Furthermore, there are a variety of texture models that have been developed that are also consistent with this observation [127,129].

Asay and Chhabildas [10] developed a self-consistent technique that allows estimation of the shear stress in the initial shocked state and the critical shear strength. Figure 2.12 summarizes the variation of these experimental quantities with shock stress, along with recent data by Al'tshuler et al. [3], who used a lateral stress gauge to determine these states independently. Ideally, τ_0 and τ_c should be equal for the elastic–plastic model, as we discussed earlier. For weak shocks (elastic–plastic two-wave structure), the experimental results [111] indicated that the elastic–plastic assumption is nearly valid, since the measured initial shear stress state was nearly equal (but not exactly, as we will see later) to the critical strength. However, for the strong shock case, i.e., over-driven elastic precursor, it was found that the initial shear stress, τ_0, produced during first shock

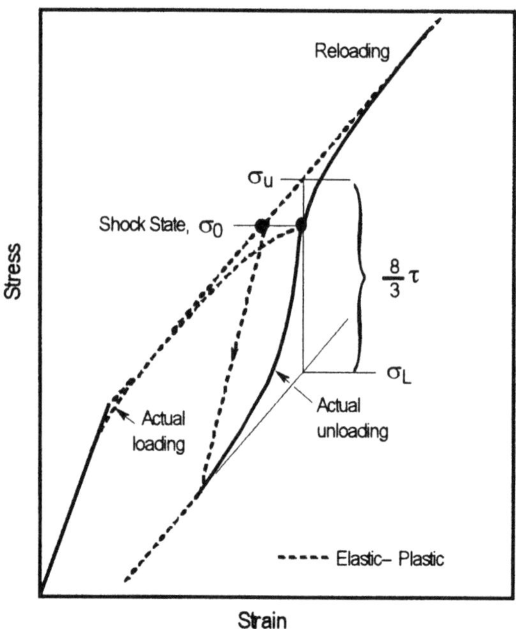

Figure 2.11. Stress-strain path implied by shock loading, unloading and reloading of aluminum [111]. The solid line is an idealized elastic–plastic model while the dashed line represents a possible interpretation of the experimental results.

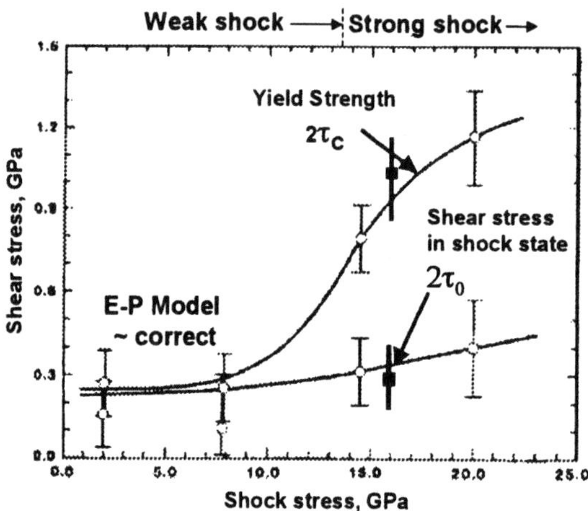

Figure 2.12. Shear stress states and shear strength versus shock stress in aluminum (open circles, [10]; solid squares, [3]). The observed deviation of the initial shear stress from the shear strength indicates a discrepancy with the elastic-plastic model.

loading was significantly less than the maximum shear strength, τ_c, in the shocked state [10]. In particular, for shock stresses approaching 20 GPa (200 kbar), the initial shear stress immediately after shock loading is only about one third of the shear strength the material is capable of supporting. This observed deviation of shear stress from shear strength, which is indicative of an initial softening effect followed by recovery later in time, suggests that the basic assumptions discussed earlier are not fully correct.

The apparent softening or decrease of the shear stress during shock compression is not restricted to polycrystalline aluminum. Experiments conducted on a variety of polycrystalline metals with varying crystal structures, including tungsten [31], beryllium, [29] and copper [28], have shown similar deviations of the initial shear stress and the critical shear strength. In addition, recent shock experiments on alumina, a high strength ceramic [139], also indicate similar effects. A basic question that must be asked concerning this discrepancy is: Is the assumption of a homogeneous and uniform state produced by the shock process correct? We will address this question in subsequent sections.

2.3.3. Heterogeneous Deformation

One of the possibilities for the observed discrepancy in the E–P model is that the deformation state is not uniform throughout the material during the shock compression process. Experimental evidence for localized deformation states produced by shock loading has been recognized for several years [4,53]. Beginning in the 1970s, several researchers [123–125,127] used 2-D numerical simulations to show that wave propagation in polycrystalline metals can be very complex, producing transient effects that result in non-homogenous deformation. In particular, these early simulations emphasized that multiple wave reflections within the polycrystal result in time-dependent stress distributions behind the shock. The reader is referred to the work of Meyers and colleagues for a discussion of these effects. More recent numerical simulations by Holian and Lomdahl [78] and Yano and Horie [164] using modern computational tools demonstrate these effects in more detail and demonstrate that localization of deformation results in dispersion of properties in the shocked state. Furthermore, there has been a significant number of metallurgical examinations on shocked materials that illustrate a rich variety of deformation features (see Curran et al. [35] and Gray [62] for recent discussions).

Lipkin and Asay [111] independently suggested a mechanism along these lines for explaining the apparent discrepancy with the E–P model by invoking a distribution of shock-induced shear stresses induced within the polycrystalline sample. They suggested that the multiple wave interactions within the polycrystalline sample could lead to a distribution of shear stress states after shock loading. They used a Gaussian probability distribution function (PDF) to represent this assumed non-uniformity of shear stress states as illustrated in Fig. 2.13a, where the gray regions within polycrystalline grains are meant to schematically represent differences in shear stress states behind the shock front. Their proposed distribution function is shown in Fig. 2.13b. This concept was implemented as a mixture-type model in a one-dimensional code [72] and used to simulate the unloading and reloading experiments on aluminum.

Figure 2.13c illustrates the results of these simulations in comparison to experiments for an initial shocked state of about 20 kbar (the initial elastic–plastic loading profile is not shown in the figure). The parameters in the PDF were arbitrarily varied until reasonable agreement was achieved between experiment and calculations as shown in Fig. 2.13c. As illustrated, relatively good agreement could be achieved between the simulations and the experimental results. This suggested that the shear stress state and hence the shocked state is not homogeneous with respect to shear stress states. These investigators and others in the United States did not continue to pursue this experimental/modeling approach because of the ad hoc nature of the model and because of the inability to independently confirm the hypothetical distribution of shear states.

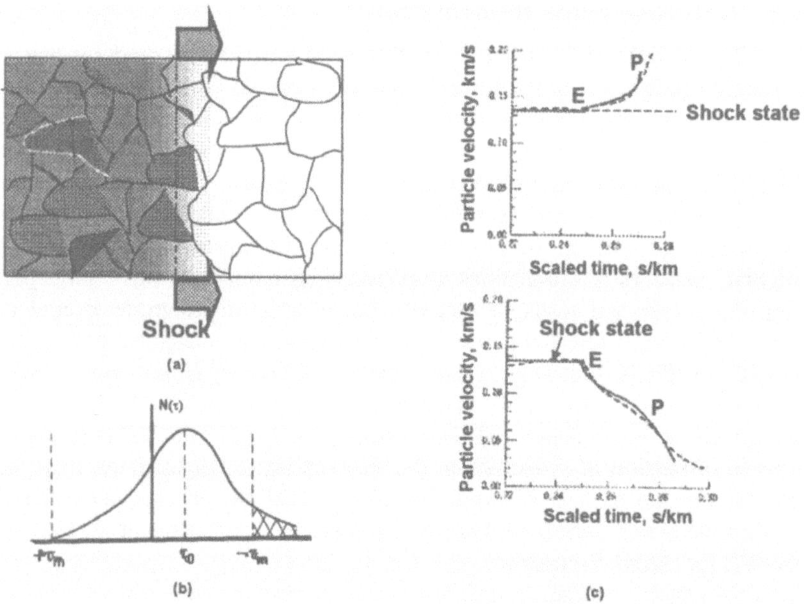

Figure 2.13. Shear stress distribution in shocked aluminum and the resulting wave profile calculations. (a) Schematic of shear stress states in a polycrystalline sample using different shades of gray to denote different shear states. (b) A probability distribution function which was assumed for the shear stress distribution. (c) Calculated (solid lines) and experimental reloading profiles (dashed lines) from an initial shocked state of about 20 kbar [111]. The initial shock profiles are not shown in this figure.

Although the existence of non-uniform stress states partially explains the principal features of the unloading and reshock experiments [111], it is not clear if this description is correct because the hypothesized shear stress distributions could not be independently determined. Without such data, the model is probably not unique and therefore has limited applicability. As an example, we present another interpretation of the high-pressure unloading/reloading experiments in aluminum. Swegle and Grady [148] used a thermal-based modeling approach to explain the anomalous recompression behavior. They assumed that localized deformation regions were produced during the risetime of the first shock wave and that the dissipative energy produced by shock loading was thermally trapped in these regions [55]. This process produces small hot spot regions, assumed to be micro-slip bands, that are initially extremely hot during shock loading but rapidly cool through thermal diffusion once the shock has passed. Because this is a time-dependent process, the risetime of the shock wave is a critical factor in modeling the local thermal deposition and diffusion processes. In the work of Swegle and Grady, the shock risetime was assumed to be stress dependent according to the fourth power relation discussed earlier [27,54,147,148]. A point to note is that their model assumed that the feature size produced during shock loading was sufficiently small that the deformation could be assumed to be homogeneous on a continuum scale. Using a mixture model to represent the localized heat deposition and diffusion and the fourth-power relation to estimate shock risetime, they were able to estimate explicitly the local temperatures in the hot spots and diffusion of heat from these regions during the time scale of shock deformation. [147,148].

The thermal trapping concept is illustrated in Fig. 2.14a along with Swegle and Grady's estimations of the shear stress states produced during shock loading, Fig. 2.14b. Over the duration of the shock risetime, the temperatures of the hot spots approached several hundred degrees Kelvin even though the calculated bulk temperature remained near ambient. This effect is shown in Fig. 14b. Because the yield stress was assumed to vary with the temperature of the local hot regions, it decreased during the peak strain rate in the shock, followed by recovery later in time due to cooling of the hot regions. The average shear stress is observed to increase to a maximum during the peak strain rate in the shock, due to a dissipative contribution, and then attains an equilibrium value late in time. Depending on the actual characteristic feature size and distribution of hot spots assumed in the model, which are not known *a priori*, Swegle and Grady found that the local temperature in the local hot regions could approach the melt temperature during peak loading rate in the shock. This was followed by relaxation to the bulk temperature calculated from thermodynamic principles as these re gions are rapidly quenched. Accordingly, the shear strength decreases to near zero in the shock wave itself followed by recovery at times sufficiently long for

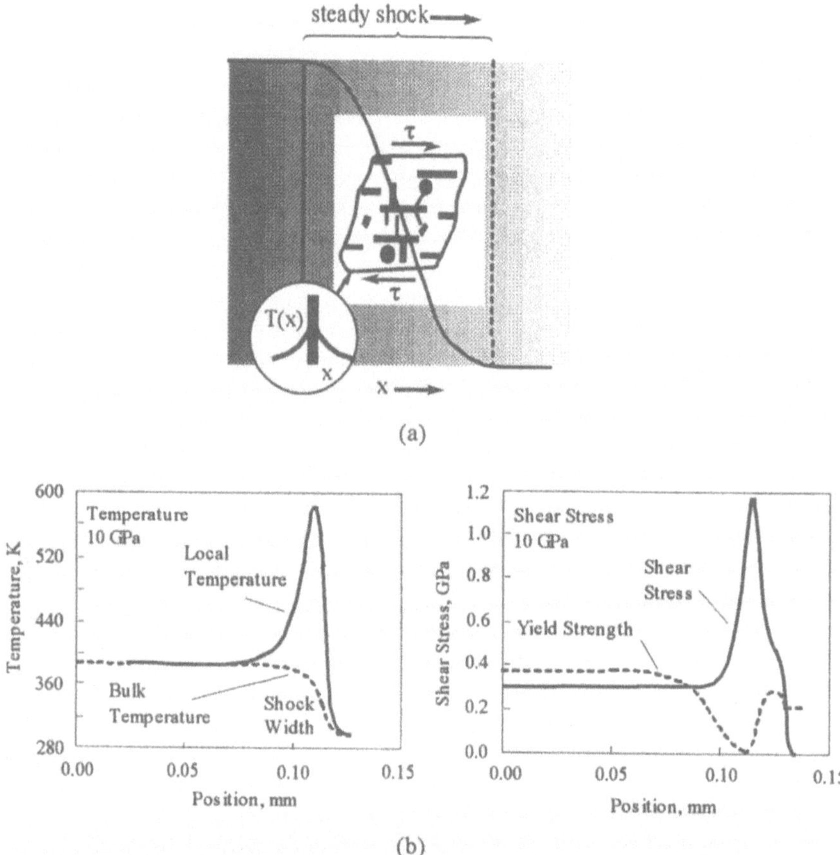

Figure 2.14. (a) Concept for thermal localization during shock compression. It is assumed that the dissipative energy produced during the shock risetime is localized in small regions of the material. (b) Upper graph shows the local and bulk temperatures calculated with a mixture model used to describe localized heating and diffusion during the shock risetime [147]. The bottom part of the graph illustrates the resulting resolved shear stress and yield strength produced during this process.

thermal equilibration. The reader is referred to Swegle and Grady's paper for a detailed discussion of the mechanical and physical effects occurring during this process.

Since the final state of shear stress is less than the yield strength immediately after shock compression in this model, an elastic wave is produced during re-shocking from the initial shocked state, as observed in the earlier work of Lipkin and Asay [111]. Figure 2.15a shows a calculated and an experimental record for an experiment conducted on aluminum at an initial shock stress of about 14 GPa [10]. As illustrated, the model of Swegle and Grady represents the experimental reshock profiles very accurately. Swegle and Grady also used the model to estimate the shear stress state and critical shear strength for different shock stress levels. Because the peak strain rate in the shock is dependent on stress to the fourth power in their model, there is a substantial effect on the local heated states. Figure 2.15b summarizes the calculated initial shear stress states, τ_0, produced by shock loading and the critical shear strength, τ_c, versus shock stress with this model. As shown in the figure, good agreement is also obtained between the model and experiment for different assumed shear band feature sizes and spacings. It is to be noted that the feature sizes used in this model were submicron, which is large compared to dislocation or microscopic descriptions and therefore represents the response at a mesoscopic scale. However, the model assumed homogeneous response on a scale "large" compared to the feature size (area of the recording instrument or about 0.1 mm) and further that one-dimensional motion of the material occurred. We shall see in the next section that both of these assumptions are not necessarily unique.

The results by Swegle and Grady suggest the important points that a shock-induced scale feature may be established at the mesoscopic scale and that mechanical behavior may be influenced by complex thermal events occurring during the risetime of the shock. Alternately, the modeling results of Lipkin and Asay indicate that shear stress distributions developed through wave interactions at the grain scale may be the dominant mechanism. There is the likely possibility that both effects play in this process. Therefore, it is important to determine the actual mechanical and thermal effects induced during shock compression in real time to evaluate these issues. However, based on the above examples we can say with confidence that the elastic–plastic model does not completely describe the response of several materials subjected to more complex loading histories.

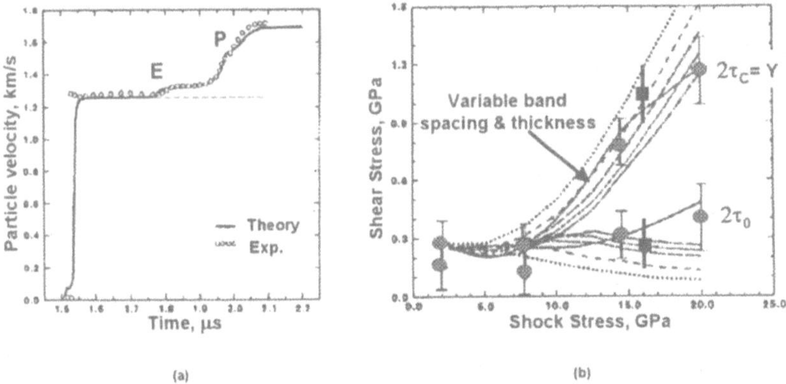

Figure 2.15. (a) Calculated recompression profile in aluminum at about 14 GPa [147] and comparison with the experimental results of Asay and Chhabildas [10]. (b) Calculated shear stress states and critical shear strength versus shock stress in aluminum.

It is to be noted that the two models that we just discussed represent deformation mechanisms occurring at a mesoscopic scale in terms of the earlier definition but there was no attempt to include deformation mechanisms at a finer scale. However, it is well known that material response at both the atomic and microscopic scales does influence shock properties at the continuum level as exemplified by the precursor decay process in LiF [65]. These should also be included for a complete representation of the dynamic response of materials [24,113].

2.3.4. Real-Time Diagnostics at the Mesoscale

It is clear that *in situ* measurements of shock-induced deformation features are needed to determine the deformation mechanisms occurring in real time during shock compression. This is one of the major challenges facing the shock wave community. Asay and Barker [8] developed an interferometer technique (VISAR) that is a first step in this direction because it allowed determination of the spread or dispersion in particle velocity during shock loading. The method was originally discovered during the investigation of shocked porous aluminum in which a significant loss of fringe contrast occurred during the shock event.

The technique is illustrated in Fig. 2.16. Figure 2.16a shows a shock wave propagating in a heterogeneous material that has a feature size smaller than the spot size, i.e., the recording area of the velocity interferometer laser signal. If the particle velocity is not uniform over this dimension, the reflected Doppler light will contain a distribution of frequencies related to the specific variation in particle velocity during the recording time of the shock measurement. Doppler broadening will produce a loss of signal contrast that depends on the interferometer sensitivity. Because the VISAR interferometer records both the phase shift produced by target velocity and the variation in signal contrast, these combined measurements can be used to determine both the average particle velocity and the dispersion in particle velocity. Asay and Barker employed a Gaussian distribution function to represent the particle velocity distribution, as illustrated in Fig. 2.16b, from which these two parameters could be determined.

Several Russian investigators have used this approach over the past fifteen years to determine the mesoscopic scale response of shocked polycrystalline metals (see for example, [40,118,119,121,122,145]). These studies demonstrate a direct correlation of average material properties, such as compressive and tensile strength and apparent strain-rate dependence to the dispersion in particle velocity produced by shock loading. Specifically, Meshcheryakov and Divakov [117] used the technique to determine particle velocity variations in several aluminum and steel alloys for elastic–plastic compression and as well as measurements of spallation for tensile loading. A result from one of these experiments is shown in Fig. 2.17. Figure 2.17a depicts the average particle velocity measured at the free surface of the planar target, which shows a typical two-

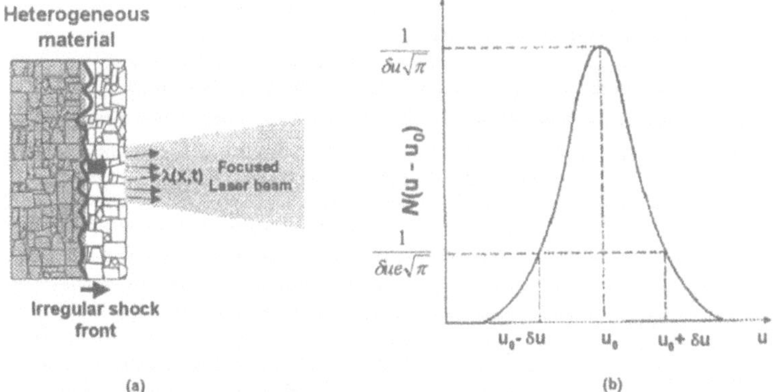

Figure 2.16. Technique for measuring particle velocity dispersion in shocked materials. (a) Concept for using an interferometer technique to measure variations in particle velocity over the recorded area of an illuminated interferometer spot. (b) Use of a Gaussian function to describe the variation in particle velocity [8]. The interferometer phase and contrast can be used to estimate the average velocity u_0 and the variance, δu.

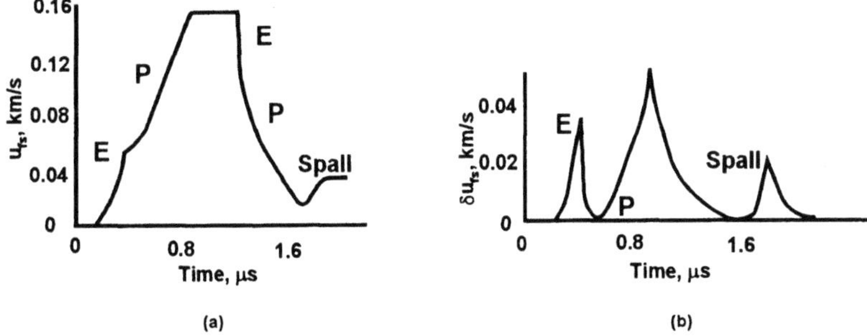

Figure 2.17. Particle velocity dispersion in metals. (a) Average free-surface velocity profile measurements in an aluminum alloy, B-95-pY-2 [121]. (b) Particle velocity dispersion corresponding to the average free-surface velocity profiles in (a).

wave structure characteristic of a weak shock. The average free-surface velocity attains an apparent constant value, followed by a decrease during unloading from the shocked state, and then spallation as evidenced by a pullback in the average free-surface velocity. The pullback signal is used to determine the spall strength of aluminum through the usual relationship:

$$\sigma^* = \frac{1}{2} \rho_0 \, c_e \, \Delta u_{fs} \,, \tag{2.11}$$

where σ^* is the spall strength, c_e is an effective wave velocity (usually assumed to be the average of the elastic and bulk sound speeds, while ignoring the wave evolution from the spall plane to the free surface), and Δu_{fs} is the pullback in free surface velocity due to spallation.

Figure 2.17b shows the corresponding dispersion in particle velocity associated with the average free-surface velocity measurements related to compression, unloading, and spallation. It is observed to become large just after elastic yielding, during peak plastic deformation, and during spallation. Note that the average velocity measurements indicate a uniform state (flat top in the free surface velocity profile) after shock compression. However, the velocity dispersion analysis shows that this is not the case. In fact, the magnitude of velocity dispersion is a rather large fraction of the average velocity, as shown in the figure. The dispersion is observed to decrease during unloading, but it becomes large again as the material undergoes tensile fracture and spallation. This effect has been observed experimentally in several polycrystalline metals [117]. These observations imply that a uniform state of motion is not produced by shock compression in a polycrystalline metal.

Meshcheryakov and coworkers have made several measurements relating material properties to particle-velocity dispersion, but these extensive results will not be discussed due to space limitations. However, one further example from their research warrants discussion because it further emphasizes limitations of the "standard" assumptions. A particularly interesting observation is the correlation between velocity dispersion and spall strength, as illustrated in Fig. 2.18. Figure 2.18a shows the free-surface velocity structure obtained from shocking a steel alloy, ST-1. Figure 2.18b gives the corresponding particle-velocity dispersion similar to the results observed for aluminum in Fig. 2.17, although the specific details of the velocity dispersion are different. The results given in Fig. 2.18c show a distinct correlation between the spall strength and the velocity dispersion. As shown in the figure, the spall strength increases with increasing free-surface velocity (or loading stress), which has been observed in previous studies [38]. However, another important observation from these experiments is the direct relationship between spall strength and velocity dispersion. The explanation offered by Meshcheryakov and Divakov for this effect is that multiple wave interactions occurring in the polycrystalline sample near the vicinity of the spall zone cause local stress reductions, necessitating an increase in applied stress to produce tensile fracture. This result is contrary to the accepted notion of nucleation and growth of voids under homogeneous loading conditions and has important implications for modeling the dynamic fracture process. This observation suggests that perhaps conventional models of spallation, such as the nucleation and growth models commonly used, should incorporate the effects of dispersion in the growth of damage during tensile loading.

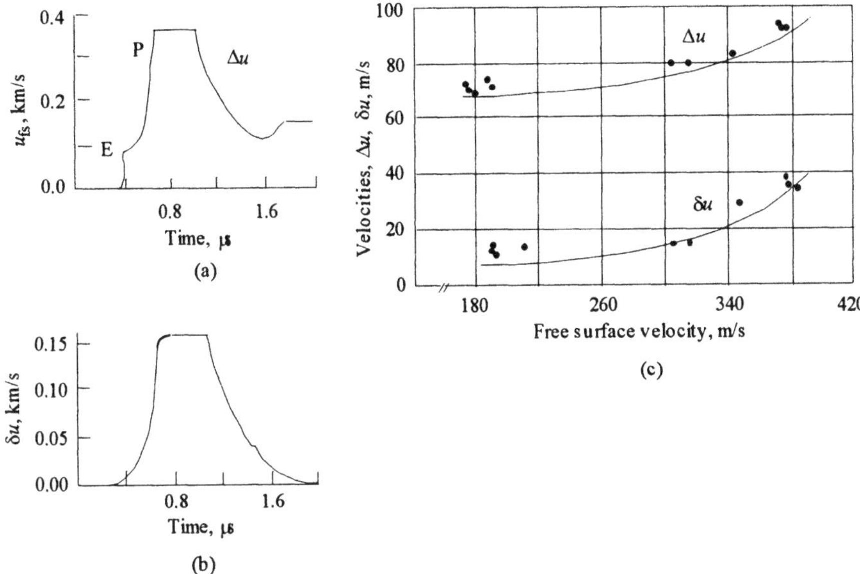

Figure 2.18. Particle velocity dispersion in polycrystalline metals. (a) Wave profile in ST-1 steel [121]. (b) Dispersion velocity corresponding to the average free surface velocity. (c) Spall strength versus velocity dispersion in steel. The top curve is the pullback velocity from free-surface velocity measurements; the bottom curve is the corresponding dispersion in velocity.

2.3.5. Numerical simulations

Recent advances in high-performance computing provide a powerful approach for probing important details about the deformation processes occurring during shock compression at the atomic and mesoscopic scales that cannot be studied experimentally. Earlier in the chapter we mentioned that several numerical studies have been and are being currently directed to the issue of wave propagation in polycrystalline materials; previous work includes [123–125, 127]. The major limitations in the earlier studies resulted from the lower memory capacity of the computers to model atomic or higher-level deformations with high resolution. With recent advancements resulting in both faster and higher storage capacity, high-resolution 3-D computations are now routinely performed. Holian notes that molecular dynamic (MD) simulations can now treat physical processes of deformation at the atomic level for large numbers of atoms and with reasonable accuracy. With further enhancements in computer capabilities it will be possible to treat yet larger groups of atoms. However, it will be some time before it will be possible to model deformations occurring from the atomic to mesoscopic or continuum scales based on a first-principles approach. With present computational capabilities, however, a practical approach is to model the collective motions of material ensembles that effectively represent mesoscopic scale deformations. In polycrystalline materials, examples could include modeling a realistic ensemble of grains with sufficient numerical resolution within grains to accurately represent sub-grain motion. A variety of numerical approaches are available for this purpose.

We will briefly discuss two numerical approaches that illustrate some of the principal features relevant to our previous discussions. One approach is to model the material elements making up individual grains and treat wave propagation in this ensemble in full three-dimensional motion. However, many of the effects we have discussed can be evaluated in two dimensions. We will discuss recent results obtained with the latter approach first. Yano and Horie [164,165] used a discrete particle code (referred to as the DM2 code) to model sub-grain deformations occurring during wave propagation in a two-dimensional array of identical-sized grains. DM2 is a two-dimensional quasi-molecular dynamics code, which represents materials as an ensemble of interacting discrete elements that make up individual grains. A two-dimensional array representing the material can then be constructed for numerical simulations of wave propagation. This approach allows simulation of deformation features that occur at a sub-grain or mesoscopic scale.

The configuration used by Yano and Horie for numerical simulations is illustrated in Fig. 2.19a, which depicts a reverse impact condition for generating shock waves in a two-dimensional array of copper grains. Upon impact, a stress front propagates upward into the unshocked array of grains that are all equal in size. The granular array illustrated in the figure represents a numerical simulation at a specific time after impact where the shock has already propagated upward to the point "A." Deformation of the grains can be observed near the impact surface. Several notable features are apparent from this simulation. As in previous studies [123–128], a highly irregular shock evolves during wave propagation. This is not readily apparent in Fig. 2.19a, but is discussed in detail by Yano and Horie [164]. In their paper, they emphasize that the deformation behind the shock includes rotational flows of material within the grains. In addition, the longitudinal particle velocity calculated by the code is not uniform over the array, but exhibits a distribution as indicated in Fig. 2.19b. Point "A" in Fig. 2.19a is the approximate distance the shock has propagated.   At this distance determination of the particle velocities on a plane perpendicular to the shock produces the variation shown in Fig. 2.19b, "A." For states well behind the shock front at distances "B" and "C," the particle velocity distributions take on a nearly Gaussian form, as indicated in Fig. 2.19b. Yano and Horie mention that these distributions are qualitatively similar to the experimental results observed by Meshcheryakov and Divakov [121]. The effects

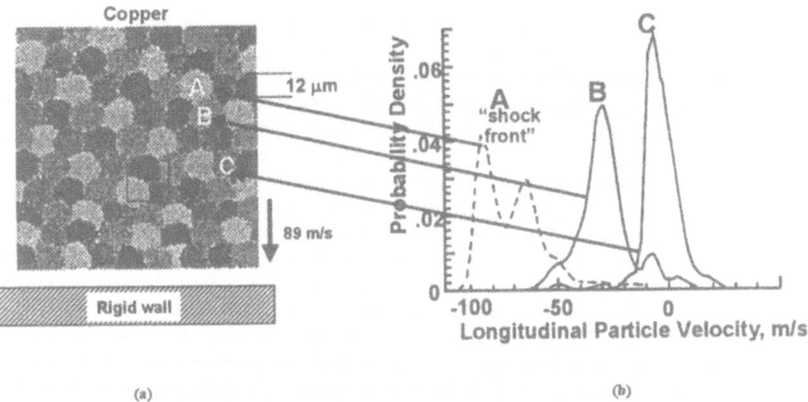

Figure 2.19. Modeling of wave propagation in copper. (a) Experimental configuration, including the calculated deformation of the granular array at a snap shot in time after the shock has progressed to approximately the level A. (b) Particle velocity distributions at different locations behind the shock at the corresponding time [164].

shown numerically are also similar to the assumptions made by Lipkin and Asay [111] for the distribution of shear stress states.

Additional numerical simulations by Yano and Horie [165] on polycrystalline iron, which are illustrated in Fig. 2.20, provide direct information about the spatial irregularity of the shock front and the dependence of local strain gradients near the front versus shock strength. A significant observation from their later calculations is that the degree of "shock front roughness," i.e., the local velocity variations and the rotational motions near the shock front appear to decrease as the amplitude of the shock stress increases. The numerical simulations further indicate that identifying a well-defined shock front at any specific location in the material is probably not realistic, since this observation will vary spatially and temporarily throughout the material. Furthermore, a measured shock risetime is probably not fully representative of atomic-based viscous mechanisms, such as the viscous damping model discussed earlier, to predict steady wave risetimes.

Yano and Horie's numerical simulations suggest that shocks produce a highly transient rotational flow and localized deformations consistent with the experimental results of Mescheryakov and coworkers. These simulations emphasize the importance of numerical calculations to identify aspects of the shock-induced flow that cannot be measured experimentally. The key results from these simulations can be summarized as follows.

- Measurements of shock risetime may be strongly dependent on the actual gauge used to make the measurement. Furthermore, risetime measurements may be dependent on the collective effects of wave

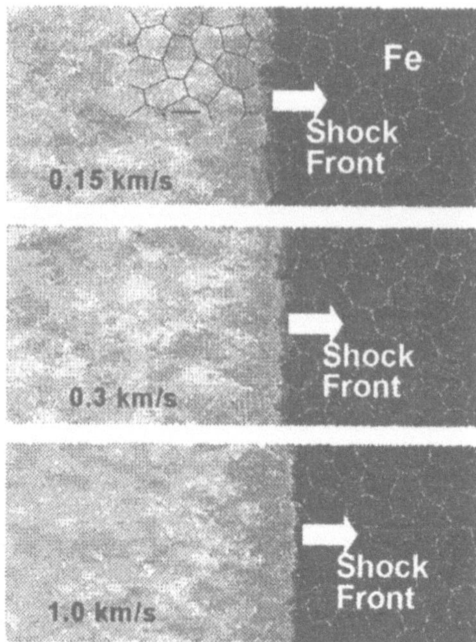

Figure 2.20. Effects of shock strength in numerical simulations of shock propagation in iron indicating smoother shock fronts at higher impact velocities [165]. The stress levels corresponding to the particle velocities of 0.15, 0.3 and 1.0 km/s are approximately 49, 104, and 433 kbar, respectively.

interactions in granular materials, rather than depending solely on physical properties such as viscous effects.

- Microscopic features of plastic deformation are likely to be important to deformations occurring at the mesoscopic scale. However, it is likely that local rotational motions and velocity distributions induced by the shock are also contributing factors and perhaps may even over-ride the microscopic effects in terms of influencing the wave structure. Until the mesoscopic scale processes are properly accounted for in realistic continuum models of wave propagation, we are not in a position to quantify the contribution from these effects.

- There is a scale effect associated with rotational motion of mass points within grains that has features resembling mesoscopic eddies or the beginnings of turbulent motions. The density of these vortices is observed to increase with shock strength, although their size decreases accordingly. A fundamental understanding of these effects

is clearly necessary to predict the mechanical and physical properties changes induced by shock waves, as discussed by Lipkin and
Asay [111] and Makarov [113]. Panin [134–136] and coworkers are
attempting to make this step by incorporating both microscopic
plasticity models of mesoscopic motions into a self-consistent
mathematical framework. With further refinement, their approach
holds promise for changing the fundamental paradigms used in generalized modeling approaches.

The combined experimental results and numerical simulations suggest the
graphical representation of shock propagation in polycrystalline materials shown
in Fig. 2.21. For planar external loading, the applied stress will cause slightly
different longitudinal wave speeds because of variable crystal orientation and
different states of yield within grains because of the applied shear stress on primary slip planes. As has been pointed out in several previous studies [123–125,
164,165], this will produce a wavy shock front that has a risetime dependent on
the internal wave interactions. There will also be local variations in particle
velocities in the shocked material that are not necessarily aligned with the applied stress. This can produce local regions that have collective motions, which
are characteristic of rotational flow or the beginnings of local vortical flow. The
variable particle velocities throughout the material can also result in the measured dispersion in these quantities at the rear free surface of the material as was
previously discussed.

Yano and Horie have noted that these local variations can produce internal
fractures and void formation at grain intersections [Yano and Horie, 1999].
Presumably, there will also be dispersion of other shock and thermodynamic
variables, such as density, pressure, and temperature throughout the shocked
material associated with local stress concentrations. In addition, the interaction
of the shock at each grain boundary and other imperfections in the material will
cause local scattering of kinetic energy into lateral directions, which may increase the total kinetic energy of the system at the expense of the internal energy. These effects could explain the apparent softening and strength recovery
effects that we previously discussed [10,111].

With present diagnostic capabilities, scattering effects resulting from grain
interactions cannot be measured *in situ* during shock compression, so the extent
to which non-translational motions occur during shock propagation in polycrystalline or even single crystals as mentioned above is not known. In addition to
the grain interactions producing the picture illustrated in Fig. 2.21, which is a
mesoscopic scale of deformation, other mechanisms occurring within grains or at
grain boundaries can also result in localized deformations and temperature regions as discussed by Swegle and Grady [147]. Thus, the shock-deformed state
can be quite complex in contrast to the simple picture illustrated in Fig. 2.2,
which depicted a shock wave resulting in a homogeneous and constant compres-

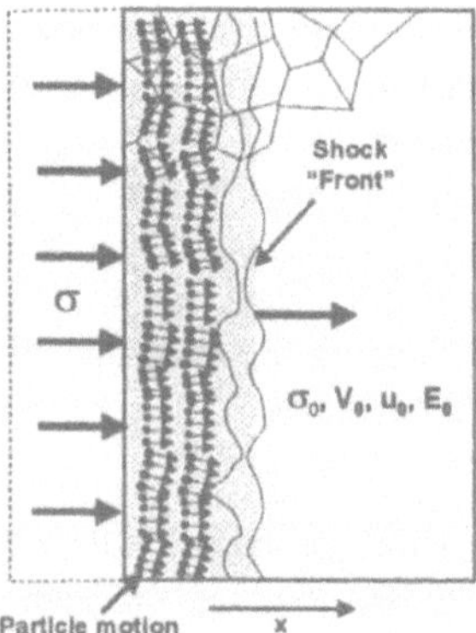

Figure 2.21. A possible scenario depicting shock propagation in polycrystalline materials. Particle velocity vectors vary throughout the shocked material in direction and magnitude. The shock front is not planar and has a risetime that may be affected by the local particle velocity gradients. Based on ideas presented by several investigators [Yano and Horie, 1999, 2002; Meyers 1975a, 1975b; Meyers and Cavalho, 1976].

sion state. Capturing this complexity in a hierarchical modeling approach covering the microscopic to macroscopic scales will be a major challenge. As Makarov [113] points out, the shocked system evolves under loading so that its microstructure is capable of self-organizing into definite deformation regions. He further emphasizes that microscopic scale analysis alone is insufficient for constructing a full-value theory capable of relating macroscopic deformation to the internal structure. Various continuum approaches are being considered to address these issues, some of which are discussed in recent publications by Makarov [113], Baer and Trott [13], Briant et al. [24], and Nesterenko [132].

The computational examples we have discussed identify certain features of the deformation process that cannot be accessed experimentally. They complement and provide additional support to the conclusions given in the earlier discussion of the experimental investigations. Although the simulations do not include all of the physics occurring at this scale, such as interface interactions, interface strength, grain boundary bonds, etc., they illustrate the complicated mechanical and thermal states which should be incorporated into multiscale models of dynamic response. In this regard, they do provide an insight into the processes occurring, even though there are many questions to be resolved before these simulations can be used to study the physical effects occurring at this scale.

We will present a final example to further illustrate the use of computational methods to design experiments that increase our understanding of mesoscopic processes. Baer and Trott [13] have studied shock propagation in granular materials, specifically porous sugar, to quantify deformation features at a mesoscopic scale. A major goal of their study is to identify scalar invariants of shock-induced deformations for use in continuum modeling. In these calculations, a shock wave was generated in a three-dimensional sample of granular sugar containing a representative distribution of grains using a 3-D hydrodynamic code simulation [13]. A reverse impact condition was used to generate the shock wave. Figure 2.22a depicts the experimental configuration for generating shocks in the 3-D sugar array and Fig. 2.22b illustrates the resulting pressure and temperature distributions induced throughout the column of sugar grains by the shock. As illustrated in the figure, considerable variation in these quantities is observed behind the shock front.

Baer has made significant progress in developing data processing methods to synthesize the output from large numerical solution sets. This allows determination of the spatial and temporal distributions for kinematic and thermodynamic variables. For the problem illustrated in Fig. 2.22, the distributions in particle velocity are represented as probability distribution functions in Fig. 2.23. In Fig. 2.23a, spatial variations in the longitudinal particle velocity, v_z, are shown at the mid-plane of the column with varying shades of gray to represent the variations at various times after impact. These are quantified in Fig. 2.23b, which shows how the longitudinal particle velocity distributions at the mid-plane develop from the initial value before shock loading to an evolution of dispersed states later in time. The major observation to note is that the longitudinal particle velocity develops a well-defined spread, essentially Gaussian, well behind the shock front. This effect was discussed in the earlier experimental and numerical examples. A more interesting effect is the development of lateral velocity components that are illustrated in Fig. 2.23c. This figure shows that the lateral particle velocity components described as $v_{x,y}$ evolve from a null result expected for uniaxial compression into considerable distributions behind the shock front. This numerical example illustrates the important point that external one-dimensional loading in granular materials does not produce strictly 1-D motion commonly accepted for analysis of these problems. This provides insight into the mechanisms for development of transverse kinetic energy and should be accounted for in continuum analyses. A challenge for the modelers and numerical analysts is to quantify the errors resulting from the assumption of strictly longitudinal motion invoked in shock wave studies.

More details of the numerical 3-D calculation can be found in the paper of Baer and Trott, [13], but it is instructive to give the results of the density and distributions produced for this simulation because density is a principal variable in standard EOS experiments. The evolution of the probability

Baer has made significant progress in developing data processing methods to synthesize the output from large numerical solution sets. This allows determination of the spatial and temporal distributions for kinematic and thermodynamic variables. For the problem illustrated in Fig. 2.22, the distributions in particle velocity are represented as probability distribution functions in Fig. 2.23. In

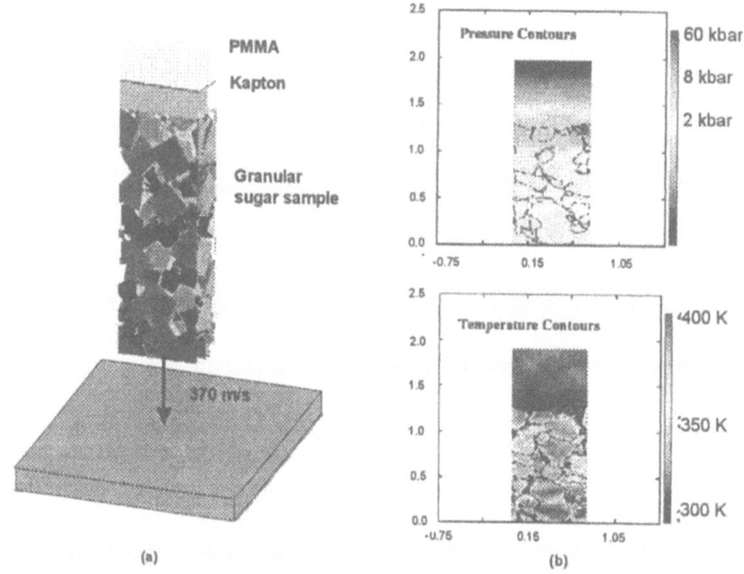

Figure 2.22. Three-dimensional numerical configuration used to induce shock propagation in granular sugar. (a) Configuration for producing shock propagation. (b) Pressure and temperature distributions induced by the shock [13].

Fig. 2.23a, spatial variations in the longitudinal particle velocity, v_z, are shown at the mid-plane of the column with varying shades of gray to represent the variations at various times after impact. These are quantified in Fig. 2.23b, which shows how the longitudinal particle velocity distributions at the mid-plane develop from the initial value before shock loading to an evolution of dispersed states later in time. The major observation to note is that the longitudinal particle velocity develops a well-defined spread, essentially Gaussian, well behind the shock front. This effect was discussed in the earlier experimental and numerical examples. A more interesting effect is the development of lateral velocity components that are illustrated in Fig. 2.23c. This figure shows that the lateral particle velocity components described as $v_{x,y}$ evolve from a null result expected for uniaxial compression into considerable distributions behind the shock front. This numerical example illustrates the important point that external one-dimensional loading in granular materials does not produce strictly 1-D motion commonly accepted for analysis of these problems. This provides insight into the mechanisms for development of distribution functions for material density at different times in the simulation are presented in Fig. 2.24. The density is observed to evolve from a Dirac delta function before impact to dispersed values after shock passage. Late in time, the distribution is nearly Gaussian, similar to that observed for particle velocity. Note that density variations persist in the shocked material for long times. Again, this is a challenge for the modelers to evaluate

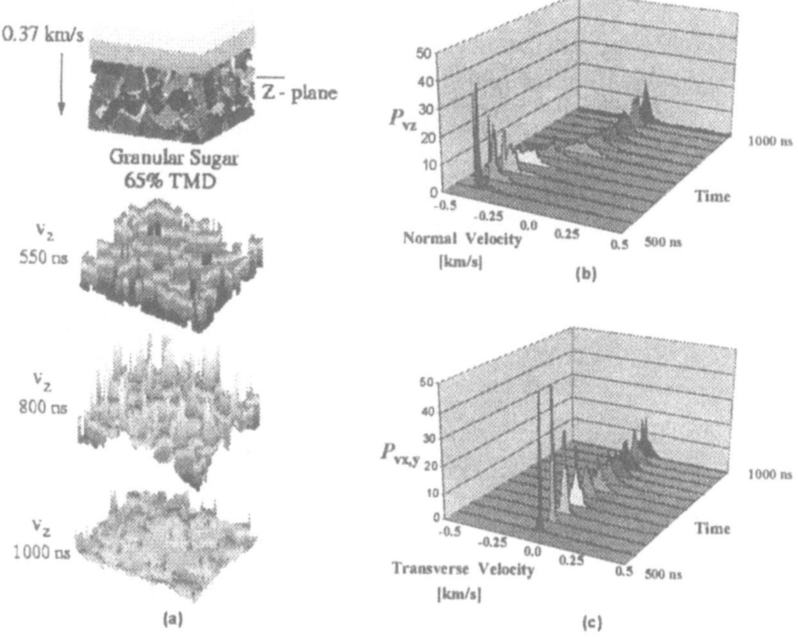

Figure 2.23. Probability distribution functions of particle velocity for shock propagation in porous sugar. (a) Graphical representation of longitudinal velocity distributions at different times after impact. (b) Probability distribution functions for longitudinal velocity as a function of time. (c) PDFs for transverse velocity as a function of time [13].

how these effects influence quoted accuracies in EOS experiments. The main point for showing these results is that, with refinements, these 3-D simulations may provide useful information to describe local temperature states that are important for driving energetic reactions or mechanical softening speculated to occur in experiments on aluminum [147]. A number of recent papers address this specific issue [12,33,97].

2.3.6. Recent Advances in Experimental Diagnostics

The experimental results and the models described above imply mesoscopic information about particle velocity distributions. However, there are different interpretations of phenomena occurring at this scale, which result in similar predicted continuum response. This emphasizes the need to quantify kinematic and thermodynamic variables associated with the deformation. Specifically, it is necessary to determine the characteristic feature size and the local velocity motions of the deformation as a function of time at the mesoscopic or smaller scales as depicted in Fig. 2.21. This will be a major challenge for developing new experimental methods.

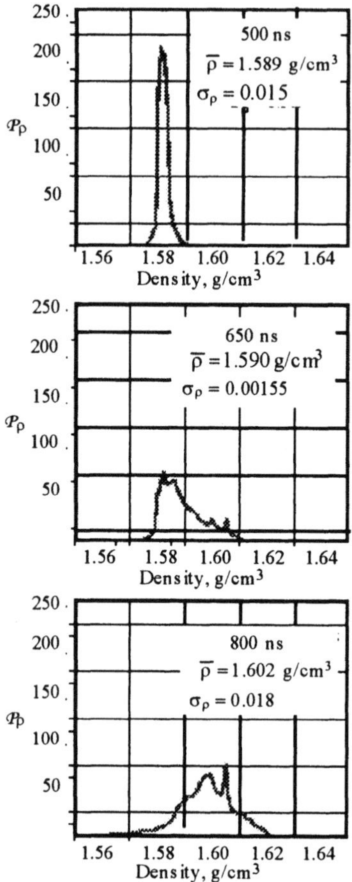

Figure 2.24. Probability distribution functions of density versus time at the mid-plane of porous sugar column for times of 500 ns, 650 ns, and 800 ns after impact [13].

An encouraging step in this direction involves a recent extension of the VISAR technique that provides simultaneous information about both the temporal and spatial distributions of longitudinal particle velocity. This technique has been alternately referred to as a Line Imaging Velocity (LIV) interferometer, a Line ORVIS, or a Microstructural Velocity Interferometer by different investigators [19,22,32,154,155]. In this technique, a continuous line of laser light is projected onto a target, in contrast to a regular single-point VISAR that uses a focused spot of about 100-μm diameter. The interferometer is slightly misaligned to produce numerous fringes, typically 10 to 40, that are recorded on a high-speed streak camera. The fringe change during shock-induced motion of the surface is proportional to the local longitudinal particle velocity at the corresponding Lagrangian position along the line. Time-resolved velocity history measurements have been made for spatial line lengths ranging from a fraction of one millimeter to over 20 mm. The spatial resolution is a function of the line length and the experimental configuration used to record data, but is about 20 μm for a 200-μm long line. A typical result [155] obtained with this technique is shown in Fig. 2.25, which depicts the experimental configuration used to measure particle velocity variations in shock experiments on granular sugar. In these experiments, a thin layer of Kapton was placed at the rear surface of the sugar sample and a polymethylmethacrylate (PMMA) window used to measure the particle velocity variations at the interface between the Kapton and PMMA with a Line Imaging VISAR. The length of the line used in this experiment was about 200 μm, resulting in a spatial resolution of about 20 μm and a time resolution of 1–2 ns. Figure 2.25b shows the variations of longitudinal particle velocity with time for positions along the line. The overall features of the velocity dispersion are in general agreement with Baer's 3-D calculations [13].

The velocity variations shown in Fig. 2.25 are observed to be not only oscillatory at a given sample position, but the magnitude of the variation persists over long periods of time. Integration of the local particle velocity variations allows the determination of the permanent non-planar local displacements. This provides a direct measure of the characteristic mesoscale dimension representative of shock compression of granulated sugar. The distinct spatial periodicity or characteristic length of the mesoscopic scale deformation observed in the experiment is finer than the size of the sugar granules and nearly time-independent. This observation is in agreement with the concept of space-averaged scalar

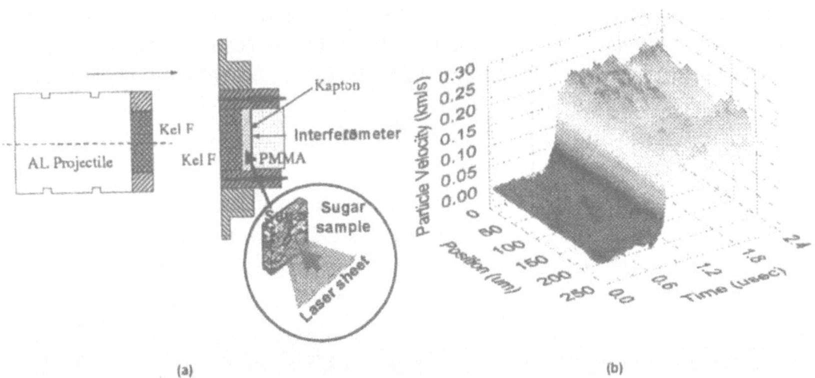

Figure 2.25. Experimental technique for Line Imaging VISAR measurements. (a) Experimental configuration used to measure particle velocity dispersion during shock loading of a porous sugar sample. The particle size for the sugar granules ranged from 150–212 μm. (b) Typical velocity result obtained on a laser line 200 microns in length at the interface between a Kapton layer and the PMMA laser window. [154].

invariants proposed by Baer [13] and would thus be valuable in developing continuum models of dynamic response.

The LIV spatial velocity profiles are also in reasonable agreement with Baer's 3-D simulations [13]. The ability to correlate computational predictions with spatially and temporally resolved experimental data in heterogeneous materials, as demonstrated by this example, presents realistic possibilities to develop a better understanding of the actual mesoscale phenomena occurring during shock deformation. Hopefully, future time- and spatially resolved studies will lead to physically based continuum models that can be used to reliably and accurately predict the mechanical responses previously observed in aluminum and other metals [111–113,118–122].

There is a final point we would like to make regarding comparisons of experimental and computational results. We can potentially learn more about the physics of shock deformation from experiments that disagree with existing continuum models than from experiments that do. The example given earlier for the dynamic shear strength effects in shocked and reshocked aluminum is one such example. Another very recent example that challenges accepted notions of shock-induced spallation was recently performed by Chhabildas and coworkers [32], who used the Line Imaging VISAR to probe localized spall regions of tantalum. These experiments were similar to those of Mescheryakov and coworkers [117] that related the spall strength of aluminum and steel alloys to the magnitude of particle velocity dispersion occurring during the spallation process. However, Chhabildas and coworkers took a significant further step by using a LIV to measure explicitly the spatial variations in velocity dispersion at different locations along the laser line. They made the assumption that the difference in volume of a spalled material compared to an idealized material that does not spall is solely due to void nucleation and growth. This allowed them to infer the *in situ* sources of void nucleation and growth at various points within the sample leading to spallation at discrete locations within the sample. In agreement with velocity dispersion measurements by Trott et al. (Fig. 2.25) and Meshcheryakov and Divakov (Figs. 2.17 and 2.18), they observed significant free-surface velocity dispersion over a 2-mm long laser line illuminated on the free surface of a polycrystalline tantalum sample. For loading stresses sufficient to produce incipient spallation, they were able to relate the local spatial displacements on the free surface to void nucleation and growth within the sample. Figure 2.26 illustrates the results of one of their experiments, which shows the growth in defect structure, characterized by a single spatial dimension, at different locations within the sample. The reader is encouraged to consult the original paper for more details, but the example demonstrates the possibilities of probing in-situ mesoscopic scale phenomena in real time to gain considerably more information about deformation processes.

2.3.7. The Challenge

The advent of numerical techniques combined with high-speed computing capabilities for studying three-dimensional deformation states of shocked materials is a significant new thrust that should provide a strong basis for developing more realistic models of dynamic material response. However, these computations can be very costly, in both time and money, as pointed out by Holian [79]. There is likewise a pressing need to develop better high-resolution advanced diagnostics in order to provide *in situ*, real-time experimental data to complement the numerical calculations. This capability will help focus simulations on the right issues, while simultaneously validating realistic models. In addition to advanced diagnostics, a dedicated experimental program is needed throughout the shock community to focus on novel experiments to further challenge the accepted norms and principles presently applied to many areas of shock physics, including mechanical, electrical, optical or phase transformation response. Russian investigators have concentrated their effort in the area of mechanical response over the past 15 years, but a more general program is needed if we are to have realistic material models. Specifically, we need to define experiments that produce results which significantly differ from the predictions of homogeneous continuum models, in addition to performing experiments that appear to "validate" these models. In addition, there is a continuing need to understand material response over the full range of atomic, microscopic, and continuum levels to tie these effects together. The biggest challenges facing the next generation of shock wave researchers is to develop the right tools, conduct the right experiments, and

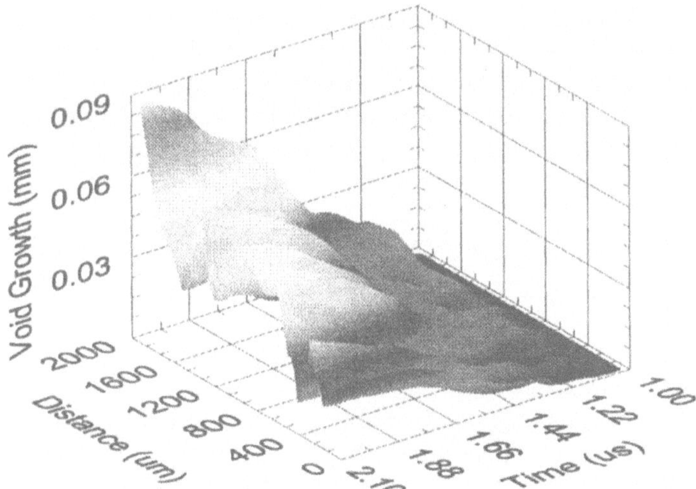

Figure 2.26. Defect growth in spalled tantalum, showing the characteristic spacing and increase in void size with time (from Chhabildas et al., [32]). Note the direction of increasing time in the graph. The width of the illuminated laser line used with the VISAR was 2 mm with a resulting spatial resolution of about 40 µm. The tantalum sample had a nominal grain size of 30–60 µm.

establish the right links between the different length scales in order to investigate or predict mesoscopic scale response.

With further refinements to the LIV technique, it may be possible to measure the shock structure occurring within individual grains in a polycrystalline sample. This information is necessary to predict the resulting defect structure induced by shock deformation. The ensuing velocity dispersion data generated with the technique have demonstrated a capability to record spatial variations on a sample surface with a resolution of 10–20 microns. With further improvements, this can likely be reduced to a few microns. Additional development of the technique could also possibly provide real-time data about local rotational motions occurring at the mesoscopic scale. This is a promising possibility based on previous work using conventional interferometer techniques to determine both the longitudinal and transverse components of motion [1,26,104]; and Chhabildas et al., 1979]. Additionally, Moiré interferometer methods may be another approach to measure localized displacements of a shocked surface. A combination of these techniques could allow measurements of real-time data about the local velocity vector at various *in situ* material points and provide confirmation of local three-dimensional deformations presently predicted by the different numerical models.

LIV measurements should also be useful for resolving the effective risetime of shock waves. As discussed earlier, planar deformation of polycrystalline metals will produce highly irregular shock fronts. Investigations by Grady [54], Swegle and Grady [147], and others have suggested that the effective risetime of shock waves in polycrystalline metals (and other material classes) varies inversely as approximately the fourth power of the final stress amplitude. However, in light of the results discussed earlier, it is not clear whether this apparent dependency is due to an inherent physical property, such as viscous dissipation occurring at the atomic or microscopic scales, or if it is due to kinematic effects arising at the mesoscopic scale.

A recent example emphasizes this point. Barker's measurements of shock risetimes that were made with a conventional velocity interferometry in polycrystalline aluminum indicate an effective risetime of about 30 ns at a shock stress of 38 kbar. However, in recent studies, Gahagan and coworkers [49], and Moore and coworkers [130] have used ultra-fast frequency-domain interferometry measurements on thin vapor-plated films of aluminum. Their results suggest that the waves were steady (although this was determined for small propagation distances) and that the effective shock risetime was about 6 ps for stress states of about 40 kbar. As discussed earlier, this apparent difference in risetime measurements could be related to the effect of intergranular wave interactions that produce in an effective risetime—the smaller crystal size of the vapor-plated samples would thus yield smaller risetimes. These conclusions are presently speculative, and frequency-domain interferometry has not been fully established for shock-profile measurements on larger sample dimensions. Therefore, an exact correlation with the risetime results based on continuum measurements should be treated with caution. Nevertheless, it is clear that this apparent discrepancy needs to be understood if we are to place confidence in shock risetime models such as the inverse fourth power relationship between shock risetime and the peak stress [54,147].

Time-resolved differential holography is another potential tool for detecting local longitudinal displacements [137] and thus local deformations in shocked materials. This technique is, however, not sensitive to transverse motions of a surface. An important attribute of holography, especially in the differential displacement mode, is that deformation features can be resolved at the optical wavelength scale, which is an order of magnitude better than that possible using the Line Imaging VISAR. However, significant further development of the technique is necessary to make it a useful tool for investigating heterogeneous deformation.

There is clearly a need for other diagnostic tools that can probe the deformation state *locally* and in *real time* during shock loading encompassing the different length and time scales appropriate to the deformation features. There are several possibilities for these measurements, including shock-induced optical emission. Optical emission experiments previously performed on shocked single crystals of quartz, $LiNbO_3$, and sapphire with fast optical framing cameras indicated light was emitted locally from discrete locations corresponding to crystallographic fracture planes for stresses above the HEL [23,70]. These experiments indicated a possible correlation between localized emission and the local deformation process itself, which is promising as a tool to probe mesoscopic scale response. However, further research is needed to identify the mechanisms of optical emissions and to correlate the spatial and temporal relationship of localized emission to the defect generation sources. If this relationship is understood, shock compression of optically transparent materials may provide *in situ* measurements of the mesoscopic scale deformation processes occurring during shock loading. It may be possible to determine local temperatures at the localized emission sites using this technique and thus evaluate the local energy deposition and thermal trapping model discussed earlier Swegle and Grady [147]. Recently developed full-field measurements of local deformations using high-speed infrared diagnostics may be another useful tool for probing localized states [63,64]. Guduru and coworkers have demonstrated the ability to measure temperature distributions around moving crack tips.

Time-resolved spectroscopy may yet be another useful tool for probing deformation features at the mesoscale. Knudson and coworkers [102,103] used picosecond time-resolved spectroscopy to study the wurtzite-to-rocksalt phase transition process in cadmium sulfide at scales ranging from the atomic to mesoscopic. Their work provided information on the transformation kinetics and the mechanisms for nucleation and growth of the daughter phase at an atomic scale. From these measurements they inferred mesoscopic scale effects that could account for the transformation kinetics. With further investigation, it should be possible to use this approach to provide physical insight into mesoscopic scale deformations occurring in dynamic phase transitions. If this capability were simultaneously combined with particle-velocity dispersion measurements, the integrated data would provide a powerful database for understanding and modeling time-dependent physical phenomena from the atomic-to-mesoscopic scale and thus provide a wealth of information for developing continuum models. This emphasizes the point that a combination of different test methodologies to study a specific phenomenon at different length scales would significantly promote an understanding of shock deformation features needed to develop physically based continuum models. As a related issue, modelers need to show what errors exist in fundamental average properties such as stress and particle velocity due to mesoscopic scale effects. These studies should address the issue of scales at which the continuum models predict measurable properties to within experimental errors. This would provide guidance for deciding how much experimental effort should be put into investigating these effects.

Time-resolved stress gauges, which have been a mainstay of shock wave continuum measurements for over forty years [30], are also useful for studying mesoscale effects. We showed earlier that significant lateral and vortical motions are likely induced as a result of shock propagation in heterogeneous materials. Presumably, these motions result from stress gradients occurring at boundaries where there is a local difference in mechanical impedance. A number of piezoresistive and piezoelectric gauges have been developed to measure longitudinal and lateral stresses, although in their current form they average over large recording surfaces and therefore could not detect local multidimensional motions [30,58]. However, with recent advances in microelectronics and microelectronic sensors [20], miniaturization of stress sensors to a size sufficient to probe *in situ* heterogeneous states becomes a possibility for investigation local stresses at the submillimeter level. Developing this next generation of stress sensors for probing mesoscopic scale phenomena will be a major challenge for both the shock wave community and the microelectronics industry. Furthermore, collaboration between researchers from diverse disciplines will also be necessary to achieve this goal.

Since the early work of Johnson and Mitchell [92,93] on LiF, real-time x-ray diffraction measurements on shocked crystals have proven to be an extremely valuable tool for inferring information about the atomic state behind the shock front. The original work by Johnson and Mitchell was pivotal in showing that crystalline states persisted behind strong shocks as high as 60 GPa in amplitude, which was not apparent at the time and contrary to the prevailing assumptions. The technique was further refined by Egorov and coworkers [46] using low-Z backing materials to approximate *in situ* x-ray diffraction measurements during shock loading. After a long hiatus, several significant advances have recently been made in x-ray diffraction experiments [68,96,143,170,173]. These recent works provide detailed information about the deformation state in real time behind the shock, including time dependence of structural transformations. The technique provides the only measurement of density or volumetric compression directly from measured lattice parameters. The technique has not been used to infer mesoscopic scale effects. However, with additional developments, it might be possible to link shock deformations at the microscopic and mesoscopic scales.

Potentially, the technique could also resolve issues of the actual stress and strain states induced by shock loading, including additivity of elastic and plastic strains. As an example, recent x-ray diffraction measurement on shocked silicon crystals by Kalantar and coworkers [96] have illustrated the ability to determine both the elastic and plastic strains behind the shock front, as well as identify the crystalline nature of the deformed lattice, using simultaneous Laue and Bragg diffraction. Further development of the technique is necessary to determine local particle velocity vectors during shock loading, and if possible estimate localized translational and rotational motions.

The use of soft recovery techniques has been extremely useful for identifying deformation mechanisms occurring under shock compression. A variety of shock-recovery techniques [59,62,105] and analysis methods have been developed for this purpose. Gray [62] provides a good summary of techniques that have been developed and the resulting metallurgical phenomena revealed with this technique. As he points out, it is essential that soft recovery methods are designed so that the recovery process itself does not adversely affect the deformation structure initially induced by the shock. In particular, it is desirable to maintain one-dimensional motion in the sample during the recovery process to minimize lateral motions that can alter the deformation state. This objective is difficult to accomplish, but several recovery methods accomplish this to a good approximation [62].

Soft recovery techniques have also been used to specifically address the issue of localized rotational motions that presumably occur during initial shock deformation. Specifically, Tomlin and coworkers used the Nematic Liquid Crystal (NLC) technique to infer local rotational motions in shocked polycrystalline metals [152]. In this technique, a thin layer of homogeneously oriented NLC material is applied to a metallic surface of optical quality prior to shock loading. Changes in the NLC structure during shock compression enable a post-shock visualization of the structural defects due to local deformations induced by the shock. These visualizations can detect local material flows at a grain and sub-grain level. As a specific example, recovery experiments conducted on copper, steel, and aluminum have identified both translational and rotational deformations that the investigators claim cannot be visualized using traditional metallurgical techniques. Their results further corroborate the numerical analyses discussed earlier, which showed that local material motions induced by planar shock loading are not strictly translational at this mesoscopic level. These results imply that the NLC method and other potentially novel metallurgical techniques need to be further developed to provide a detailed understanding of localized shock-induced deformation features.

The limitations of all soft recovery measurements are that they provide time-integrated information about deformation mechanisms occurring in the shock. As mentioned above, this is a critical requirement to understand transient effects occurring during shock deformation. Even with these real restrictions, shock recovery techniques have been one of our principal tools for probing the deformation features produced by shock compression. Continued progress of the technique should include identification of deformation mechanisms resulting in local lateral and rotational flows induced by the shock [5,101,118,122,152].

Finally, we reemphasize the observation that modern computational techniques are playing an increasingly important role in promoting an understanding of shock-deformation features. Traditionally, modeling of shock wave profiles produced in planar impact experiments has been limited to using 1-D to reproduce experimental continuum wave profiles, as illustrated in Figs. 2.7 and 2.8. The dramatic advances recently made in computer power as discussed by Holian [79] make multidimensional first-principles calculations a viable "experimental" tool for investigating shock deformation processes. These tools provide a systematic method for studying the sensitivity of various material effects to the development of distributions in kinematic and thermodynamic variables during the shock event. This information would have been extremely valuable in the early investigations of Lipkin and Asay [111] for quantifying *a priori* the functional form of the shear stress distributions needed for modeling. These computational techniques are also useful in identifying new experimental diagnostics to probe mesoscopic scale response or to design new experimental configurations to better characterize or measure this response. A related challenge in this area is to develop "data mining" techniques for 3-D numerical simulations, as proposed by Baer [13] to obtain the essential information, such as PDFs discussed earlier, from the large volumes of data generated in these simulations.

There are other several other examples of shock-induced material property effects that cannot be easily explained using the standard assumptions of shock compression that were not discussed in this paper because of space limitations. These include phase transitions [45,50,78,164,165], porous materials [8,11,13, 38,106], optical [23,60,61,70], and electrical properties [60,61]. The heterogeneous deformations observed in single crystal materials [23,60,61,70] provide additional evidence that these effects are not limited to granular or polycrystalline materials and may be a pervasive effect of dynamic yielding. Furthermore, it has been speculated for many years that the reactivity of energetic materials is strongly correlated with localized deformations. There are many examples of heterogeneous or cellular structures in energetic materials during shock propagation, including the classic example of cellular structure in shocked nitromethane [13,33,157–162]. Further examples of heterogeneous deformation are apparent in ceramics [114]. Understanding the importance of these effects at the different length scales and their relation to mesoscopic and continuum scales is essential for developing accurate material descriptions.

2.3.8. What is a Shock?

The topics we discussed in this chapter were motivated by the question: What is a shock? The answer to this question is not simple and depends to some extent on the degree of detail desired. In terms of the classical continuum definition of shock propagation applicable in many situations, a steady planar shock wave propagates without change in structure, produces one-dimensional, translation motion, and results in a uniform deformation state behind the structured wave. In fluids, these assumptions are probably very good except for motions at an atomic scale. At this level, atomic vibrations will induce local displacements that are not strictly one-dimensional on the time scale of 10^{-12} s or less but the average shock-induced motion in the continuum limit is presumed to be simply translational in the direction of shock propagation. Thus, the use of shock conservation equations to determine average thermodynamic properties in the shocked state over times scales of 10^{-6} s during shock compression should result in a good approximation of homogeneous and uniform response. In this case, the concept of a steady shock wave described by conventional mathematical formalism is probably realistic.

Shock propagation in single-crystal solids results in the next level of complexity because the scale of the structural deformation involves generation of defects such as dislocations, twins, and vacancies that cause deviations from the idealized picture of purely translational motion. If these defects are uniformly distributed and have feature dimensions small compared with the time and spatial measurement system, then the continuum assumptions of steady wave propagation should also provide a reasonable description of steady wave evolution. However, there are many cases of shock propagation in single crystals where this assumption breaks down. A particular example includes dynamic yielding of high-strength ceramics where highly localized motions on millimeter scales have been observed. This result is clearly a departure from the assumptions of uniform and translational motion. Although continuum theories have been used to describe "plastic" waves in these materials, the concept of a "plastic" shock wave described by continuum equations is highly questionable and the degree of accuracy in using these models should be carefully investigated

The "standard" assumptions become even more problematical for shock loading of polycrystalline or granular materials. As discussed in this paper, there is experimental and computational evidence, which indicates that one-dimensional loading can produce significant departure from pure translational motions. Localized transient material rotations are observed that have scale features much larger than the microscopic scale of dislocations and approaching grain scales. Furthermore, the shock "front" in these materials is not planar, and material properties in the shocked state are not single-valued but dispersed over space and time. Because of these effects, the pressure, density, and temperature states are not homogeneous or uniform on the time scales typically assumed for equilibrium behind "steady" shock waves. The observed scale size for these effects (tens to hundreds of microns) is a significant fraction of the gauge dimensions used to record shock motions and is therefore defined as mesoscopic. In this case, the definition of a single steady "shock wave" becomes ambiguous because the scales of the inhomogeneity and shock irregularities are comparable to the shock transition thickness. Furthermore, it is not clear if a single wave motion can be accurately described through continuum formalisms or whether the wave should be described by an ensemble of wavelets, as suggested by several investigators. The question of whether a shock wave in polycrystalline materials is really steady for this case should also be addressed with more accuracy than previous "steady wave" experiments and in the context of the diagnostics used to measure the shock. That is, the temporal and spatial resolution of the measuring system may influence the measurement of the shock wave. In addition, the accuracy of treating the shock motion with the "standard" assumptions should be critically evaluated, both experimentally and theoretically, with metrics developed to quantitatively describe errors resulting from these assumptions. Perhaps by critically evaluating these issues, we will have a better answer to the question of "What is a shock?"

2.4. Closure

In this chapter, we have attempted to address the question of "what is a shock" from a slightly different perspective. We have chosen not to focus on a review of recent accomplishments and a comprehensive survey of deformation processes produced by shock waves. Instead, we have taken the approach of examining the effect of shock compression on measurable physical or mechanical properties. With this goal in mind, we examine if shock-induced material changes can be interpreted in terms of traditional assumptions of shock propagation. Our emphasis has been on shock propagation in polycrystalline and granular materials in which non-homogeneous and non-equilibrium effects appear to play a dominant role. The examples presented in the chapter highlight several deviant or "anomalous" results that cannot be readily explained in terms of the simple assumptions and material models often used to interpret shock wave experiments.

A specific example concerns the compressive yield properties of aluminum under shock loading. This example demonstrates an apparent time dependence of yield stress after shock passage, which is not expected from existing elastic–plastic models. Interpretation of these results includes: (1) postulation of a distribution of shear stress states produced during initial shock compression, or (2) shock-induced local deformations that create local transient high temperatures during initial shock loading, which causes a temporary loss of material strength. In either case, the yield stress becomes a time-dependent function, which is not predicted with elastic–plastic models and raises questions about the fundamental assumptions used to derive these models.

Experimental results are also presented for the shock compression of other polycrystalline metals, which indicate significant dispersion of the particle velocity induced by shock compression. This effect is also not expected from existing models and raises further questions about the uniformity and homogeneity of shock compression. Mescheryakov and coworkers [116–121] have performed extensive experiments on the dispersion of particle velocity in shocked materials. Their results demonstrate that velocity dispersion is a common phenomenon during plastic shock compression of polycrystalline metals, particularly immediately after elastic yielding, during large plastic deformation and during spallation. They also show that the dispersion velocity can be correlated with spall strength [121], which is inconsistent with commonly accepted assumptions of dynamic failure.

We have also highlighted recent two- and three-dimensional numerical simulations that provide further insight into deformation mechanisms that may be responsible for the observed experimental results. These simulations illustrate that deviations in shock and thermodynamic variables are observed to persist for relatively long times after shock compression in polycrystalline metals [13,165]. These numerical examples further elucidate the fundamental assumptions used to analyze shock wave experiments. They also help to identify additional experimental diagnostics, particularly *in situ* diagnostics, that will be needed to understand these effects. Unless experimental diagnostics are developed to probe the actual deformation mechanisms occurring at different length scales during the shock process, our knowledge of deformation mechanisms will remain extremely limited.

For the most part, shock wave experiments have been limited to planar loading conditions. This general class of experiments will not yield information about the full anisotropic stress tensor, especially if the shock gauges respond primarily to one-dimensional response. If we are to develop the next generation of dynamic material models for describing shock compression of solids, we must define experiments that determine the multidimensional stress-components as well [67,98,141]. Complex loading paths (such as shock followed by unloading or reloading, ramp loading, ramp loading followed by release, etc.) are also valuable for critically testing material deformation models, as we have discussed. For evaluating the question of "what is a shock?," the additional complexities in material response introduced by these loading conditions are useful for probing the deformation state produced in the initial loading process.

Continuum wave profile measurements have been very useful in the development of dynamic material models over the past five decades. They continue to be the mainstay of shock physics research and usually provide a first step in understanding the compression shock behavior. However, it is increasingly clear that existing material models based on continuum measurements are far from predictive. Combining continuum property measurements with other mesoscopic scale information, such as real-time probability distribution functions for making critical physical property measurements is vital for understanding the shock deformation processes. Developing the right experimental and computational tools to accomplish this goal is one of the most challenging tasks facing the shock wave community. As modeling capabilities and simulation tools improve, it will become feasible to include microscale and even atomic scale effects at the mesoscopic scale for developing better continuum models. It is essential that the community address these issues in a timely fashion. Otherwise, we will still have to rely on material models developed several decades ago.

Although the experimental and theoretical challenges we face are formidable, we feel that they present exciting opportunities to the next-generation of researchers who are willing to deal with these complex physics, chemistry and mechanical problems. The development of predictive dynamic material models hinges on the success of this endeavor.

Acknowledgments

We would like to acknowledge the valuable discussions we've had with several researchers about their thoughts and concerns regarding the shock deformation processes. In particular, Mel Baer, John Gilman, Bob Graham, Brad Holian, and Yuki Horie, provided many useful perspectives. We are especially appreciative to Bob Graham for his constant interest and for providing quotations by several previous investigators who have given considerable thought to this issue. We are also indebted to Jerry Forbes and Marlin Kipp for providing a critical review of this manuscript and for valuable suggestions to improve the paper. We are also grateful to Mel Baer who provided several original graphics (regarding 3-D shock propagation in granular materials) for use in this publication.

References

[1]   Abou-Sayed, A.S., R.J. Clifton, and L. Herman (1976), "The oblique-plate impact experiment," *Exp. Mech.* **16**, pp. 127–132.

[2]   Al'tshuler, L.V. (1978), "Phase transitions in shock waves (review)," J. *Appl. Mech. Phys.* **18**, pp. 496–505.

[3]   Al'tshuler, L.V., M.N. Pavlovskii, V.V. Komissarov, and P.V. Makarov (1999), "Shear strength of aluminum in shock waves," *Combustion, Explosion and Shock Waves*, **35**, pp. 92–96.

[4]   Ananin, A.V., O.N. Breusov, A.N. Dremin, S.V. Pershin, and V.F. Tatsii, (1974), "The effect of shock waves on silicon dioxide 1. quartz," *Combustion, Explosion, and Shock Waves* **10**, p. 372.

[5]   Atroshenko, S.A., and N.I. Zhigacheva (1996), "The method of visualization of dynamic deformation modes of metals," in *Shock Compression of Condensed Matter—1995* (eds. S.C. Schmidt and W.C. Tao), Amer. Inst. of Phys., NY.

[6]   Asay, J.R., G.R. Fowles, G.E. Duvall, M. H.Miles and R.F Tender (1972), "Effects of point defects on elastic precursor decay in LiF," *J. Appl. Phys.* **43**, pp. 2132–2145.

[7]   Asay, J.R., D.L. Hicks, and D.B. Holdridge (1975), "Comparison of experimental and calculated elastic-plastic wave profiles in LiF," *J. Appl. Phys.* **46**, pp. 4316–4322.

[8]   Asay, J.R. and L.M. Barker (1974), "Interferometric measurement of shock-induced internal particle velocity and spatial variations of particle velocity," *J. Appl. Phys.* **45**, pp. 2540–2546.

[9]   Asay, J.R., and L.C. Chhabildas (1980), "Shear strength of shock-loaded polycrystalline tungsten," *J. Appl. Phys.* **51**, pp. 4774–4783.

[10]   Asay, J.R. and L.C. Chhabildas (1981), "Determination of the shear strength of shock compressed 6061-T6 aluminum," in *Shock Waves and High-Strain-Rate Phenomena in Metals: Concepts and Applications*, (eds. Marc A. Meyers and Lawrence E. Murr), Plenum Publishing Corporation, NY.

[11]   Asay, J.R. and G.I. Kerley (1987), "The response of materials to dynamic loading," *Int'l. J. Impact Engineering* **5**, pp. 66–99.

[12]   Baer, M.R., M.E. Kipp, and F. van Swol (2001), "Microstructural mechanical modeling of heterogeneous energetic materials," in: Proceedings of the Eleventh International Detonation Symposium, in press.

[13]   Baer, M.R. and W.M. Trott (2001), "Mesoscale descriptions of shock-loaded heterogeneous porous materials," Bulletin of the Amer. Phys. Soc. **46**, p. 102.

[14]   Bancroft, D., E.L. Peterson, and S. Minshall (1956), "Polymorphism of Iron at High Pressure," *J. Appl. Phys.*, **27**, pp. 291–298.

[15]   Band, W. (1960), "Studies in the theory of shock propagation in solids," *J. Geophys. Res.* **65**, pp. 695–719.

[16]   Barker, L.M. (1968), "Fine Structure of Compressive and Release Wave Shapes in Aluminum Measured by the Velocity Interferometer Technique," in: *Behavior of Dense Media Under High Dynamic Pressures*, Gordon & Breach, N.Y, p. 483.

[17]   Barker, L. M. (1971), "A Model for Stress Wave Propagation in Composite Materials," *J. Composite Materials* **5**, 140–162.

[18]   Barker, L.M., and R.E. Hollenbach (1974) "Shock wave study of the $\alpha \Leftrightarrow \varepsilon$ phase transition in iron," *J. Appl. Phys.* **45** (11), pp. 4872–4887.

[19]   Baumung, K., H.J. Bluhm, B. Goel, P. Hoppe, H.U. Karow, D. Rusch, V.E. Fortov, G.I. Kanel, S.V. Razorenov, A.V. Utkin, and O.Yu. Vorobjev (1986), in: *Laser and Particle Beams*, **14**, Cambridge Univ. Press, pp. 181–209.

[20]   Bishop, D., P. Gammel, and C.R. Giles (2001), "The little machines that are making it big," in *Physics Today* **54**, pp. 38–44.

[21]   Bland, D.R. (1964), "Dilatational waves and shocks in large displacement isentropic dynamic elasticity, " *J. Mech. Phys.* **12**, pp. 245–267.

[22]   Bloomquist, D.D., and S.A. Sheffield (1983), "Optical recording interferometer for velocity measurements with subnanosecond resolution," *J. Appl. Phys.* **54** pp. 1717–1722.

[23]   Brannon, P.J., C. Konrad, R.W. Morris, E.D. Jones and J.R. Asay (1983), "Studies of the Spectral and Spatial Characteristics of Shock-Induced Luminescence from x-cut Quartz, " *J. Appl. Phys.* **54**, pp. 6374–6381.

[24]   Briant, C., B. Morris, T. Tombrello, and S. Yip (2001) "Report on the Bodega Bay Workshop on Multi-Scale Modeling of Materials, " Oct. 7–10.

[25]   Bridgman, P.W. (1956), "High-pressure polymorphism of iron," *J. Appl. Phys.* **27**, p. 659.

[26]   Chhabildas, L.C. H.J. Sutherland, and J.R. Asay (1979), "A velocity interferometer technique to determine shear-wave particle velocity in shock-loaded solids," *J. Appl. Phys.* **50**, pp. 5196–5201.

[27]   Chhabildas, L.C., and J.R. Asay, (1979), "Rise-time measurements of shock transitions in aluminum, copper and steel," *J. Appl. Phys.* **50**, pp. 2749–2756.

[28]   Chhabildas, L.C., and J.R. Asay (1982), "Time-Resolved Wave Profile Measurements in Copper to Megabar Pressures," in: *High Pressure in Research and Industry* **V1** (eds. C.M. Backman, T. Johanisson, and L. Tegner), pp. 183–189.

[29]   Chhabildas, L.C., J.L. Wise, and J.R. Asay (1982), "Reshock and Release Behavior of Beryllium," in: *Shock-waves in Condensed Matter—1981,* (eds. W.J. Nellis, L. Seaman, and R.A. Graham, American Institute of Physics, New York, pp. 422–426.

[30]   Chhabildas, L.C. and R.A. Graham (1987), in: *Techniques and theory of stress measurements for shock wave applications,* AMD **83** (eds. R.R. Stout, F.R. Norwood, and M.E. Fourney), ASME New York, pp. 1–18.

[31]   Chhabildas, L.C., J.R. Asay, and L.M. Barker (1988) *Shear Strength of Tungsten under Shock- and Quasi-Isentropic Loading to 250 GPa*, Sandia Report SAND88-0306 (unpublished).

[32]   Chhabildas, L.C., W.M. Trott, and W.D. Reinhart (2001), "Incipient Spall Studies in Tantalum – Microstructural Effects," *Bulletin of the Amer. Phys. Soc.* **46**, p. 25.

[33]   Conley, P.A., D.J. Benson, and P.M. Howe (2001), "Microstructural effects in shock initiation," in: *Proceedings of the Eleventh International Detonation Symposium*, in press.

[34]   Courant, R., and K.O. Friedrichs (1948), *Supersonic Flow and Shock Waves*, Interscience, New York, 1991,

[35]   Curran, D.R., L. Seaman, and D.A. Shockey (1987), "Dynamic Failure Of Solids," *Phys. Reports* **147**, p. 253.

[36]   D'Almeida, T., and Y.M. Gupta (2000), "Real-time x-ray diffraction measurements of the phase transition in KCl shocked along [100]," *Phys. Rev. Lett.* **85**, pp. 330–333.

[37] Davis, J.P. (2001), "Formation of shocks from ramp input," Sandia National Laboratories, private communication.

[38] Davison, L. and R.A. Graham (1979), "Shock Compression of Solids," *Physics Reports* **55**, pp. 257–379.

[39] Davison, L. (2002), "Traditional analysis of nonlinear wave propagation in solids," *This volume,* Chapter 1.

[40] Divakov, A.K., L.S. Kokhanchik, Yu.I. Mescheryakov, and M.M. Myshlyaev (1987), "Micromechanics of dynamic deformation and failure," Translated from Zhurnal *Prikladnoi Mekhaniki i Tekhnicheskoi Fiziki*, pp. 135–144.

[41] Dremin, A.N., and O.N. Breusov (1968), "Processes Occurring in Solids Under the Action of Powerful Shock Waves," *Russian Chemical Reviews* **37**, pp. 392–402.

[42] Dhere, A.G., and M.A. Meyers (1982), "Correlation between texture and substructure of conventionally and shock-wave-deformed aluminum," *Materials Science and Engineering* **54**, pp. 113–120.

[43] Dunn, J.E. (1990), "Implications and origins of shock structure in solids," in: *Shock Compression of Condensed Matter —1989* (eds. S.C. Schmidt, J.N. Johnson, and L.W. Davison), Elsevier Science Publishers B.V., Amsterdam, pp.21–32 (1990).

[44] Duvall, G.E., and G.R. Fowles (1963), "Shock Waves," *High Pressure Physics and Chemistry 2*, Chap. 9 (ed. R.S. Bradley), Academic Press, New York.

[45] Duvall, G.E., and R.A. Graham (1977), "Phase transitions under shock-wave loading," *Rev. Modern Phys.* **49**, pp. 523–579.

[46] Egorov, L.A., A.I. Barenboim, V.V. Mokhova, V.V. Dorohin, and A.I. Samoilov (1997), "X-ray diffraction studies of structures of Be, Al, LiF, Fe +3%Si, Si, SiO_2, KCl under dynamic pressures from 2 GPa to 20 GPa," *Journal de Physique IV*, **7** (C3), pp. 355–360.

[47] Fowles, G.R. (1961), "Shock Wave Compression of Hardened and Annealed 2024 Aluminum," *J. Appl. Phys.* **32**, 1475–1487.

[48] Fowles, G.R. (1972), "Experimental technique and Instrumentation," in: *Dynamic Response of Materials to Impulsive Loading* (eds. P.C. Chou and A.K. Hopkins), pp. 405–480.

[49] Gahagan, K.T., D.S. Moore, D.J. Funk, R.L. Rabie, S.J. Buelow, and J.W. Nicholson (2000), "Measurement of shock wave rise times in metal thin films," *Phys. Rev. Lett.* **85**, pp. 3205–3208.

[50] Germann, T.C., B.L. Holian, P.S. Lomdahl (2000), "Orientation dependence of molecular dynamic simulations of shocked single crystals," *Phys. Rev. Lett.* **84**, pp. 5351–5354.

[51] Gilman, J.J. (2001), "Elastic-plastic impact (some persistent mis-conceptions)," in: *Fundamental Issues and Applications of Shock-Wave and High-Strain-Rate Phenomena* (eds. K.P. Staudhammer, L.E. Murr, and M.A. Meyers), Elsevier Sciences Ltd., New York.

[52] Gillman, J.J. (2001), private communication.

[53] Grady, D.E., W.J. Murri, and P.S. DeCarli, (1975) "Hugoniot Sound Velocities and Phase Transformation in Two Silicates," *J. Geophys. Res.* **80**, p. 4857.

[54]   Grady, D.E. (1981), "Strain-rate dependence of the effective viscosity under steady-wave shock compression," *Appl. Phys. Lett.* **38**, pp. 825–827.

[55]   Grady, D.E. and J.R. Asay (1982), "Calculation of thermal trapping in shock deformation of aluminum," *J. Appl. Phys.* **53**, p. 7350.

[56]   Grady, D.E. (1998), "Scattering as a mechanism for structured shock waves in metals," *J. Mech. Solids* **10**, pp. 2017–2032.

[57]   Graham, R.A. (1979). "Shock-induced electrical activity in polymeric solids. A mechanically induced bond scission model," *J. Phys. Chem.* **83**, pp. 3048–3056.

[58]   Graham, R.A., and J.R. Asay (1978), "Measurements of wave profiles in shock-loaded solids," *High-Temperatures-High Pressures* **10**, pp. 355–390.

[59]   Graham, R.A. and D.M. Webb (1984), "Fixtures for Controlled Explosive Loading and Preservation of Powder Samples," in: *Shock-waves in Condensed Matter—1983* (eds. J.R. Asay, R.A. Graham, and G.K. Straub), Elsevier Science Publishers, Amsterdam, pp. 211–214.

[60]   Graham, R.A. (1993a), *Solids Under High-Pressure Shock Compression: Mechanics, Physics, and Chemistry*, Springer–Verlag, New York.

[61]   Graham, R.A., (1993b), "Bridgman's Concern," *High Pressure Science and Technology, Part 1*, (eds. S.C. Schmidt, J.W. Shaner, G.A. Samara, and M. Ross), American Institute of Physics, New York, pp. 3–12.

[62]   Gray, G.T. III (1993), "Influence of shock-wave deformation on the structure/property behavior of materials," in *High-Pressure Shock Compression of Solids* (eds. J.R. Asay and M. Shahinpoor), Springer-Verlag, New York, pp. 187–213.

[63]   Guduru, P.R., A.J. Rosakis, and G. Ravichandran (2001a), "Dynamic shear bands: an investigation using high speed optical and infrared diagnostics," *Mech. of Materials* **33**, pp. 371–402.

[64]   Guduru, P.R., A.T. Zehnder, A.J. Rosakis, and G. Ravichandran (2001b), "Dynamic full field measurements of crack tip temperatures," *Eng. Frac. Mech.* **68**, pp. 1535–1556.

[65]   Gupta, Y.M. (1975), "Stress dependence of elastic-wave attenuation in LiF, " *J. Appl. Phys.* **46**, pp. 3395–3401.

[66]   Gupta, Y.M. (1977), "Effect of crystal orientation on dynamic strength of LiF, " *J. Appl. Phys.* **48**, pp. 5067–5073.

[67]   Gupta, Y.M., D.D. Keough, D. Henley, and D.F. Walter (1980), " Measurement of lateral compressive stresses under shock loading," *Appl. Phys. Lett.* **37**, pp. 395–397.

[68]   Gupta, Y.M., K.A. Zimmerman, P.A. Rigg, E.B. Zaretsky, D.M. Savage, and P.M. Bellamy (1999), "Experimental developments to obtain real-time x-ray diffraction measurements in plate impact experiments," *Rev. Sci. Instr.* **10**, pp. 4008–4014.

[69]   Gupta, Y.M. (2000) "Shock wave experiments at different length scales: recent achievements and future challenges," *Shock Compression of Condensed Matter—1999*, (eds. M.D. Furnish, L.C. Chhabildas, and R.A. Hixson), Amer. Inst. Phys., pp. 3–10.

[70]   Hare, D. (2001), "Optical emission of sapphire shock-loaded to 255 GPa," *Bulletin of the Amer. Phys. Soc.* **46**, p. 22.

[71]   Herrmann, W (1974), "Development of a high strain rate constitutive equation for 6061-T6 aluminum," Sandia National Laboratories Report SLA-73 0897, unpublished.

[72]   Herrmann, W., and R.J. Lawrence (1978), *J. Eng. Mat. Tech.* **100** (84).

[73]   Hodowany, J.G. Ravichandran, A.J. Rosakis and P. Rosakis (1998), "On the partition of plastic work into heat and stored energy in metals; part I: Experiments," California Institute of Technology Report SM Report 98-7.

[74]   Holian, B.L., and G.K. Straub (1979), "Molecular dynamics of shock waves in three-dimensional solids: transition from nonsteady to steady waves in perfect crystals and implication for the Rankine–Hugoniot conditions," *Phys. Rev. Lett.* **43**, pp. 1598–1600.

[75]   Holian, B.L. and G.K. Straub (1980), "Shock-wave structure via nonequilibrium molecular dynamics and Navier-Stokes continuum mechanics," *Phys. Rev. A* **22**, pp. 2798–2808.

[76]   Holian, B.L. (1988), "Modeling shock-wave deformation via molecular dynamics," *Phys. Rev. A* **37**, pp. 2562–2568.

[77]   Holian, B.L. (1995), "Atomistic computer simulations of shock waves," *Shock Waves* **5**, pp. 149–157.

[78]   Holian, B.L. and P.S. Lomdahl (1998), *Science* **280**, p. 2085.

[79]   Holian, B.L. (2002), "What is a shock wave? — The view from the atomic scale," this volume, Chap. 4.

[80]   Hoover, W.G. (1979), "Structure of a shock wave in a liquid," *Phys. Rev. Lett.* **42**, pp. 1531–1534.

[81]   Horie, Y. (1970a), "Characteristics of compressible fluids and the effects of heat conduction and viscosity," *Amer. J. Phys.* **38**, pp. 212–215.

[82]   Horie, Y. (1970b), "An aspect of energy transfer in axial impact of rods," TR 70-6, North Carolina State University, September 1970, Raleigh, NC, prepared for ONR under contract N00014-68-A-0187.

[83]   Horie, Y. (1971), "Adiabatic theory of plane steady shock profiles in solids," *J. Appl. Phys.* **42**, pp. 2925–2933.

[84]   Horie, Y. and H.L. Chang (1972), "Analysis of plane shock structures in 6061-T6 aluminum," *J. Appl. Phys.* **43**, pp. 3362–3368.

[85]   Horie, Y. (1973), "Classification of steady-profile shocks in liquids," *J. Appl. Phys.* **45**, pp. 759–764.

[86]   Horie, Y. (1976), " Thermal-energy relaxation in shock-compressed solids," *J. Mech. Phys. Solids* **24**, pp. 361–379.

[87]   Hugoniot, P.H. (1889), "On the propagation of movement in bodies, in particular in ideal gases," *J. de l'Ecole Polytechnique* **57**.

[88]   Johnson, J.N. (1993), "Micromechanical considerations in shock compression of solids," in *High-Pressure Shock Compression of Solids* (eds. J.R. Asay and M.Shahinpoor) Springer-Verlag, New York, pp. 222–240.

[89]   Johnson, J.N. and L.M. Barker (1969), "Dislocation dynamics and steady plastic wave profiles in 6061-T6 aluminum," *J. Appl. Phys.* **40**, pp. 4321–4334.

[90]   Johnson, J.N., O.E. Jones, and T.E. Michaels (1970), "Dislocation dynamics and single-crystal constitutive relations: shock-wave propagation and precursor decay," *J. Appl. Phys.* **41**, pp. 2330–2339.

[91]   Johnson J.N., R.S. Hixson, and G.T. Gray III (1994), "Shock-wave compression and release of aluminum/ceramic composites," *J. Appl. Phys.* **76**, pp. 5706–5718.

[92]   Johnson, Q., A. Mitchell, R.N. Norris, and L. Evans (1970), "X-ray diffraction during shock-wave compression," *Phys. Rev. Lett.* **25**, pp. 1099–1101.

[93]   Johnson, Q., and A.C. Mitchell (1972), "First x-ray diffraction evidence for a phase transition during shock-wave loading," *Phys. Rev. Lett.* **29**, p. 1369.

[94]   Jones, O.E., and J.D. Mote (1969), "Shock-induced dynamic yielding in copper single crystals," *J. Appl. Phys.* **40**, pp. 4920–4928.

[95]   Kadau, K., T.C. Germann, P.S. Lomdahl, and B.L. Holian (2002), "Microscopic view of structural phase transitions due to shock waves," submitted to *Science*.

[96]   Kalantar, D.H., B.A. Remington, J.D. Colvin, K.O. Mikaelian, S.V. Weber, L.G. Wiley, J.W. Wark, A. Loveridge, A.M. Allen, A.A. Hauer, and M.A. Meyer (2000), "Solid-state experiments at high pressure and strain rate," *Phys. Plasmas* **7**, pp. 1999–2006.

[97]   Kerley, G.I. (2000), "XRB: A new reactive burn model for heterogeneous and homogenous explosives," Kerley Publishing Co. Report KPS00-1.

[98]   Kanel, G.I., G.E. Vakhitova, and A.N. Dremin, (1978), "Metrological characteristics of manganin pressure pickups under conditions of shock compression and unloading," *Combustion, Explosion, and Shock Waves* **14**, pp. 244–248.

[99]   Kanel', G.I., M.F. Ivanov, and A.N. Parshikov (1995), "Computer simulation of the heterogeneous materials response to the impact loading," *Int. J. Impact. Eng.* **17**, pp. 455–464.

[100]   Khantuleva, T.A. (2000), "Non-local theory of high-rate straining followed by structure formations," *J. Phys. IV, France* **10**, 485–490.

[101]   Khantuleva, T.A., and Yu.I. Meshcheryakov (1999), "Non-local theory of the high-strain-rate processes in structured media," *International J. Solids and Structures* **36**, pp. 3105–3129.

[102]   Knudson, M.D. (1999), "Transformation mechanism for the pressure-induced phase transition in shocked CdS," *Phys. Rev. B* **59**, pp. 11704–11715.

[103]   Knudson, M.D. and Y.M. Gupta (2002), "Transformation kinetics for the shock wave induced phase transition in cadmium sulfide crystals," submitted to *Phys. Rev. Lett.*

[104]   Kim, K.S., R.J. Clifton, and P. Kumar (1977), "A combined normal- and transverse-displacement interferometer with an application to impact of y-cut quartz," *J. Appl. Phys.*, **48**, pp. 4132–4139.

[105]   Kumar, P. and R.J. Clifton (1977), "Star-shaped flyer for plate-impact recovery experiments," *J. Appl. Phys.* **48**, pp. 4850–4852.

[106]   Kipp, M.E., L.C. Chhabildas, W.D. Reinhart, and M.K. Wong (2000), "Polyurethane foam impact experiments and simulations," in: *Shock Waves in Condensed Matter—1999* (eds. M.D. Furnish, L.C. Chhabildas, and R.S. Hixson), American Institute of Physics, New York, pp. 313–316.

[107]   Kormer, S.B. (1968), "Optical study of the characteristics of shock-compressed condensed dielectrics," *Sov. Phys. Uspekhi* **11**, pp. 229–254.

[108] Larsson, J., Z. Chang, E. Judd, P.J. Schuck, R.W. Falcone, P.A. Heimann, H.A. Padmore, H.C. Kapteyn, P.H. Bucksbaum, M.N. Murnane, R.W. Lee, A. Machacek, J.S. Wark, X. Liu, and B. Shan (1997), "Ultrafast x-ray diffraction using a streak camera detector in averaging mode," *Optics Letters* **22**, pp. 1012–1014.

[109] Larsson, J., P. A. Heimann, A. M. Lindenberg, P. J. Schuck, P. H. Bucksbaum, R.W. Lee, H.A. Padmore, J.S. Wark, and R.W. Falcone (1998), "Ultrafast structural changes measured by time-resolved x-ray diffraction," *Appl. Phys. A* **66**, pp. 587–591.

[110] Lindenberg, A.M., I. Kang, S.L. Johnson, T. Missalla, P.A. Heimann, Z. Chang, J. Larsson, P.H. Bucksbaum, H.C. Kapteyn, H.A. Padmore, R.W. Lee, J.S. Wark, and P.W. Falcone (2000), "Time-resolved x-ray diffraction from coherent phonons during a laser-induced phase transition," *Phys. Rev. Lett.* **84**, pp. 111–114.

[111] Lipkin, J. and J.R. Asay (1977), "Reshock and release of shock-compressed 6061-T6 aluminum," *J. Appl. Phys.* **48**, p. 182.

[112] Makarov, P.V. (1992) "Microdynamic theory of plasticity and failure of structurally inhomogeneous media," Translated from *Izvestiya Vysshikh Uchebnykh Zavedenii, Fizika* **4**, pp. 42–58.

[113] Makarov, P.V. (1998), "Physical mesomechanics approach in simulation of deformation and fracture processes," *Physical Mesomechanics* **1**, pp. 57–75.

[114] Mashimo, T., and M. Uchino (1997), "Heterogeneous free-surface profile of B_4C polycrystal under shock compression," *J. Appl. Phys.* **81**, pp. 7064–7066.

[115] McQueen, R.G., S.P. Marsh, J.W. Taylor, J.N. Fritz and W.J. Carter (1970), "The equation of state of solids from shock wave studies," in: *High-Velocity Impact Phenomena* (ed. Ray Kinslow), Academic Press, New York, pp. 293–417.

[116] Meshcheryakov, Yu.I. (1983), "Relaxation of submicrosecond pressure pulses in a solid," Translated from *Zhurnal Prikladnoi Mekhaniki i Tekhnicheskoi Fiziki*, **2**, pp. 102–109.

[117] Meshcheryakov, Yu.I., and A.K. Divakov (1985), "Particle velocity variations in a shock wave and fracture strength of aluminum," *Sov. Phys. Tech. Phys.* **30**, pp. 348–350.

[118] Meshcheryakov, Yu., S.A. Atroshenko, V.B. Vasilkov and A.I. Chernyshenko (1992), "Criteria of transition from translation to rotational motion of media under shock loading," in: *Shock Compression of Condensed Matter—1991* (eds. S.C. Schmidt, R.D. Dick, J.W. Forbes and D.G. Tasker), Elsevier Sciences Publ., Amsterdam, pp. 407–41.

[119] Meshcheryakov, Yu.I., and S.A. Atroshenko (1992), "Dynamic rotations in crystals," Translated from *Izvestiya Vysshikh Uchebnykh Zavendenii, Fizika* **4**, pp. 105–123.

[120] Meshcheryakov, Yu.I., N.A. Mahutov, and S.A. Atroshenko (1994), "Micromechanisms of dynamic fracture of ductile high-strength steel," *J. Mech. Phys. Solids* **42**, pp. 1435–1457.

[121] Meshcheryakov, Yu. I., and A. K. Divakov (1994), "Multi-scale kinetics of microstructure and strain-rate dependence of materials," *DYMAT Journal* **1**, 271–287.

[122] Meshcheryakov, Yu., and S.A. Atroshenko (1995), "Dynamic recrystallization in shear bands," in: *Metallurgical and Materials Applications of Shock-Wave and High-Strain-Rate Phenomena* (eds. L.E. Murr, K.P. Staudhammer and M.A. Meyers), Elsevier Science B.V., Amsterdam.

[123] Meyers, M.A. (1975a), "A "wavy wave" model for the shocking of polycrystalline metals," in: *Proc. 5th International Conf. On High-Energy Rate Fabrication*, Denver Research Institute, June.

[124] Meyers, M.A., (1975b), The effect of surface conditions on shock hardening," *Scripta Metallurgica* 9, pp. 667–669.

[125] Meyers, M.A. and M.S. Carvalho, (1976), "Shock-front irregularities in polycrystalline metals," *Mat. Sci. and Eng.* 24, pp. 131–135.

[126] Meyers, M.A. (1977), "A model for elastic precursor waves in the shock loading of polycrystalline metals," *Materials Science and Engineering* 30, pp. 99–111.

[127] Meyers, M.A., L.E. Murr, C.Y. Hsu, and G.A. Stone (1983), "The effect of polycrystallinity on the shock wave response of Fe-34.5wt.%Ni and Fe-15wt%Cr-15wt.%Ni," *Materials Science and Engineering* 57, pp. 113–126.

[128] Meyers, M.A., K.-C. Hsu, and K. Couch-Robino (1983), "The attenuation of shock waves in nickel: second report," *Materials Sci. and Engineering* 59, pp. 235–249.

[129] Mogilevsky, M.A. (1981), "Mechanisms of deformation under shock loading," in: *Shock Waves and High-Strain-Rate Phenomena in Metals: Concepts and Applications* (eds. Marc. A. Meyers and Lawrence E. Murr), Plenum Press, New York, pp. 531–546.

[130] Moore, D.S., K.T. Gahagan, J.H. Reho, D.J. Funk, S.J. Buelow, and R.L. Rabie (2001), "Ultrafast nonlinear optical method for generation of planar shocks," *Appl. Phys. Lett.* 78, pp. 40–42.

[131] Murri, W. (1974), *Advances in High-Pressure Research, 4* (ed. R.H. Wentorf, Jr.), Academic Press, London and NY.

[132] Nesterenko, V.F. (2002), *Dynamics of Heterogeneous Materials*, Springer-Verlag, New York, 2001.

[133] Nunziato, J.W., E.K. Walsh, K.W. Schuler, and L.M. Barker (1974), "Wave Propagation in Nonlinear Viscoelastic Solids," in: *Handbuch der Physik*, Vol. VIa/4, (ed. S. Flügge), Springer-Verlag, Berlin, pp. 1–108.

[134] Panin, V., Yu.V. Grinyaev, T.F. Elsukova, and A.G. Ivanehin (1982), "Structural levels of deformation in solids," *Izvestiya Vysshikh Uchebnykh Zavedenii, Fizika*, 6, pp. 5–27.

[135] Panin, V.E. (1997), "Physical mechanisms of mesomechanics of plastic deformation and fracture of solids," *Acta Metallurgica Sinica* 33, pp. 187–197.

[136] Panin, V.E. (1998), "Foundations of physical mesomechanics," *Physical Mesomechanics* 1, pp. 5–20.

[137] Perry, F.C., and D.D. Noack (1986), "Heterogeneous response of aluminum and lithium niobate to shock waves," *Shock Waves in Condensed Matter* (ed. Y.M. Gupta), Plenum Press, New York, pp. 655–660.

[138] Persson, Per-Anders and Gunnar Persson (1972), "High-resolution photography of transverse wave effects in the detonation of condensed explosives," *Proceedings of 6th Symposium on Detonation*, (ed. D.J. Edwards), Office of Naval Research, Arlington, VA. pp. 414–425.

[139] Reinhart, W.D., L.C. Chhabildas, D.E. Grady, and T. Mashimo (2002), "Shock compression and release properties of Coors AD995 alumina," to be published in the proceedings of the American Ceramic Society PACRIM – Conference, November 3–6, 2001.

[140] Rosakis, P., A.J. Rosakis, G. Ravichandran, and J. Hodowany (1998), "On the partition of plastic work into heat and stored energy in metals; Part II: Theory," California Institute of Technology Report SM Report 98-8.

[141] Rosenberg, Z. and Y. Partom (1985), "Lateral stress measurement in shock-loaded targets with transverse piezoresistance gauges," *J. Appl. Phys.* **58**, pp. 3072–3076.

[142] Rice, M.H., R.G. McQueen and J,M, Walsh (1958), "Compression of solids by strong shock waves," in: *Solid State Physics* **1** (eds. F. Seitz and D. Turnbull), Academic, New York, pp. 1–63.

[143] Rigg, P.A., and Y.M. Gupta (1998), "Real-time x-ray diffraction to examine elastic-plastic deformation in shocked lithium fluoride crystals," *J. Appl. Phys. Lett.* **73**, pp. 1655–1657.

[144] Rigg, P.A. and Y.M. Gupta (2001), "Multiple x-ray diffraction to determine transverse and longitudinal lattice deformation in shocked lithium fluoride," *Phys. Rev B* **63**, p. 094112.

[145] Savenko, G.G., Yu.I. Mescheryakov, V.B. Vasil'kov and A.I. Chernyshenko (1990), "Grain vibrations and development of the turbulent nature of plastic deformation during high-velocity interaction of solid bodies," Translated from *Fizika Goreniya I Vzryva*, **26**, pp. 97–102.

[146] Swan, G.W, G.E. Duvall and C.K. Thornhill (1973), "On steady wave profiles in solids," *J. Mech. Phys. Solids* **21**, pp. 215–227.

[147] Swegle, J.W., and D.E. Grady (1985), "Shock viscosity and the predictions of shock wave risetimes," *J. Appl. Phys.* **58**, pp. 692–701.

[148] Swegle, J. W. and D. E. Grady (1986), "Calculation of thermal trapping in shear bands," in: *Metallurgical applications of shock-wave and high-strain-rate phenomena*, (eds. Lawrence E. Murr, Karl P. Staudhammer, and Marc A. Meyers), NY.

[149] Taylor, J.W. (1965), "Dislocation dynamics and dynamic yielding," *J. Appl. Phys.* **36**, pp. 3146–3150.

[150] Taylor, J.W. (1984), "Thunder in the mountains," in: *Shock Waves in Condensed Matter —1983* (eds. J.R. Asay, R.A. Graham, and G.K. Straub), North-Holland, New York, pp. 3–15.

[151] Teller, E. (1962), "On the speed of reactions at high pressures," *J. Chem. Phys.* **36**, pp. 901–903.

[152] Tomlin, M.G., Yu.I. Meshcheryakov, S.A. Atroshenko, N.I. Zhigacheva, and S.I. Vavilov (1994), "LC vision: the dynamically induced structure of deformation modes in metals," *Mol. Cryst. Liq. Cryst.* **251**, pp. 343–349.

[153] Trott, W.M., J.N. Castaneda, J.J. O'Hare, M.R. Baer, L.C. Chhabildas, M.D. Knudson, J.P. Davis, and J. R. Asay (2000), "Examination of the mesoscopic scale response of shock compressed heterogeneous materials using a line-imaging velocity interferometer," in: *Shock-Wave and High-Strain-Rate Phenomena*, (eds. K.P. Staudhammer, L.E. Murr, and M.A. Meyers), pp. 647–654.

[154] Trott, W.M., J.N. Castaneda, J.J. O'Hare, M.R. Baer, L.C. Chhabildas, M.D. Knudson, J.P. Davis, and J.R. Asay (2000), "Dispersive velocity measurements in heterogeneous materials," Sandia National Laboratories report SAND2000-3082.

[155] Trott, W.M., L.C. Chhabildas, M.R. Baer, J.N. Castaneda, J.J. O'Hare (2001), "Investigation of dispersive waves in low-density sugar and HMX using line-imaging velocity interferometry," *Bulletin of the Amer. Phys. Soc.* **46**, p. 31.

[156] Tonks, D.L. (1988), "Relation between shock strength and strain-rate plasticity at maximum deviatoric stress," *Shock Waves in Condensed Matter—1987*, (eds. S.C. Schmidt and N.C. Holmes), Elsevier Science Publishers B.V., Amsterdam.

[157] Urtiew, P.A. and A.K. Oppenheim (1966), "Experimental observations of the transition to detonation in an explosive gas," *Proc. Royal Soc.* **295**, pp. 13–28.

[158] Urtiew, P.A. (1970), "Reflections of wave interactions in marginal detonations," *Astronautica Acta* **15**, pp. 335–343.

[159] Urtiew, P.A..and A.S. Kusubov (1970), "Wall traces of detonation in nitromethane-acetone mixtures," presented at the 5th Symposium on Detonation, Pasadena, CA, Aug. 18–21.

[160] Urtiew, P.A., A.S. Kusubov, and R.E. Duff (1970), "Cellular structure of detonation in nitromethane," *Combustion and Flame* **14**, pp. 117–122.

[161] Urtiew, P.A. (1975), "From cellular structure to failure waves in liquid detonations," *Combustion and Flames* **25**, pp. 241–245.

[162] Urtiew, P.A. (1976), "Idealized two-dimensional detonation waves in gaseous mixtures," *Acta Astronautica* **3**, pp. 187–200.

[163] Urtiew, P.A. and C.M. Tarver (1981), "Effects of cellular structure on the behavior of gaseous detonation waves under transient conditions," in *Gasdynamics of Detonations and Explosions* (eds. J.R. Bowen, N. Manson, A.K. Oppenheim, and R.I. Soloukhin), in: Progress in Astronautics and Aeronautics Vol. 75.

[164] Yano, K. and Y. Horie (1999), "Discrete-element modeling of shock compression of polycrystalline copper," *Phys. Rev. B* **59**, p. 13672.

[165] Yano, K., and Y. Horie (2002), "Mesomechanics of the $\alpha-\varepsilon$ transition in iron," submitted to *Int. Journal of Plasticity*.

[166] Wallace, D.C. (1980a), "Flow processes of weak shocks in solids," *Phys. Rev. B*, **22**, pp. 1495–1502.

[167] Wallace, D.C. (1980b), "Irreversible thermodynamics of flow in solids," *Phys. Rev. B* **22**, pp. 1477–1486.

[168] Wallace, D.C. (1981), "Irreversible thermodynamics of overdriven shocks in solids," *Phys. Rev. B* **24**, pp. 5597–5606.

[169] Wark, J.W. (1996), "Time-resolved X-ray diffraction," *Contemp. Phys. (UK)*, **36**, pp. 205–218.

[170] Whitlock, R.R., and J.S. Wark (1995), "Orthogonal strains and onset of plasticity in shocked LiF crystals," *Phys. Rev. B* **52,** pp. 8–11.

[171] Widmann, K., G. Guethlein, M.E. Foord, R.C. Cauble, F.G. Patterson, D.F. Price, F.J. Rogers, P.T. Springer, and R.E. Stewart (2001), "Interferometric investigation of femto-second laser-heated expanded states," *Physics of Plasmas* **8**, pp. 3869–3872.

[172] Zel'dovich, Ya.B., and Yu.P. Raizer (1966), *Physics of Shock Waves and High-Temperature Hydrodynamic Phenomena*, **1** (eds. W.D. Hayes and R.F. Probstein) Academic Press, New York and London.

[173] Zaretskii, E.B., G.I. Kanel, P.A. Mogilevskii, and V.E. Fortov (1991), "Device for investigating x-ray diffraction on shock-compressed material," translated from *Teplofixika Vysokikh Temperatur* **29**, pp. 1002–1008.

[174] Zaretskii, E.B., G.I. Kanel, P.A. Mogilevskii, and V.E. Fortov (1991), "X-ray diffraction study of phase transition mechanism in shock-compressed KCl single crystal," *Sov. Phys. Dokl.* **36**, pp. 76–78.

CHAPTER 3

The Universal Role of Turbulence in the Propagation of Strong Shocks and Detonation Waves

John Lee

3.1. Introduction

A shock wave is a steep compression wave that propagates at supersonic speeds relative to the medium ahead of it. In gases, the thickness of the shock transition zone ranges from the molecular mean free path (10^{-9} m) in a perfect gas to millimeters (10^{-3} m) when there are long relaxation and chemical processes involved in reaching the final equilibrium state of the shocked medium. Thus, the shock thickness can cover a spectrum of length scale that differs by six orders of magnitude. For a strong shock, irreversible changes, i.e., plastic yield, fracture, phase transformations, molecular dissociations and ionizations, and chemical reactions, have affected the material behind the shock. The chemical reactions may be exothermic, producing a net increase in the internal energy of the reaction products. For a sufficiently exothermic reaction, the shock wave can be self-propagating, supported by the expansion work of the reaction products. Such self-propagating shock waves are referred to as detonation waves, and media capable of supporting a detonation wave are usually highly exothermic and the reaction rates are also sufficiently fast for the shock generated to be maintained by the expanding products. There is no fundamental difference between strong shocks and detonations except for the self-sustained nature of detonation waves.

Our understanding of shock waves is based mainly on macroscopic experiments where the shock velocity and the post-shock equilibrium states are measured (e.g., particle velocity and pressure). In a macroscopic description, the resolution is based on a length scale that is large compared to the shock thickness and a time scale that is long compared to the transition time across the shock transition zone. Because the shock thickness is generally very thin (10^{-9} m to 10^{-3} m) and the shock velocity is usually high (of the order of a few km/s), the transition time scale involved is very short, of the order of nanoseconds to microseconds. Thus, most experiments are of a macroscopic nature. In a macroscopic description, the shock can be treated as a steady, one-dimensional wave (see Fig 3.1). Losses are sufficiently small in general across the thin transition zone.

The one-dimensional steady conservation laws can be written across two equilibrium planes that bound the shock transition zone, i.e.,

$$\rho_0 D = \rho_1 (D - u_1)$$

$$p_1 - p_0 = \rho_0 D u_1$$

$$e_0 + \frac{p_0}{\rho_0} + \frac{D^2}{2} = e_1 + \frac{p_1}{\rho_1} + \frac{(D - u_1)^2}{2}.$$

The so-called Hugoniot properties of a shock wave can be readily derived from the conservation laws, e.g., the transition from downstream to upstream states follows the straight Rayleigh line on the $p-v$ diagram, i.e.,

$$(\rho_0 D)^2 = j^2 = \frac{p_1 - p_0}{v_0 - v_1},$$

whereas the locus of upstream states is represented by the Hugoniot curve

$$e_1 - e_0 = \frac{1}{2}(p_1 - p_0)(v_0 - v_1),$$

which can also be plotted on the $p - v$ plane if the equation of state $e(p,\rho)$ is known.

With the conservation laws, it suffices to measure one other upstream state variable in addition to the shock velocity to determine all the rest of the post-shock state variables, e.g., if one measures the particle velocity u_1, then the pressure p_1 can be found from the momentum equation, the density ρ_1 from the conservation of mass, and the internal energy e_1 from the energy equation. A macroscopic description does not tell us anything about the propagation mech-

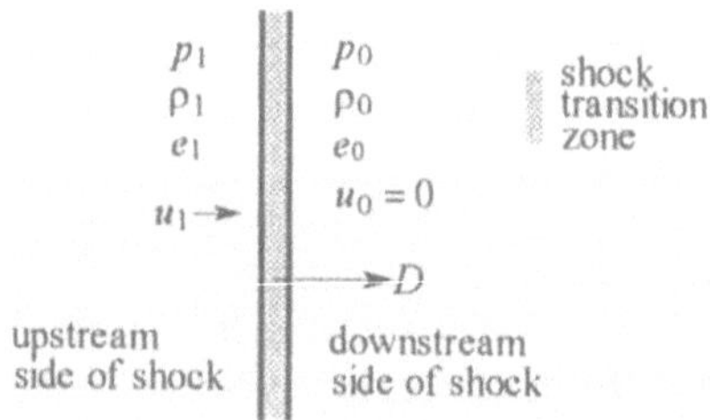

Figure 3.1. Shock wave propagating at velocity D. The pressure is p, the mass density is ρ, the internal energy density is e, and the particle velocity relative to the laboratory coordinate frame is u.

anism of the shock wave. By propagation mechanism, we refer to the physical and chemical processes that are responsible for effecting the transition across the shock itself.

To learn something about the propagation mechanism, we must do higher resolution experiments so that the shock structure can be resolved. A mesoscopic description on a length scale that is small as compared to the shock thickness and a time scale that is short relative to the transit time across the shock transition zone is required to describe the shock structure. This presents a difficult experimental challenge in view of the fine space and time resolutions required. However, recent advances in experimental diagnostics have now revealed more and more detail of the shock structure. Macroscopic experiments of the past five decades have strongly reinforced the notion that shock waves are steady-state, one-dimensional waves sufficiently thin to be treated as discontinuities. Hence, when a description of the relaxation processes occurring behind the leading boundary of the shock is required, a "laminar" one-dimensional model is usually adopted. This model includes a leading normal shock where translational equilibration is first achieved and then a region in which the slower relaxation processes of intramolecular energy (e.g., vibrational) equilibration, dissociation, and chemical reactions occur. However, there is sufficient experimental evidence to indicate that, on a mesoscopic level, the shock structure is highly nonsteady and three dimensional, with large hydrodynamic fluctuations superimposed on the mean flow quantities. Instead of a normal planar front, the leading boundary of the shock takes on a "corrugated" or "cellular" structure with the cell boundaries defined by the ensemble of transverse shock waves sweeping across the leading front. The transition zone itself is highly turbulent with a collection of eddies as well as interacting transverse shock waves, shear layers, and density interfaces

Hence, the relaxation and chemical processes are now occurring in this highly complex turbulent zone. However, if one takes the upstream plane to be sufficiently far behind the transition zone to ensure the decay of the hydrodynamic fluctuations, then the global one-dimensional "steady" conservation laws can still be applied to preserve the Hugoniot properties of the shock wave (as derived from the conservation laws). However, the steady one-dimensional conservation laws are no longer valid inside the turbulent transition zone where large fluctuations and a strong departure from local equilibrium occurs. Note that the steady conservation laws do not contain any physics of the shock transition process. Hence, many of the practical parameters of interest cannot be obtained unless one resolves the complex turbulent shock transition zone itself. For example, understanding deformation and fracture mechanisms, the critical initiation pressure and run-up distances for the transition from deflagration to detonation in an explosive medium, etc., all require a description of the nonequilibrium turbulent shock structure. Current models for the transition zone are generally based on one-dimensional steady flow concepts; hence, they are incapable of describing the turbulence mechanisms. Thus arbitrary empirical constants, fit to experimental data, are generally used to render these models of practical value.

It is important to define what is meant by "turbulence" in the context of this chapter. The general notion of turbulence is derived from the study of aerodynamics where compressibility effects are negligible in general. Turbulence in an incompressible flow involves mainly velocity fluctuations, and the turbulent mechanism is essentially one of vortex–vortex interactions with vorticity being generated in the shear flows at the boundaries. The spectrum of length scales goes from the smallest dissipative Kolmogorov scale to the largest eddy size, i.e., from a fraction of a millimeter for high Reynolds number flows to tens of centimeters. When compressibility effects are important, strong pressure fluctuations are present in addition to the velocity fluctuations. An ensemble of strong pressure waves or shock waves will be generated as a result of the strong pressure fluctuations. Apart from vortex–vortex interaction, we now have shock–shock and shock–vortex interactions. Vorticity production can result from shock–shock (Mach) interactions as well as from pressure and density gradient fields, i.e., the baroclinic mechanism $\vec{\nabla} p \times \vec{\nabla}(1/\rho)$. Instability of density interfaces (Richtmyer–Meshkov) also gives rise to strong turbulence production. Thus, in a compressible medium, the turbulent structure is far more complex and includes a wider spectrum of length scales ranging from the shock thickness, the spacing of interacting shocks, the Kolmogorov scale, and the spectrum of eddy sizes. It should be noted that turbulence may not necessarily be random or chaotic, but an orderly structure can be realized quite often (i.e., coherent structure). In the present discussion of shock waves, we are dealing with a highly compressible flow and hence must adopt a broader view of turbulence.

3.2. Turbulent Structure of Gaseous Detonations

It is of interest to first illustrate the experimental observations of the unstable turbulent structure of shock waves in various media. Perhaps the most extensively studied turbulent structure is that of gaseous detonations. In gaseous detonations, the scale of the transition zone can be increased to the order of centimeters by a reduction in the initial pressure or by dilution of the explosive mixture with an inert gas. Diagnostics with a time resolution of the order of microseconds are thus sufficient to reveal the internal structure of the detonation wave. Furthermore, the detonation products are also transparent, permitting a variety of optical methods to be used (e.g., interferometry, schlieren, etc.). Detailed studies of the gaseous detonation structure began in the late 1950s. In Fig. 3.2, examples of interferograms of gaseous detonations taken by White [55] are illustrated. The unstable turbulent structure with the scale of the density fluct-

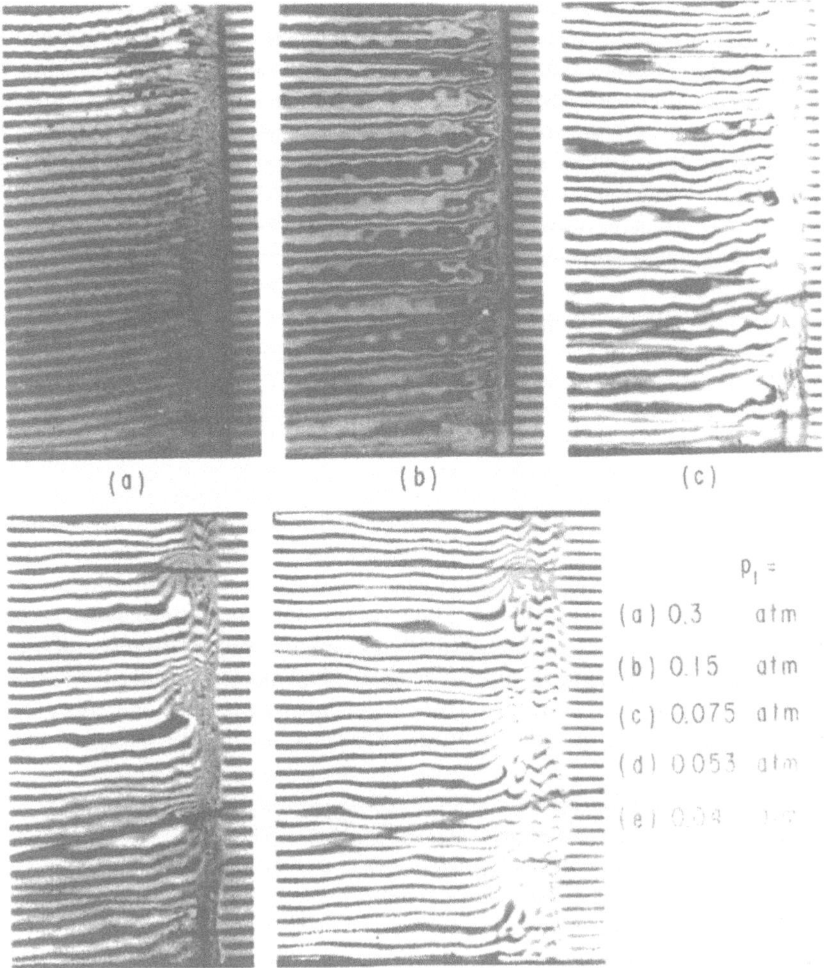

Figure 3.2. Interferograms of gaseous detonations in $2H_2 + O_2 + 0.92$ Xe mixtures at various initial pressures illustrating their turbulent structure (Courtesy of D.R. White, [55] This and the following figure reprinted with permission from the Annual Review of Fluid Mechanics, V. 16 © 1984 by Annual Reviews, www.annualreviews.org).

uations increasing with decreasing initial pressure of the explosive mixture is clearly indicated. The interferograms of White showed that the classical one-dimensional laminar ZND model of the detonation is not realized experimentally. White was also the first to call the detonation structure "turbulent." However, his description was never adopted by the community because turbulence mechanisms had always been associated with those of incompressible aerodynamics. Shock waves and a certain degree of large-scale coherence were not the general notion of turbulent flows at that time. Head-on observation of the detonation revealed a cellular pattern where the cell boundaries are defined by the intersections of the transverse shock waves with the leading front. Figure 3.3 shows the imprint on a soot-covered surface subsequent to the reflection of the detonation wave off the surface. The cellular structure can also be observed by the direct photography of the self-luminous pattern [45]. It appears that the luminosity is the highest at the cell boundaries. Using the impedance-mirror method, Presles et al. [46] also obtained data on the cellular structure of gaseous detonations almost identical to the imprint on soot foils. In the impedance-mirror technique used by Presles, a thin Mylar foil is coated with a thin reflecting layer of aluminum. When the detonation reflects off the foil, the deformation of the foil will conform to the non-uniform pressure distribution on the detonation front. An instantaneous photograph of the foil can be taken when it is illuminated by a bright flash. The impedance-mirror method shows that the cell boundaries also correspond to local high-pressure regions in the detonation front. A schematic sketch of an idealized cellular detonation front at various positions is shown in Fig. 3.4. The "corrugations" of the front correspond to the intersections of the transverse shocks with the leading front. The velocity of the shock front fluctuates from about $1.5V_{CJ}$ to about $0.6V_{CJ}$ within a cycle of collision of the transverse shocks. Thus, locally, the front propagates in a pulsating manner with an average velocity V_{CJ} as determined from equilibrium consideration of the global conservation laws. The trajectories of the triple shock intersections map out a characteristic "fish scale" pattern on the soot-coated surface as illustrated in Fig. 3.5. In this sequence of framing schlieren photo-

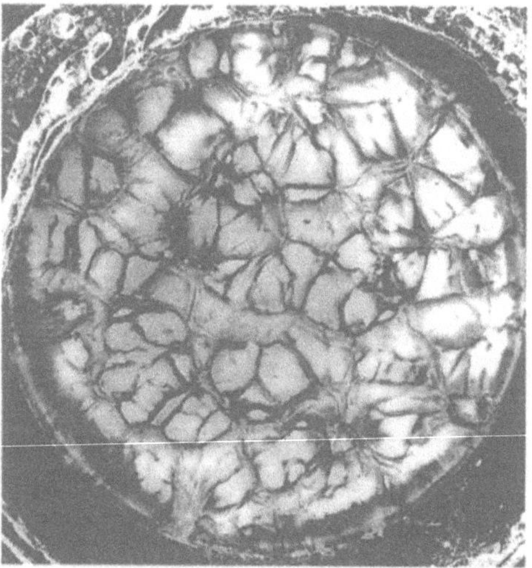

Figure 3.3. End on reflection of a gaseous detonation on a smoked glass plate (stoichiometric acetylene–oxygen mixture).

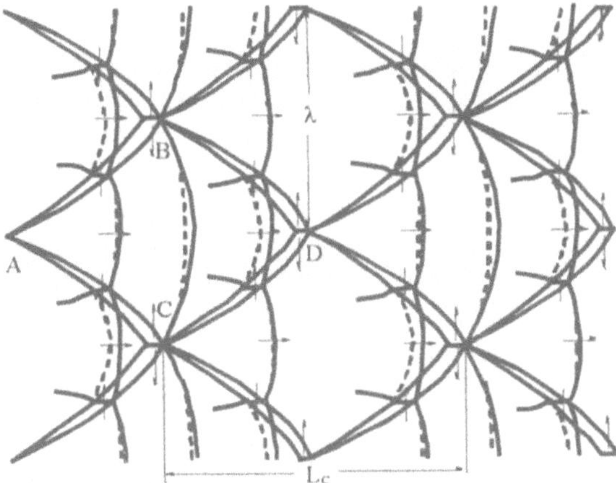

Figure 3.4. Schematic sketch of a cellular detonation illustrating the Mach interactions at the front and trajectories of the triple points.

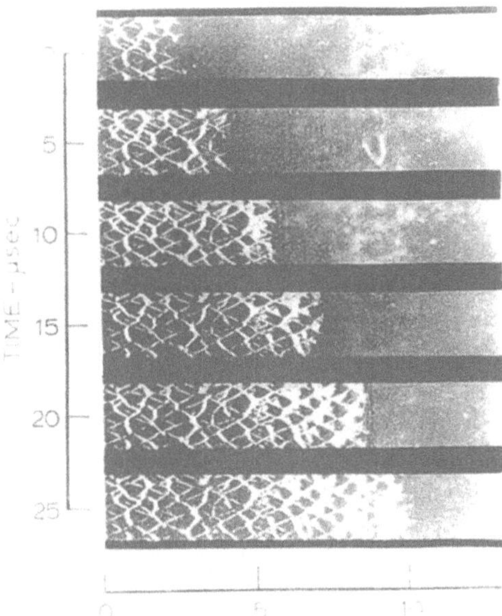

Figure 3.5. Simultaneous schlieren framing photographs and smoked foil record of the trajectories of the triple points (Courtesy of A.K. Oppenheim, 1969).

graphs, one of the side glass windows of the detonation tube was coated with soot. Thus, apart from instantaneous schlieren pictures of the turbulent detonation front, the trajectories of the "triple points" can also be recorded because the soot is erased by the shear layer associated with the triple Mach shock interaction. The average cell size can generally be determined from the average spacing of the triple point trajectories. A variation of cell sizes is generally obtained in a soot foil and, quite often, the soot pattern is complicated and it is difficult to determine a single value of the dominant cell size. Turbulent cellular detonations are found to be a rule rather than an exception, and only in special cases (e.g., an acetylene–oxygen mixture highly diluted with argon) is the detonation sufficiently stable so that the transverse shocks become essentially weak Mach waves that play a minor role (if any) in the shock wave propagation. In such stable detonations, the structure is laminar and is in accord with the one-dimensional ZND model.

3.3. Turbulent Structure in Condensed Phase Detonations

For condensed explosives, the scale of the turbulence is much smaller than in gaseous explosives and hence much more difficult to observe, requiring much higher time and space resolutions. The most extensively studied cellular structure in condensed phase explosives is in nitromethane, where the scale of the turbulence can be increased by dilution with acetone. Figure 3.6 is a direct head-on photograph of the detonation front in nitromethane [19] and a non-uniform cellular pattern is clearly demonstrated. Using the impedance mirror technique, extensive observations of the cellular structure of nitromethane detonations had been made by Mallory [37,38]. Figure 3.7 shows a head-on view of the cellular pattern formed by the pressure inhomogeneities of the detonation on the mirror front. A time-resolved picture of the pressure fluctuation at a diametrical slit of the impedance mirror is shown in Fig. 3.8, indicating the duration of the pressure fluctuation in the reaction zone. A fish-scale pattern of the trajectories of the triple points can also be obtained on the walls of the confining metal tube [53] and on a copper witness plate [25]. These observations of cellular detonation fronts in nitromethane suggest that the propagation mechanism in condensed phase liquid explosives may be similar to that of homogeneous gaseous explosives. We may conclude also that turbulence is required for self-sustained propagation in homogeneous liquid explosives.

For solid explosives, the fine structure of the detonation front in a cast TNT/RDX (35/65) explosive charge had been reported by Held [27,28]. Held used a transverse impedance mirror technique similar to that used by Mallory for liquid explosives. Held found that the detonation front in solid TNT/RDX charges also contains surface undulations indicating a similar cellular structure as in nitromethane. However, the perturbations are of much smaller amplitudes (of the order of $50\,\mu m$) than in nitromethane. A linear dependence of amplitude

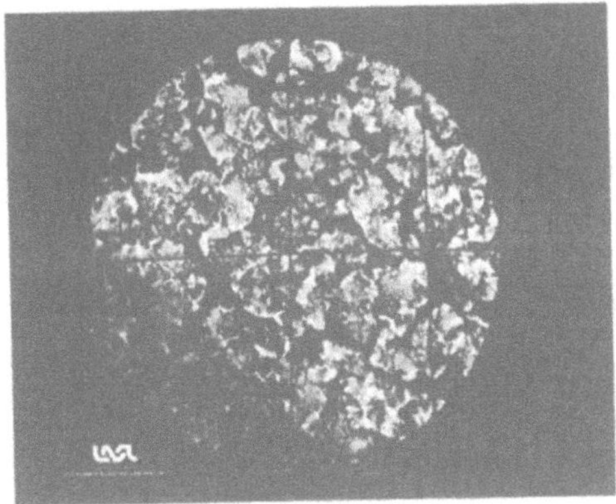

Figure 3.6. End on view of a detonation front in nitromethane (Courtesy W. Davis).

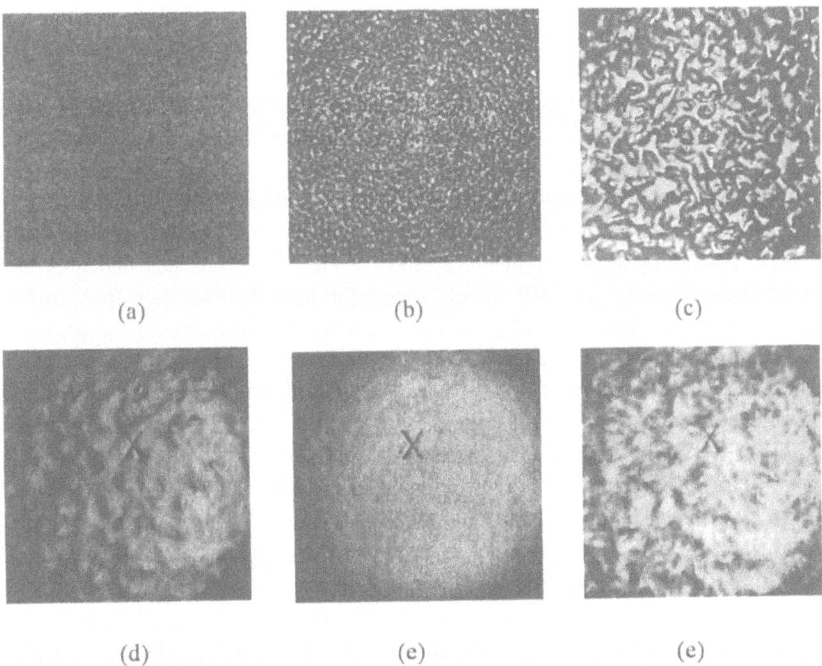

Figure 3.7. Framing camera photographs of impedance mirror patterns of NM/acetone detonation (a) 75/25; (b) 55/45; (c) 45/55; (d) 35/65; (e) 30/70; (f) 35/65 (Courtesy H.D. Mallory [37]).

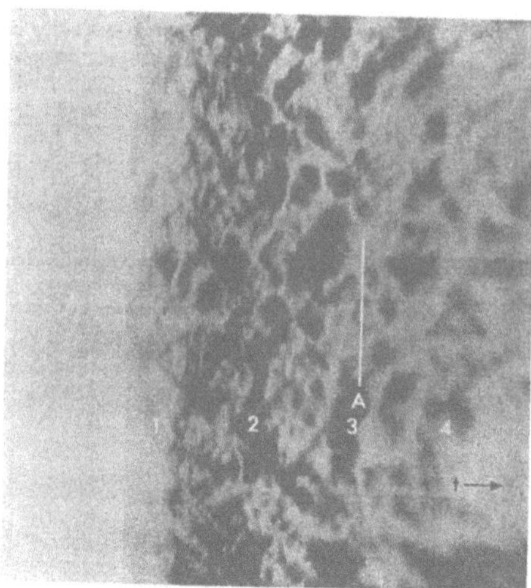

Figure 3.8. Time resolved impedance mirror record of an overdriven detonation wave in 35/65 nitromethane/acetone. Line A indicates the approximate end of the reaction zone (Courtesy of H.D. Mallory [38]).

of the undulation with the grain size of the explosive was observed by Held, indicating that the origin of those perturbations may be due to the nonuniformities of the heterogeneous explosive medium. In a homogeneous medium, the perturbations have to arise from instability of the front itself, with the scale of the cellular front being governed by the most unstable mode. An interesting pattern of transverse waves at the surface of cast TNT charges had also been reported by Howe et al. [30]. In the technique used by Howe, a thin uniform layer of argon (of the order of a millimeter thick) is introduced on the surface of the rectangular explosive charge and the argon layer is confined by a thin plastic sheet. To avoid luminosity from the atmospheric air due to the transmitted shock, a layer of heavy polyatomic gas is placed on top of the argon layer, thus isolating it from the atmospheric air. The air shock transmitted through the polyatomic gas layer is strongly attenuated to produce no significant luminosity in the air when it emerges from this polyatomic gas layer. The high heat capacity of the polyatomic gas layer serves to lower the shock temperature and hence the luminosity of the shocked gases. Thus, one may be able to observe only the luminosity pattern in the argon layer on top of the explosive charge. For the other three sides of the rectangular explosive charge, confinement is provided by brass walls. The luminosity pattern produced in the thin, shocked argon layer as the detonation front propagates through the TNT charge is recorded by open shutter photography. Figure 3.9 [29] shows a typical open shutter photograph of the luminosity pattern of the shocked argon layer. A remarkable similarity to the transverse wave pattern (recorded on smoked foil) of gaseous detonation fronts

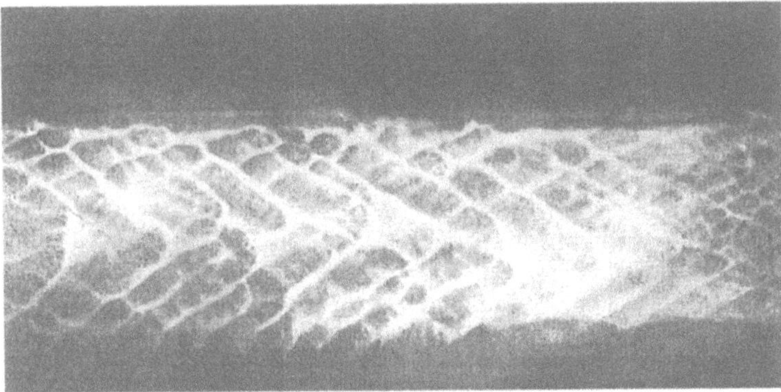

Figure 3.9. Luminosity pattern in a thin Argon layer on the surface of a TNT charge generated by a traversing detonation front. (Courtesy of P.M. Howe, et al. [30])

can be observed. It should be noted that the scale of these transverse wave patterns observed on the surface of solid explosive charges using the argon layer technique is orders of magnitude greater than the scale of the surface undulations on the detonation front observed by Held, using the impedance mirror technique. This is probably due to the highly attenuated curved detonation near the charge boundary, resulting in a much increased reaction zone thickness locally. Corresponding to the larger reaction time, the frequency of the instability is thus lowered. Hence, the observations by Held and Howe appear to suggest that a turbulent cellular structure may also exist in solid condensed explosive detonation fronts. However, solids are intrinsically heterogeneous and thus the turbulent structure may be a direct consequence of this heterogeneity of the granular solid media. Irrespective of the origin of the turbulent fluctuations (i.e., from hydrodynamic instability or from the inherent heterogeneity of the medium itself), the physical effects of the turbulent fluctuations on the propagation of the detonation wave should be the same.

3.4. Turbulent Structure of Non-Reacting Shocks

Shock waves (non-self-propagating waves that require a continuous external energy input to maintain steady propagation) can also be unstable. The structure may also take on a corrugated leading front with turbulent flow in the transition zone in the wake of the front. Instability may arise from different sources. For example, the equation of state can render steady solutions of the one-dimensional conservation laws impossible or the relaxation processes in the transition zone itself may also be unstable. This occurs when strong coupling between the energetics of the relaxation processes and the fluid mechanics is present. The stability of shock waves has been extensively studied for the five decades since the pioneering work of Bethe [4], who investigated the existence of steady shock waves in media with an arbitrary equation of state. Since then, the stability of shock waves has been investigated by numerous researchers in both the U.S. and Russia [10,17,20–22,31,32,49,56]. The nature of shock stability can readily be seen from the basic requirements for a steadily propagating shock, i.e., i) the shock must be supersonic relative to the medium ahead of it and subsonic with respect to the medium behind and ii) the entropy must always increase across the shock. The necessity for the first requirement is clear; the shock must be supersonic so that flow perturbations generated by the shock cannot advance beyond the front and attenuate its steady structure. The downstream subsonic requirement is necessary to permit a communication with the external energy input (i.e., piston) to maintain steady shock propagation. The entropy requirement is necessary to distinguish the compression shock from the rarefaction shock solution because the conservation laws across the shock are symmetrical to both solutions. The condition $\Delta s \geq 0$ (where s is the entropy) for a compression shock is required to avoid violating the Second Law of Thermodynamics. The positive entropy increase requirement leads to the so-called "convexity condition", which requires that the Hugoniot have positive curvature, i.e.,

$$\left(\frac{\partial^2 p}{\partial v^2} \right)_{\mathrm{H}} > 0 . \tag{3.1}$$

The above condition can readily be obtained by expanding the Hugoniot equation about its initial state for a weak shock transition, i.e.,

$$\Delta s \cong -\frac{1}{12T} \left(\frac{\partial^2 p}{\partial v^2} \right)_{\mathrm{H}} (v - v_0)^3 + \cdots .$$

Since $T > 0$, where T is the temperature, and $(v - v_0) < 0$ for a compression shock, we thus obtain the positive curvature requirement of Eq. 3.1 for $\Delta s > 0$. Bethe [4] first showed that all pure substances in a single phase satisfy the above convexity requirement. However, there are situations when the equation of state of the media fail to satisfy the "convexity constraint", e.g., solid undergoing plastic deformation or phase changes, vapor with very high heat capacities (seven or more atoms to a molecule) and in a state near the curve of phase equilibrium with the vapor state near the critical point, ionizing and dissociating shocks, etc. The convexity constraints can only tell us whether steady compression shock solutions of the steady one-dimensional conservation laws are possible or not. To investigate in more detail the nature of the instability, we must analyze the condition for the growth of small perturbations, both longitudinal and transverse to the direction of shock propagation. Longitudinal instability leads to the destruction of the steady shock structure, i.e., splitting of the shock into two shocks or shock and rarefaction, moving either in the same or opposing directions. Longitudinal instability may also be due to the amplification of perturbations reflecting off the shock and a back boundary such as a piston surface.

Spontaneous emission of perturbations from the shock front is indicated by an infinite reflection coefficient. Amplification of transverse perturbations leads to a corrugated or cellular shock front. Finally, the shock structure, i.e., the flow in the relaxation zone upstream of the leading front may also be hydrodynamically unstable. The unsteady flow perturbations then lead to fluctuations in the shock front itself as it conforms to the boundary conditions in its wake. Thus, the turbulent, unstable structure of a shock front can originate from a variety of mechanisms. The stability-limit conditions may be different for the different kinds of instability mechanisms and may also overlap. Irrespective of the mechanisms, instability implies that under certain conditions a turbulent mechanism is required to effect a shock transition process.

It is of value to summarize the results of the various shock stability studies. The early result from the linear stability analysis of Dyakov [10] gave the limits for shock stability as

$$-1 \leq j^2 \left(\frac{\partial v}{\partial p} \right)_H \geq (1 + 2M_1),$$ (3.2)

where $j^2 = (p - p_0)/(v_0 - v_1)$ is the negative slope of the Rayleigh line and $M_1 = (D - u_1)/c_1$ is the upstream Mach number (relative to the shock front). The same results were obtained by Erpenbeck [17] and later by Swan and Fowles [49]. The stability limits (Eq. 3.2) define the so-called corrugation instability to describe the growth of transverse perturbations on the shock surface. Gardner [22] demonstrated that the limits for corrugation stability also correspond to the conditions for stability against the one-dimensional breakup (or shock splitting) analyzed by Bethe. A more restrictive limit was found by Kontorovich [32] who gave the upper limit as

$$j^2 \left(\frac{\partial v}{\partial p} \right)_H \geq \frac{1 - M_1^2 \left(1 - \dfrac{v_0}{v_1} \right)}{1 - M_1^2 \left(1 + \dfrac{v_0}{v_1} \right)}$$ (3.3)

instead of $(1 + 2M_1)$ given by Eq. 3.2. Since $M_1 \leq 1$ and $(v_0/v_1) > 1$, the above inequality is smaller than that given by Eq. 3.1 and is therefore the controlling one. When the lower limit of Eq. 3.1 is not satisfied, the shock splits into two waves moving in the same direction [54] whereas the upper limit defines the shock stability to multidimensional corrugations.

Experimental observation of shock instability was reported by Glass et al. [23,24] and Griffiths et al. [26] for strong ionizing shocks in the noble gases and dissociating shocks in carbon dioxide. Figure 3.10 shows typical interferograms of the turbulent structure of ionizing shocks in argon that are strongly reminis-

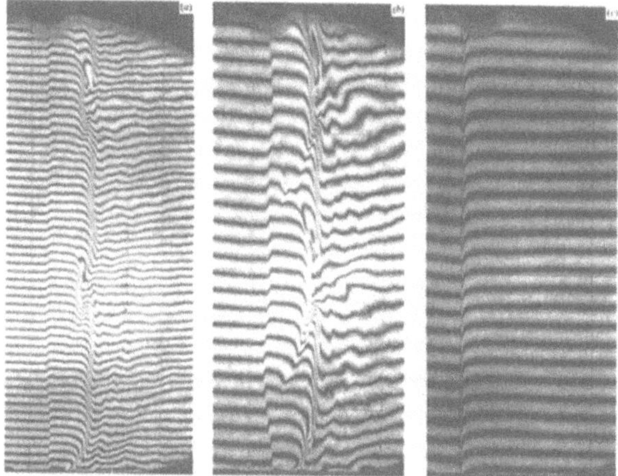

Figure 3.10. Interferograms of strong ionizing shock waves in noble gases (Courtesy of I.I. Glass, [23,24]).

cent of the turbulent structure of unstable gaseous detonations illustrated in Fig. 3.2. The equilibrium Hugoniots for shocked argon and CO_2 were computed by Griffiths et al. [26] with the experimental results for stable and unstable shocks represented on the Hugoniots (Fig. 3.11). The regions on the Hugoniot where unstable shocks were observed experimentally are shown as full (dark) circles. As can be observed in Fig. 3.11, instability occurs in the region where $(\partial p / \partial v)_H$ is positive in accord with the prediction from stability. However, the experimental results are not in quantitative agreement with the stability theory as given by Eq. 3.3. Typical experimental observations of the limiting values of $j^2(\partial v / \partial p)_H$ were found to be about an order of magnitude smaller than the theoretical values. The onset of instability and turbulence in strong shock waves in gases is still not fully understood.

Shock waves that cause plastic deformation in solid media are also found to have a non-planar front. The plastic deformation flow behind the front is also found to be of a "turbulent" nature. In a polycrystalline solid, the shock front may become irregular as a result of the dependence of the wave velocity on the crystallographic orientation. As a result of the velocity anisotropy, the shock shape becomes irregular and the irregularities amplify as the shock penetrates into the media [44]. The irregularities also change the stress state at the shock front and hence influence the plastic strain and the residual structure. The plastic flow of the shocked material may be one-dimensional when viewed on a macroscopic level. However, it is found to be turbulent when viewed on a finer mesoscopic level as a result of the non-uniform stress field. Mescheryakov and co-workers [39–43,48] have studied in detail the shock deformation processes

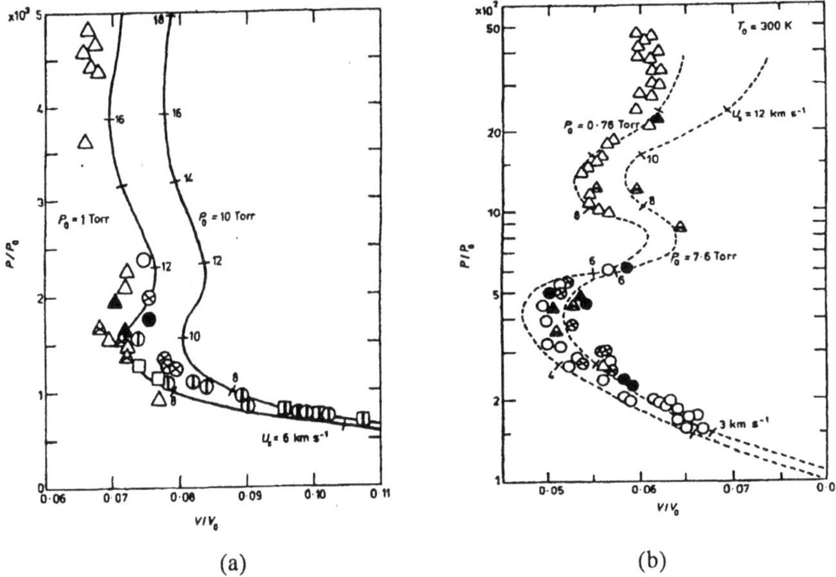

Figure 3.11: (a) Stability results and normalized equilibrium Hugoniot curves for Argon; (b) Stability results for shock waves in CO_2. Solid, crossed and open symbols denote cases classed as unstable, doubtful and stable shock respectively (Courtesy of H. Hornung, et al., 1975).

in planar targets of different materials (copper, ductile steel, aluminum and its alloys, titanium) due to high velocity impact. In their experiments, the impactor plates (the same material as the target) were mounted on the front face of the projectile, which was accelerated in a gas gun to an impact velocity in the range of 50–500 m/s. The free surface velocities of the targets were measured by an interferometer with the laser beam focused to a diameter of 100–120 μm, corresponding to 2–10 grain sizes of the polycrystalline material. This resolution is intermediate between the mesoscopic level (1–10 μm) and the macroscopic level (> 1 mm). Besides the velocity history, the interference method permits the particle velocity distribution width to be obtained (i.e., the square root of the particle velocity dispersion) by measuring the interference signal contrast. The particle velocity dispersion indicates that adjacent micro-volumes change their position relative to each other resulting in a slip of the micro-volumes relative to each other. If the duration of the slip process is sufficiently long, then shear bands of a non-crystallographic nature occur in the direction of the shock wave propagation. Post-shock analysis of the microstructure using optical and electron microscopy also revealed evidence of a rotational motion of the plastic flow similar to turbulent motion in hydrodynamics. Numerical simulations can provide valuable qualitative information on the detail of the complex non-equilibrium processes in shock waves in solids. Of particular interest is the recent numerical study of shock compression of crystalline copper by Yano and Horie [57]. They used a two-dimensional, discrete element code (DM2) in which the medium is considered to be made up of a collection of interacting particles (or elements). These particles or elements have a dimension of the order of microns, thus the scale of the numerical model is much larger than that of molecular dynamics. However, it is not a continuum model either, since the motion of the particles is governed by classical dynamics rather than the usual conservation equations of continuum mechanics. The interaction between the elements consists of a central potential force, a shear force, a dry friction force, a central damping force as well as a tangential damping force. In an attempt to simulate the impact experiment of Meshcheryakov with the DM2 discrete element code, Yano and Horie first created the target specimen by slow compression of copper particles of 12 µm diameter to 95% theoretical density to eliminate voids. The particles then become the grains of the specimen after compression. The particles themselves are comprised of elements of 1 µm diameter and each particle has on the average 113 elements. The orientation of the element packing in each particle was assigned randomly to create local anisotropy. Grain boundary bond strength is derived from the interaction of neighboring particles that come into contact. The initial geometry of the model specimen is shown in Fig. 3.12. The specimen is assumed to impact a rigid wall located at the bottom of the specimen. The velocity field at 14 µs after impact is shown in Fig. 3.13. It is clearly shown that the shock wave front is non-planar. The velocity field of the plastic deformation flow behind the shock wave is illustrated in Fig. 3.14. It can be observed that the flow is rotational or vortical, similar in nature to turbulent fluid motion. This turbulent motion is more clearly shown in Fig. 3.15 at 18 µs after impact. The profiles of the velocity distribution for three particles, A, B, and C, defined in Fig. 3.12 are shown in Fig. 3.16. The Reynolds number on the grain

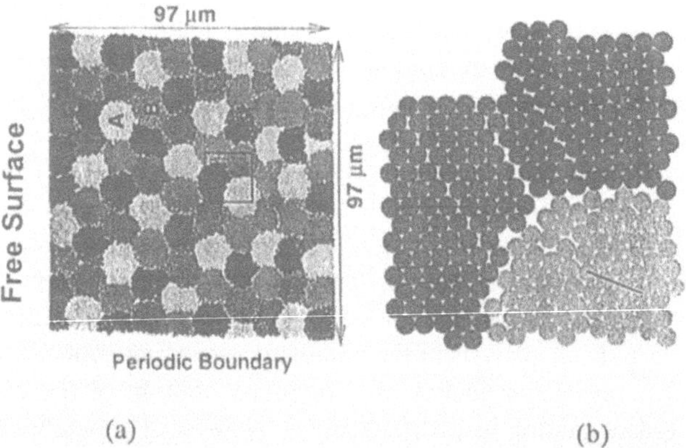

Figure 3.12. (a) Initial geometry of the model copper specimen; (b) Enlarged triple poir grain boundary in the square region in (a) (Courtesy of Yano and Hori, [57]).

size level computed from the particle velocity fluctuation is of the order of unity. This is analogous to the Kolmogorov dissipative length scale in hydrodynamics. The numerical simulation reproduced the essential qualitative features of Mescheryakov's experiments and indicates that, at high impact velocity, the structure of the shock wave is also of a turbulent nature.

The above examples indicate that non-reacting shock waves in gas and condensed phase media under certain conditions have non-planar fronts and the flow behind the front can also exhibit a turbulent-like structure.

3.5. Why Shocks Have a Turbulent Structure

Although stability analysis and experimental observation can tell us, in general, the condition when a certain phenomenon becomes unstable, the more profound question as to why it is unstable is often more difficult to answer. For example, we can predict via stability theory the critical Reynolds number beyond which laminar pipe flow or a boundary layer becomes unstable. Experimentally, we can also measure this critical Reynolds number when laminar flow breaks down. However, neither theory nor experiment informs us of the reason for the breakdown of laminar flow. To find out why the laminar flow has to become turbulent beyond a certain Reynolds number, we must examine in detail the basic mechanisms of turbulent flow, and by comparing them with those of laminar flow, we

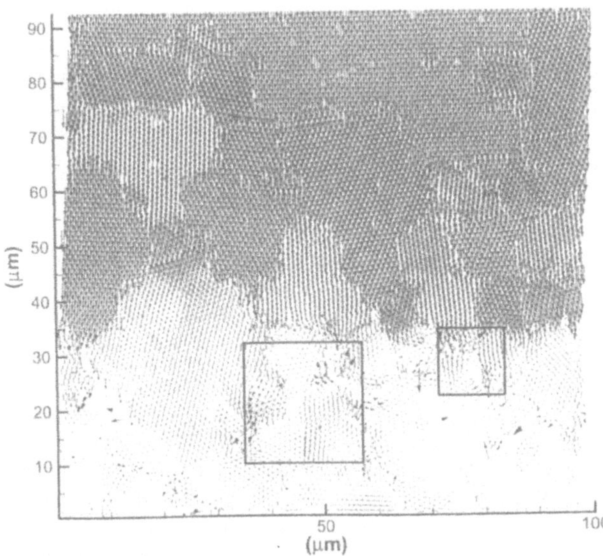

Figure 3.13. Velocity field in the model specimen at 14 µs after impact at 89 m/s (Courtesy of Yano and Horie [57]).

138   John Lee

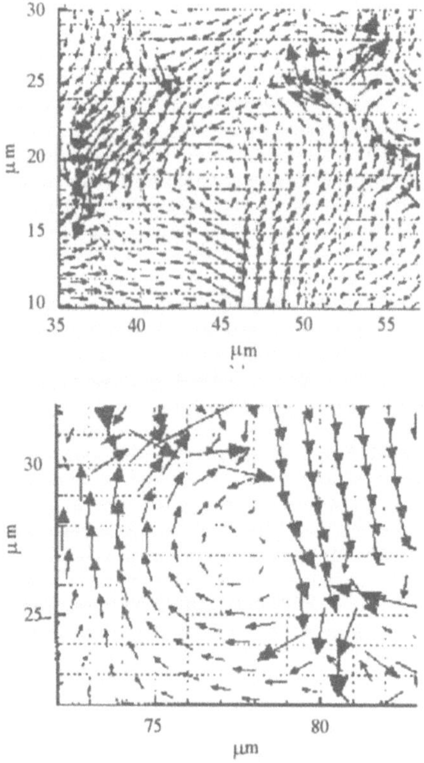

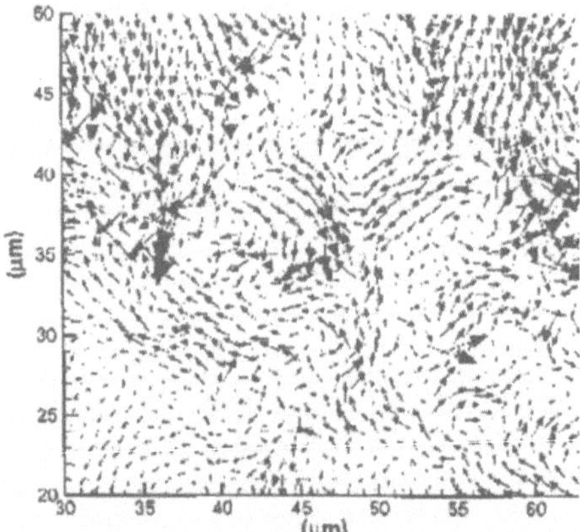

Figure 3.14. Enlarged view of the velocity field marked by the large (a) and small (b) squares of Fig. 3.13 (Courtesy of Yano and Horie [57]).

Figure 3.15. Eddy-like velocity fields behind shock in copper specimen 18 μs after impact (Courtesy of Yano and Horie [57]).

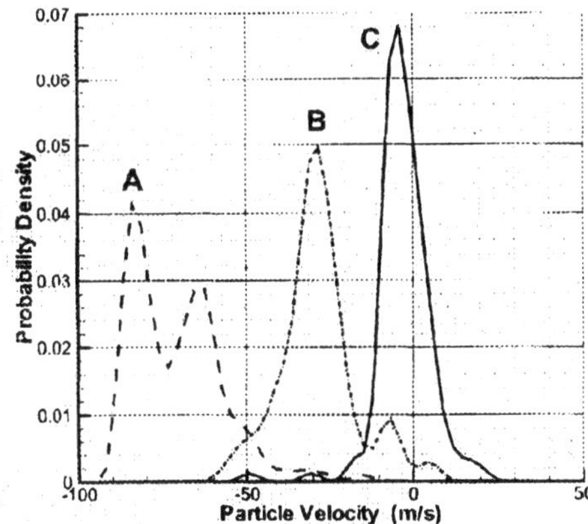

Figure 3.16. Particle velocity distribution function within three grains indicated by A, B, and C in Fig. 3.12 (Courtesy of Yano and Horie [57]).

may speculate as to the reasons why nature prefers to go from one type of flow to the other. For pipe flows, we may say that the onset of turbulence permits viscous dissipation to be more effective. We can readily see from the conservation laws that viscous effects diminish with increasing Reynolds number. The onset of turbulence reduces the characteristic length scale (from the pipe diameter to the mean eddy size) and thus maintains the role played by viscosity. For gaseous detonations, turbulence must also be required to effect the detonation process under certain conditions, i.e., when the effective activation energy of the chemical reaction rate exceeds a certain critical value. The propagation of a detonation wave depends on an ignition mechanism where the chemical reactions can be initiated and a self-sustaining mechanism where the chemical energy released can be fed back to maintain steady propagation of the shock. In the classical laminar one-dimensional ZND model, the ignition mechanism is via adiabatic shock heating. Molecules are dissociated into active radical species to initiate the chemical reactions. The chemical energy released is then fed back to maintain steady propagation of the leading shock front via the expansion work of the combustion products. Radical species are derived from normal shock heating via thermal dissociation of the reactant molecules. Note that radical species are present in abundance in the reaction zone as well as in the products. However, molecular diffusion is too slow for the transport of these radical species ahead of the reaction zone to effect ignition in a detonation wave. Thus, we have to rely on the shock front to provide these radical species in a laminar detonation. For a turbulent detonation, the transport rate is enhanced considerably, thus permitting the radical species in the reaction zone and products to be mixed with the reactant to initiate chemical reactions rather than relying solely on shock dissociation as the source of the radical species. It should be pointed out that, in compressible turbulence, shock waves play a dominant role in vorticity production via shock–shock (Mach interaction), shock–vortex, and shock–density interface interactions, as well as the baroclinic mechanism [33]. In fact, when the transverse shock fluctuations were eliminated from a cellular detonation, the subsequent "one-dimensional" wave failed, with the reaction zone decoupling from the leading shock front and the detonation wave velocity decreasing to about one half of the C–J value. Figure 3.17 illustrates the failure of a cellular turbulent detonation when the transverse shocks are damped out by their reflections off the porous wall of the detonation tube [52]. This conclusively demonstrates the necessity for the detonation structure to be turbulent for steady propagation. Also, the importance of shock waves in mixing enhancement in compressible turbulence is illustrated.

Apart from mixing enhancement, a turbulent cellular detonation structure also leads to the formation of local high-temperature regions (hot spots) from the transverse shock interactions. Mach stem temperatures can be considerably higher

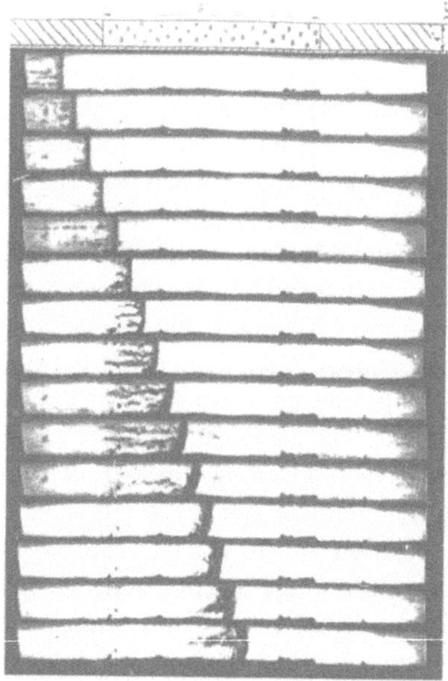

Figure 3.17. Failure of turbulent detonation when transverse shocks in the reaction zone are damped out by a porous wall.

than the bulk shock temperature based on the averaged shock speed (i.e., C–J velocity). Thus, these "hot spots" can also serve as local ignition sources where chemical reactions are first initiated and subsequently spread to neighboring unreacted, shock-heated regions. So, turbulence in gaseous detonation can also provide "hot spots" to facilitate the initiation of chemical reactions in addition to the role turbulence plays in the rapid mixing and transport of radical species from the reaction zone to the unreacted mixture. Thus, we may argue that a turbulent structure facilitates the initiation of chemical reactions and increases the reaction rate over a laminar detonation structure.

In addition to "hot spots," turbulence also results in very high shear rates that could lead directly to the mechanical initiation of chemical reactions in condensed explosives. There is strong evidence that shear and mechanical deformation can lead directly to the initiation of the chemical processes without going through the thermal path (i.e. conversion of mechanical work to heat and subsequent thermal initiation of chemical reaction) in condensed explosives. Experiments on the shock initiation of a monocrystal of PETN have been carried out by Dick and co-workers [7–9,58] and they observed that the shock sensitivity to detonation initiation depends strongly on the orientation of the crystal relative to the direction of propagation of the shock wave. If the initiation process is thermal, then the sensitivity to initiation should not be dependent on the crystal orientation. The results of Dick's experiment was explained in terms of an anisotropic shear flow in the uniaxial strain of the planar shock wave. In a crystal, the shear flow occurs on crystallographic planes oriented at approximately 45° to the plane of the shock. In molecular crystals, the molecules are able to pass by each other without interference on some planes, but present obstacles to each other's passage in other planes in the shear flow. This effect is referred to as "steric hindrance" to shear and, as a result, the orientations that have strong hindrance will have a higher sensitivity (especially at low stresses). Orientations that have a low hindrance are thus less sensitive. For PETN crystals, it was found that the [100] and [101] orientations are relatively insensitive whereas the [110] and [001] are more sensitive. These results have also been corroborated by Soulard [50] who observed considerably more sensitivity for the [001] orientation than the [111] orientation. Further evidence that a bulk thermal process via shock heating cannot explain the shock initiation process in monocrystals is the discovery of an anomalous sensitivity at low stresses (also reported by Dick). It was found that the "run-up distance" at 4.2 GPa is shorter than at 8.5 GPa for the [110] orientation, a result that contradicts the thermal ignition theory. This anomalous sensitivity was credited to the restraint generated by the elastic wave that renders the steric hindrance to shear even more effective in causing the initiation of the explosive chemistry. These experiments demonstrate the important role of mechanical stress on the initiation of chemical reactions.

It is of interest to note that as early as 1940, Bridgman had already demonstrated that mechanical stresses of compression and shear could result in the initiation of the explosive decomposition of energetic molecules [5]. In Bridgman's experiment, a number of explosives were subjected to high axial compression in a hydraulic press. In some experiments, the sample is further subjected to a shear subsequent to axial compression by the rotation of one of the anvils. Bridgman observed that a number of the explosives tested could detonate under this action of mechanical stresses at isothermal conditions. However, Bridgman stated that these preliminary results were not conclusive and he indicated the possibility of friction sparks that could have been generated during the rotation of the anvil in the shearing process. In attempting to explain the explosive decomposition of sugar crystals under high compression and shear, Edward Teller discussed various possible mechanical initiation mechanisms and postulated that mechanical stresses can result in a lowering of the effective activation energy of the reactant molecules [51]. The mechanical work associated with the compression and shear is assumed to contribute toward the production of an appropriate activated state that forms the lowest barrier between the reactant and product molecules. Thus, the decomposition of the reactant molecule is greatly facilitated as a result of this reduced activation potential. In recent years, a number of Russian researchers [12–16] have carried out similar studies in diamond anvils on the mechanical initiation of a number of organic and inorganic compounds. They confirmed the original observations of Bridgman that detonations can indeed be initiated under high mechanical stresses alone.

The "non-thermal" initiation of the explosive decomposition of energetic molecules has led to the formulation of chemical models that recognize the role of mechanical stresses explicitly. For example, Benderskii et al. [2,3] have proposed a phenomenological reaction model where the overall activation energy is taken to be the difference between the thermal activation energy and mechanical energy of deformation. Thus, an increase in the mechanical deformation energy would lower the thermal requirement leading to the existence of a critical pressure (at normal temperatures) at which explosive decomposition can take place. Chemical reaction rate models, where mechanical parameters are explicitly involved, are being used to model experiments and are beginning to find more general acceptance nowadays. For example in the recent numerical study of Baer on heterogeneous explosives [1], the chemical reaction is assumed to be triggered by a critical pressure threshold as opposed to the normal temperature criteria of the conventional "thermal model." Recognizing the role of mechanical initiation, turbulent shear stress may well be incorporated along with thermodynamic parameters in chemical models.

In solid explosives, heterogeneities are inherent in the structure of the media. Thus, turbulence can arise from the random orientation of the grains. However, for homogeneous liquids, turbulence has to arise from instability of the detonation front. This is well illustrated in the case of cellular detonations in nitromethane. One can also control and induce the onset of the turbulent structure by the addition of solid particles or glass microballoons into the liquid. It is of interest to note that the introduction of hot spots via cavities in the end plate of a shock initiation experiment can reduce the critical shock strength for detonation initiation in nitromethane from 8.0 GPa to about 5.5 GPa [18]. Addition of glass spheres and glass microballoons can also lead to a significant reduction in the critical diameter of nitromethane [6,11,25,34,35,47]. These experiments demonstrated that artificially inducing a non-uniform turbulent structure by mechanical means can sensitize and promote detonation in the explosive without modifying the chemistry of the reaction. The addition of the heterogeneities is generally considered to create "hot spots", i.e., small regions of higher temperature as compared to the bulk temperature for the von Neumann state behind the shock. The higher local temperature then promotes ignition and higher reaction rates at these local "hot spots". However, these heterogeneities also create intense shear and turbulence which can also promote ignition mechanically. The strong dependence of the shock sensitivity of nitromethane to initial temperature also suggests that mechanical stresses may play a stronger than expected role in the initiation of the explosive reaction. From the point of view of the thermal initiation mechanisms, a few tens of degree variation in the initial temperature would not significantly influence the shock temperature. However, the viscosity of a liquid has a strong dependence on temperature and can change by orders of magnitude for a small change in the temperature. Thus, the strong dependence of the shock sensitivity of nitromethane may suggest that viscosity influences the onset of instability and turbulence. This can also provide further evidence of the important role played by mechanical stresses at the shock in the initiation of the chemical reactions.

Thus, we may conclude that, as in the case of gaseous detonations, a turbulent structure also can provide a more "efficient" means to effect the initiation of the chemical reactions in a condensed phase detonation front as well. In particular, if mechanical shear deformation plays an important role in the ignition processes, then a turbulent structure can facilitate the decomposition process and hence facilitate the propagation of the detonation wave.

3.6. Conclusion

Future advances in shock wave physics would require experimental and numerical studies on the mesoscopic and microscopic levels. That would undoubtedly reveal the complex three-dimensional turbulent structure of shock waves more as rule than an exception. The orders of magnitude increase in the amount of information that results from higher resolution experiments and computations require the development of physical models to permit the data obtained to be interpreted. The formulation of appropriate physical models of the phenomenon requires an understanding of the mechanisms involved in the shock transition process. Thus, we must speculate as to the reason why a turbulent structure is preferred over the simple laminar one. Unfortunately, the answer is not apparent from the conservation equations. Normal mode stability analysis yields the conditions where steady "laminar solutions" cannot be realized but do not answer the question "why." The alternate formulation based on thermodynamic stability theory simply shifts the blame onto entropy production and the second law. It appears that the preference of a turbulent structure for shock (and detonation) waves is associated with the ease in which the transition across the shock takes place. A turbulent detonation structure permits formation of local "hot spots" that facilitate the initiation of chemical reactions in homogeneous gaseous and liquid explosives. Turbulence also promotes mixing and permits radical species from the reaction zone and products to be transported rapidly to the unburnt mixture to effect ignition. If we acknowledge the mixing mechanism, then the chemical model would have to be different from the standard Arrhenius rate models based on a purely thermal mechanism. A chemical model that recognizes the turbulent mixing mechanism would have to incorporate the turbulence parameters. Similarly, in condensed explosives, if compression and shear were considered as appropriate mechanisms for the initiation of chemical reactions, then the chemical models would have to include the mechanical stress parameters (in addition to the thermodynamic parameters). It should be noted that current models are mostly based on thermal explosion theory. There are sufficient empirical parameters to fit any chemical model proposed even if the mechanism in the model is incorrect. Since these empirical parameters were fitted from experiments, agreement with experiment is not an indication of the correctness of the model used. As more information on the macroscopic and microscopic levels become available, the role of turbulence in shock transition process may become more apparent and permit correct physical and chemical models to be proposed to describe the shock transition process on the global macroscopic level. In other words, detailed study of the mesoscopic and microscopic levels are now required for the formulation of better models for the macroscopic description of shock and detonation phenomena. Thus, we may propose that the frontier in shock wave research lies in the study of the turbulent structure of the shock wave.

Acknowledgments

The author acknowledges the critical and editing reading of the manuscript by Andrew Higgins and the preparation of the final manuscript by Della Maharajh.

References

[1]   Baer, M.R. (1999), "Computational Modeling of Heterogeneous Reactive Materials at the Meso-Scale," in *Shock Compression of Condensed Matter* (ed. M.D. Furnish, L.C. Chhabildas, and R.S. Hixson), The American Institute of Physics, New York, pp. 27–33.

[2]   Benderskii, V.A., P.Q. Filipov, and M.A. Ovchinmikov (1989), "Ratio of Thermal and Deformation Ignition in Low Temperature Solid Phase Reactions," *Dokl. Akad. Nauk. SSSR* **308** [57], pp. 401–405.

[3]   Benderskii, V.A., E.Ya. Misochko, A.A. Ovchinmikov, and P.G. Filipov (1982), "A Phenomenological Model of Explosion in Low Temperature Chemical Reactions," *Sov. J. Chem. Phys.*, pp. 117–1182.

[4]   Bethe, H.A. (1942), *The Theory of Shock Waves for an Arbitrary Equation of State*, Office of Scientific Research and Development No. 545, Serial No. 237.

[5]   Bridgman, P. (1947), "The Effect of High Mechanical Stress on Certain Solid Explosives," *J. Chem. Phys.* **15**, pp. 311–313.

[6]   Campbell, A.W., W.C. Davis, and J.R. Travis (1961), "Shock Initiation of Detonation in Liquid Explosives," *Phys. Fluids* **4**, pp. 498–510.

[7]   Dick, J.J. (1995), "Shock-Wave Behavior in Explosive Monocrystals," *J. de Physique IV* **5**, pp. C4-103–106.

[8]   Dick, J.J. (1997), "Anomalous Shock Initiation of Detonation in Pentaerythritol Tetranitrate Crystals," *J. Appl. Phys.* **81**, pp. 601–612.

[9]   Dick, J.J. and J.P. Ritchie (1994), "Molecular Mechanics Modeling of Shear and the Crystal Orientation Dependence of the Elastic Precursor Shock Strength in Pentaerythritol Tetranitrate," *J. Appl. Phys.* **76**, pp. 2726–2737.

[10]   Dyakov, S.P. (1954), "On the Stability of Shock Waves," *J. Exp. Theor. Phys. (U.S.S.R.)*, **27**, p. 288.

[11]   Engelke, R., (1983), "Effect of the Number Density of Heterogeneities on the Critical Diameter of Condensed Explosives," *Phys. Fluids* **26**, pp. 2420–2424.

[12]   Enikolopyan, N.S. (1985), "Some Aspects of Chemistry and Physics of Plastic Flow," *Pure & Appl. Chem.* **57**, pp. 1707–1711.

[13]   Enikolopyan, N.S., A.A. Mkhitaryan, A.S. Karagezyan, and A.A. Khzardzhyan (1987), "Critical Events in the Explosion of Solids under High Pressure," *Dokl. Akad. Nauk. SSSR* **292**, pp. 887–890.

[14]   Enikolopyan, N.S., A.A. Mkhitaryan, and A.S. Karagezyan, (1987), "Explosive Chemical Reactions of Metals and Oxides of Salts in Solids," *Dokl. Akad. Nauk. SSSR* **294**, pp. 912–915.

[15]   Enikolopyan, N.S., A.A. Khzardzhyan, E.E. Gasparyan, and V.B. Vol'eva (1987), "Kinetics of Explosive Chemical Reactions in Solids," *Dokl. Akad. Nauk. SSSR* **294**, pp. 1151–1154.

[16]   Enikolopyan, N.S., V.B. Vol'eva, A.A. Khzardzhyan, and V.V. Ershov (1987), "Explosive Chemical Reactions in Solids," *Dokl. Akad. Nauk. SSSR* **292**, pp. 1165–1169.

[17]   Erpenbeck, J.J. (1962), "Stability of Step Shocks," *Phys. Fluids* **5**, pp. 1181–1187.

[18]   Fauquignon, C. and H. Moulard (1978), "Shock Sensitivity of Nitromethane with Well Defined Hot Spots Distribution," *Acta Astronautica* **5**, pp. 1035–1040.

[19]   Fickett, W. and W.C. Davis (1979), *Detonation*, University of California Press, Berkeley.

[20]   Fowles, G.R. (1976), "Conditional Stability of Shock Waves—A Criterion for Detonation," *Phys. Fluids* **19**, pp. 227–238.

[21]   Fowles, G.R. (1981), "Stimulated and Spontaneous Emission of Acoustic Waves from Shock Fronts," *Phys. Fluids* **24**, pp. 220–227.

[22] Gardner, C.S. (1963), "Comment on 'Stability of Step Shocks,'" *Phys. Fluids* **6**, pp. 1366–1367.

[23] Glass, I.I. and W.S. Liu (1978), "Effects of Hydrogen Impurities on Shock Structure and Stability in Ionizing Monatomic Gases. Part 1. Argon," *J. Fluid Mech.* **84**, pp. 55–77.

[24] Glass, I.I., W.S. Liu, and F.C. Tang (1977), "Effects of Hydrogen Impurities on Shock Structure and Stability in Ionizing Monatomic Gases: 2. Krypton," *Canadian Journal of Physics* **55**, pp. 1269–1279.

[25] Góis, J.C., J. Campos, and I. Plaksin (2001), "Effect of GMB on Failure and Reaction Regime of NM/PMMA–GMB Mixtures," in Shock Compression of Condensed Matter—2001 (eds. M.D. Furnish, N.N. Thadhani, and Y. Horie), American Institute of Physics, Melville, NY, pp. 898–901.

[26] Griffiths, R.W., R.J. Sandeman, and H.G. Hornung (1975), "The Stability of Shock Waves in Ionizing and Dissociating Gases," *J. Phys. D: Appl. Phys.* **8**, pp. 1681–1691.

[27] Held, M. (1973), "Struktur der Detonationsfront in Verschiedenen Gas-Luftgemischen," *Auszug aus dem Bericht der Jahrestagung 1973 des Institute für Chemie der Treib-und Explosivstoffe (ICT)*, pp. 279–297.

[28] Held, M. (1972), "Method of Measuring the Fine Structure of Detonation Fronts in Solid Explosives," *Astronautica Acta* **17**, pp. 599–607.

[29] Howe, P.M. (1969), "Transverse Waves in Detonating Liquids," *Appl. Phys. Lett.* **15**, pp. 197–198.

[30] Howe, P., R. Frey, and G. Melani (1976), "Observations Concerning Transverse Waves in Solid Explosives," *Combustion Science and Technology* **14**, pp. 63–74.

[31] Iordanskii, S.V. (1957), "About Stability of a Plane Shock Wave," *Prikl. Mat. Mekh.* **321**, pp. 465–471.

[32] Kontorovich, V.M. (1957), "Concerning the Stability of Shock Waves," *J. Exp. Theor. Phys. (U.S.S.R.)* **33**, pp. 1525–1526.

[33] Lee, J.H.S. (1997), "Shock–Vortex Interaction: Its Role in Compressible Turbulence and Detonation Structure," in *Proceedings of the 2nd International Workshop on Shock Wave/Vortex Interaction*, Shock Wave Research Center, Institute of Fluid Sciences, Tohoku University, Sendai, Japan, pp. 15–30.

[34] Lee, J.J., D.L. Frost, J.H.S. Lee, and A. Dremin, (1995), "Propagation of Nitromethane Detonation in Porous Media," *Shock Waves* **5**, pp. 115–120.

[35] Lee, J.J., M. Brouillette, D.L. Frost, and J.H.S. Lee (1995), "Effect of Diethylene-triamine Sensitization on Detonation of Nitromethane in Porous Media," *Combustion and Flame* **100**, pp. 292–300.

[36] Lee, J.H.S., R. Soloukhin, and A. Oppenheim (1969), "Current Views on Detonations," *Astronautica Acta* **14**, pp. 565–584.

[37] Mallory, H.D. (1967), "Turbulent Effects in Detonation Flow: Diluted Nitromethane," *J. Appl. Phys.* **38**, pp. 5302–5306.

[38] Mallory, H.D. (1976), "Detonation Reaction Time in Diluted Nitromethane," *J. Appl. Phys.* **47**, pp. 152–156.

[39] Meschcheryakov, Yu.I. and A.K. Divakov (1985), "Particle Velocity Variations in a Shock Wave and Fracture Strength of Aluminum," *Sov. Phys. Tech. Phys.* **30**, pp. 348–350.

[40] Mescheryakov, Yu.I., A.K. Divakov, and S.A. Atroshenko (1989), "Role of Shock-Induced Particle Velocity Distribution in Processes of Localization During Spallation," in *Shock Compression of Condensed Matter—1989* (ed. S.C. Schmidt, J.N. Johnson, and L.W. Davison), North-Holland, Amsterdam, pp. 449–453.

[41] Mescheryakov, Yu.I. and S.A. Atroshenko, (1992), "Multiscale Rotations in Dynamically Deformed Solids," *Int. J. Solids Structures* **29**, pp. 2761–2778.

[42] Mescheryakov, Yu.I., N.A. Mahutov, and S.A. Atroshenko (1994), "Micromechanisms of Dynamic Fracture of Ductile High-Strength Steel," *J. Mech. Phys. Solids* 42, pp. 1435–1457.

[43] Mescheryakov, Yu.I., A.K. Divakov, and N.I. Zhigacheva (2000), "Shock-Induced Phase Transformation and Vortex Instabilities in Shock Loaded Titanium Alloys," *Shock Waves* **10**, pp. 43–56.

[44] Meyers, M.A. and M.S. Carvalho (1976), "Shock-Front Irregularities in Polycrystalline Metals," *Mater. Sci. Engineering* **24**, pp. 131–135.

[45] Murray, S.B. and J.H.S. Lee (1985), "The Influence of Yielding Confinement on Large Scale Ethylene–Air Detonations," in *Dynamics of Shock Waves Explosions and Detonations*, (ed. J.R. Bowen, N. Manson, A.K. Oppenheim and R.I. Soloukhin), *Progress in Astonautics and Aeronautics* **94**, AIAA, New York, pp. 80–103.

[46] Presles, H.N., P. Bauer, C. Guerraud, and D. Desbordes, (1987), "Study of Head On Detonation Wave Structure in Gaseous Explosives," *J. de Physique* **48**, pp. 119–124.

[47] Presles, H.N., J. Campos, F. Heuzé, and P. Bauer, (1989), "Effects of Microballoons Concentration on the Detonation Characteristics of Nitromethane–PMMA Mixtures," *Proc. 9th Symp. (Intl.) on Detonation*, Office of Naval Research, Dept. of the Navy, Arlington, Virginia, pp. 925–929.

[48] Savenko, G.G., Yu.I. Mescheryakov, V.B. Vasil'kov, and A.I. Chernyshenko (1990), "Grain Vibrations and Development of the Turbulent Nature of Plastic Deformation During High-Velocity Interaction of Solid Bodies," *Fizika Goreniya i Vzryva* **26**, pp. 97–102.

[49] Swan, G.W. and G.R. Fowles, (1975), "Shock Wave Stability," *Phys. Fluids* **18**, pp. 28–35.

[50] Soulard, L. (1990) "Etude du Moncristal de Pentrite Soumis a un Choc Plan," thesis, L'Universite de Haute-Alsace; L. Soulard & F. Bauer in *Shock Compression of Condensed Matter—1990* (ed. S.C. Schmidt, J.N. Johnson and L.W. Davison), p. 817.

[51] Teller, E. (1962), "On the Speed of Reactions at High Pressures," *J. Chem. Phys.* **36**, pp. 901–903.

[52] Teodorczyk, A.,and J.H.S. Lee (1995), "Detonation Attenuation by Foams and Wire Meshes Lining the Walls," *Shock Waves* 4, pp. 225–236.

[53] Urtiew, P.A., A.S. Kusubov, and R.E. Duff (1970), "Cellular Structure of Detonation in Nitromethane," *Combustion and Flame* 14, pp. 117–122.

[54] Wackerle, J. (1962), "Shock Compression of Quartz," *J. Appl. Phys.* **33**, pp. 922–937.

[55] White, D.R. (1961), "Turbulent Structure of Gaseous Detonations," *Phys. Fluids* **4**, pp. 465–480.

[56] Weyl, H. (1949), "Shock Waves in Arbitrary Fluids," *Commun. Pure Appl. Math.* **2**, pp. 103–122.

[57] Yano, K. and Y. Horie (1999), "Discrete-Element Modeling of Shock Compression of Polycrystalline Copper," *Phys. Rev. B* **59**, pp. 13672–13680.

[58] Yoo, C.S, N.C. Holmes, P.C. Souers, C.J. Wu, and F.H. Ree (2000), "Anisotropic Shock Sensitivity and Detonation Temperature of Pentaerythritol Tetranitrate Single Crystal," *J. Appl. Phys.* **88**, pp. 70–75.

CHAPTER 4

What is a Shock Wave?
—The View from the Atomic Scale

Brad Lee Holian

It is easy to say that everything starts at the time and distance scale of atoms. Strong shock waves provide the most appropriate conditions under which to study processes at the atomistic level on the computer. In the last three decades, molecular-dynamics (MD) simulations have been applied to shock waves in gases, liquids, and solids. In the case of solids, the problem becomes more complicated because of defect structures, which have an intrinsically larger length scale than that of the mean atomic spacing. In sufficiently strong shocks, defects can be produced homogeneously. For weak shocks, they can be triggered as the wave interacts with pre-existing defects that serve as inhomogeneous nucleation sites.

In all cases, the critical feature of nonequilibrium MD (NEMD) simulations is the time it takes to attain a steady wave, beginning at the time of impact. As larger and larger samples have become feasible with large-scale, massively parallel NEMD, the level of complexity has increased dramatically, to include the possibility of structural phase transformation, as well as plastic deformation, and chemical reaction for molecular systems.

The biggest challenge for the future is the approach to modeling truly mesoscale features, such as exhibited in weak shocks in polycrystalline samples with grain sizes above the micrometer scale, and the simulation of detonations for complex molecules with large reaction zones. However, similar to expected refinements in experimental techniques, advances in atomistic simulations will allow us to probe even further the limits to our fundamental understanding of shockwave phenomena.

4.1. Introduction

The time and distance scales of shock-wave processes, caused by high-velocity impacts, are ideal for investigation at the atomistic level. Molecular-dynamics (MD) simulations, where Newton's equations of motion for large numbers of strongly interacting atoms are solved on the computer, have been employed to study shock waves for over thirty years [1], but most successfully in the last decade, when massively parallel computers have become available.

Strong shock waves in dense fluids can exhibit the steepest rises in density, velocity, pressure, and energy, with widths as little as a couple of mean atomic spacings (nearest-neighbor distances in solids and dense liquids are typically of the order of 0.3 nm). Because shock velocities are supersonic in the uncompressed medium, and because sound speeds are of the order of 6,000 m/s, the risetime of the shock can be as short as a vibrational period in the solid (or mean collision time in the fluid), which is of the order of 0.3 ps. These are extremely tiny time and distance scales, which are dictated by the puny sizes of atoms. These facts of life were known a century ago, that is, that matter is made up of atoms, and that atoms are incredibly tiny.

Avogadro's number of atoms (about 10^{24}) in a typical metal is contained in a cube whose side length is about 3 cm, the scale of a typical laboratory shock-wave experimental sample at a firing site. It is no wonder that engineers doing continuum-mechanical simulations of such experiments have long felt that atomistic simulations could never possibly have any relevance to the "real world." The gap between the continuum, where matter is treated as though it is infinitely divisible and smooth, and the atomistic, which is as grainy and discontinuous as matter can get (beyond subatomic particles), is worth quantifying, so that we can see just how far apart these world views really are.

At the atomistic level, it has been historically evident, time and time again, that the continuum presses in upon atoms at surprisingly small time and distance scales. When, in the late 1950s, Berni Alder [2] first carried out equilibrium MD simulations for hard disks in two dimensions (2D) and hard spheres in 3D, many scoffed at his notion that a few tens of atoms, or at most a hundred, could exhibit anything resembling the thermodynamic limit, thought to be something far closer to 10^{24} atoms or molecules. Later, at the end of the 1960s, Alder demonstrated that hydrodynamic behavior of dense fluids [3] could be ascribed to a hundred hard spheres, on a time scale of a few mean collision times, and on a distance scale of a couple of neighbor shells, or mean atomic spacings. In fact, the fundamental assumption that Boltzmann was forced to make, because he did not have computers available in 1900, was "molecular chaos," the exponential loss of a particle's memory of its collisional history. Molecular chaos made it possible to solve the problem of gas dynamics for the velocity distribution function, which came to be known as the Boltzmann Equation [4]. Molecular chaos was shown by Alder to be wrong precisely because atoms "remember" their previous velocity history much better than Boltzmann assumed, by a power-law decay, rather than exponential. The path to that discovery was the demonstration of hydrodynamic behavior at the atomic scale.

Near the beginning of the 1970s, Bill Hoover (who had come to the Livermore Lab to work as a postdoc of Berni's) pioneered the method of nonequilibrium molecular dynamics (NEMD) [5]. NEMD applies experimental driving forces and constraints, similar to those imposed in the laboratory, but at the atomic scale, in order to induce nonequilibrium steady-state flows of mass, momentum, and energy. Naturally, the tiny scales of the samples that could be investigated by NEMD meant that the gradients or strain rates studied were enormous compared with the ones to which laboratory experimenters were accustomed.

Because of the influence of Alder and Hoover, it is no wonder that shock-wave propagation would quickly become a natural object of study for NEMD. (In the 1950s at Livermore, Alder was the group leader of the shock physics group, both experiment and theory. The group morphed into a division from which Hoover has only recently retired—though neither Hoover nor Alder has actually retired yet, in the full sense of the word.) The gradients in shock waves are grudgingly exempted from the complete scorn of experimentalists, who might otherwise claim that NEMD is somehow inappropriately constrained to the "wrong time and distance scales."

Nevertheless, although NEMD and shock waves are ideally well married, the window of opportunity is not without constraints, and I will attempt to delineate the limits, at least as they presently exist.

First of all, the samples that can be subjected to shocks in NEMD must be sufficiently long that steady waves can be achieved, for that is the first criterion of a planar shock wave. Planar shock waves are studied in laboratory experiments, not because most shock waves in the real world are planar—most are fully 3-D, either divergent or convergent—but because planar waves are most easily analyzed, because they are 1-D in space. They become steady in time by virtue of the transport of energy and momentum from the direction of propagation into the transverse dimensions. In fluids, the transverse flow is viscous, while in solids, transverse displacement is either reversible elastic distortion or irreversible plastic flow. This dissipation therefore necessitates 2-D or 3-D simulations; otherwise, in order to become steady, strictly 1-D systems require some kind of "artificial viscosity" to mimic transverse degrees of freedom* [6].

In order to accommodate the transverse flow, sufficient size in the cross section of the simulation is required. In real experiments, the planar impact of a flier plate upon a target is designed to minimize the influence of release from the

---

* The usual *linear* viscous damping force opposing each atom's velocity is inappropriate for shock waves, since a penalty is exacted for fluid flow along with the piston. This tends to struggle against the shock-wave and leads to zero temperature behind the shock front, regardless of shock strength—an unphysical final state. Therefore, a more appropriate form for atomistic artificial dissipation, which may mimic transverse motion in 1-D, is nonlinear viscous damping of the kind introduced by Firsov for electronic stopping power, namely, a force opposing an atom's velocity relative to the local velocity of atoms in its neighborhood.

edges, so that diagnostics are taken in the center of the sample. In NEMD simulations, the uniaxial strain is accomplished most efficiently by imposing periodic boundary conditions (PBCs) transverse to the shock propagation direction, where the universe is made up of a periodic checkerboard of images. Thus, when a particle leaves one side of the computational cell, its periodic image enters on the opposite side.

4.1.1. Comparison of Atomistic and Continuum Simulations

In order to see how atomistic simulations stack up to continuum ones, we can now apply some simple estimates. Suppose that we wish to carry out a truly "superheroic" NEMD calculation by today's standards, namely, one billion atoms, starting the shock-wave at one end, propagating it to the other, and letting the rarefaction wave sweep back to the origin. One thousand atoms on a side gives a total side length of 0.3 micrometers; this is not much stuff: a cube this tiny would not be enough dirt under your thumbnail to be felt. The sound traversal time, back and forth, would be just 0.1 ns; and this isn't much time, either! Since the classical equations of motion require the vibrational period of 0.3 ps to be divided up into at least 30 time steps, in order to have an accurate solution, the time step would be 10 fs, and the NEMD calculation of 0.1 ns would necessitate 10,000 time steps. (Rather than 10^9 atoms, routinely performed NEMD shock wave simulations reduce each cross-sectional dimension by an order of magnitude, so that the total size is "only" 10 million atoms.)

The goal of such an atomistic shock-wave simulation is to achieve some measure of continuum-like behavior. Bigger size is therefore better; longer (in time) is likewise better. On the other hand, the goal of a continuum calculation is to resolve wave interactions on finer and finer distance scales. Hence, one strives, if possible, to reduce the mesh size until convergence is achieved. Thus, a macroscopic chunk of metal might be an inch on a side (about 3 cm, say, or one thumb knuckle thick). A behemoth continuum simulation of this size sample, containing one billion cells, would correspond to a mesh size of $30\,\mu m$, which is 100 times bigger than the biggest NEMD calculational cell. If the continuum calculation were conducted at the maximum Courant-condition time step (mesh size divided by sound speed), the time step would be 5 ns, which is 50 times the longest NEMD calculation time.

The gap of 100 in distance scale and 50 in time scale between superheroic NEMD and behemoth engineering calculations is precisely the realm of the mesoscale. The continuum mesh size, where sub-grid physics occurs (whose effects are needed for physically based engineering calculations), is on the order of the size of grains in a polycrystalline metal. In order to model this sub-grid mesoscale world, information from atomistic simulations will be required, just as quantum-mechanical information from the electronic structure at atomic distance scales is required in order to generate the interatomic potentials used in NEMD.

4.1.2. Essential Ingredients for NEMD Simulations

Now that we see how NEMD fits into the hierarchy of physical time and distance scales—micro, meso, and macro—it is appropriate to give a brief description of the method of molecular dynamics. I wish to emphasize a fundamental difference between continuum engineering and atomistic simulations: At the continuum level, sub-grid physics and chemistry is introduced by means of constitutive models—plastic flow in solids, viscous flow in fluids, heat conductivity, and failure mechanisms for fracture and fragmentation. At the atomistic level, however, all of these behaviors are not introduced as input models, but rather, are simply the outcome of cooperative motion of large numbers of atoms moving under the influence of forces from neighboring atoms, whose interaction is a fixed initial input. Apart from the force laws most commonly used—pairwise-additive and many-body potentials, such as the embedded-atom method (EAM) appropriate to metals [7]—the other atomistic inputs are simply the initial conditions (atomic positions and velocities at time zero); boundary conditions, which can include external forces applied in special reservoir regions distinct from the sample, or else applied homogeneously throughout the sample; periodic boundaries (possibly moving ones); or rigid boundaries. All of these atomistic inputs require justification, usually inspired by experiment, in order not to be viewed as "smoke and mirrors" or "knobs". In the next section, I will review the fundamental equations of motion, initial and boundary conditions, and interaction potentials that we use in shock-wave NEMD simulations.

4.2. Tricks of the NEMD Trade

The most straightforward approach to shock-wave simulations at the atomistic scale is NEMD, using Newton's equations of motion, or more usefully, Hamilton's equations for positions $\mathbf{r}$ and velocities $\mathbf{u}$:

$$\frac{d\mathbf{r}}{dt} = \mathbf{u}$$

$$\frac{d\mathbf{u}}{dt} = \frac{\mathbf{F}}{m},$$

where the forces $\mathbf{F} = -\partial\Phi/\partial\mathbf{r}$ are derived from the total potential energy Φ, which is obtained by summing up interactions of all atoms with each other, whether pairwise-additive or many-body.

An atomistic shock wave can be generated by imposing a combination of initial and boundary conditions. The boundary conditions are: (1) a specularly reflecting "momentum mirror", or a stationary piston at the origin $x = 0$ (a particle's collision with the mirror is detected when its x-position at the end of a time step is found to be negative, whereupon the x-velocity is reversed, and the x-coordinate is replaced by the absolute value); (2) transverse (y and z) PBCs (image particles enter with the same velocity as the exiting particle, but at the opposite "wall"). Initial conditions are: particle positions and initial Maxwell–Boltzmann distribution velocities corresponding to temperature T_0 and augmented by a drift velocity $-u_p$ toward the mirror (see Fig. 4.1). As soon as the "target" material collides with the stationary, infinitely smooth piston, a shock wave begins moving to the right at velocity $u_s - u_p$. By Galilean invariance, this is equivalent to a piston moving at velocity u_p into stationary material, with the shock wave moving out in front of the piston at the shock velocity u_s. For purposes of visualization, the former frame of reference has the advantage that the shocked material is now stagnated against the immobile piston with average velocity equal to zero. There are other schemes to generate shock waves [1i], but the momentum mirror is the simplest, and for solids, it's *almost* foolproof, particularly when phase transformations or plasticity give rise to split, two-wave shock structures.

The most faithful rendition of the experimental generation of planar shock waves is the symmetric impact, where a flyer plate of atoms is launched to collide with a target plate [8] (see Fig. 4.1). Since two oppositely running shock waves are generated from the impact surface, this approach costs twice as much as the mirror piston method. (Note that transverse to the shock propagation direction, we always impose static PBCs.)

Similar to the symmetric impact, if we impose *shrinking* PBCs in the direction of the shock [9], with the boundaries moving toward each other at $\pm u_p$, a pair of shock waves move toward each other at $\pm u_s$, again at the expense of a factor of two compared with the mirror piston (see Fig. 4.1). Shrinking PBCs are better for describing fluid shocks, because free surfaces are absent, and no special care need be taken for ensuring that atoms do not evaporate from the sample.

To be completely fair and open about the momentum mirror method, however, the possibility exists that the material near the mirror can heat up in a way that is distinct from symmetric impacts and shrinking PBCs: the latter two methods have identical material on either side of the interface, while the momentum mirror is asymmetric—on one side is shocked material; on the other is nothing, vacuum, interstellar space. The material near the mirror is trapped: it seeks desperately to escape from the shocked, higher temperature, higher pressure, higher density material, but the momentum mirror keeps it in check, and heats it up in an extraordinary way by making the atoms near it bang their heads continuously against the "window". For this reason, in computing averages over the shocked state, we tend to disregard material near the mirror. We also have to

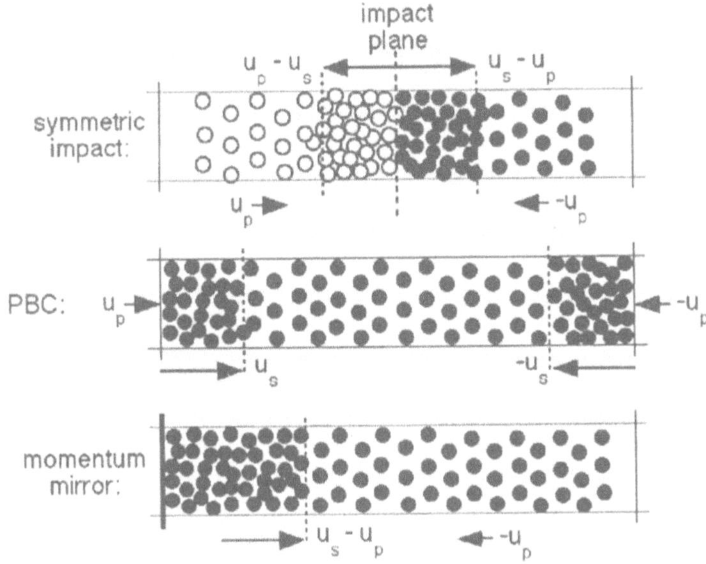

Figure 4.1. Illustration of three methods of generating shock waves in NEMD: symmetric impact, shrinking periodic boundaries (PBC), and the momentum mirror (piston velocity u_p, shock velocity u_s).

equilibrate the surface at the opposite free end of the sample, so that no rarefaction or compressive pulse begins propagating backward toward the mirror. The bottom line is that there is no really free lunch when it comes to setting up shockwave simulations by NEMD [10].

Apart from initial and boundary conditions, the only other input to the NEMD simulation of a shock wave is the interaction potential between atoms. Having chosen the interatomic potential, be it pairwise-additive or many-body in character, the resulting behavior that is observed is completely specified. The rest of the story is told by Newton (or Hamilton, if you prefer). Of course, the time step and the system size must be chosen with some care, but in any event, dynamics determines the outcome. The material we choose to simulate depends on the potential that is supposed to characterize it, and the minimal properties that must be fitted are the normal density at zero pressure and temperature, the cohesive energy, and the bulk modulus. Once the nearest-neighbor distance r_0 and energy scale (such as the bond energy ε_0) are set, along with the atomic mass m, the unit of time is determined, namely, $t_0 = \sqrt{(\varepsilon_0/m)}$; since this unit of time is not far from the Einstein period of the crystal, the computational time step δ is typically on the order of 0.01—less for shocks.

In the case of EAM many-body potentials for metals, an additional defect or surface property, such as the vacancy formation energy, must be incorporated into the fitting procedure. For shock-wave processes, where high-pressure data are important, a fifth parameter, the third-order anharmonicity, must also be included, as represented by the pressure derivative of the bulk modulus [7]. The anharmonicity determines, among other properties, the slope of the essentially linear Hugoniot relationship between u_s and u_p. In order to obtain a reliable value of the anharmonicity, one must resort to a fit of equation-of-state data (a low-temperature pressure–volume isotherm, for example), either from experiment or else from quantum-mechanical calculations [11].

The range of interactions, whether pairwise-additive or many-body EAM, is usually taken to be a radius r_{max} of only a few neighbor shells in the uncompressed lattice. There is a good physical reason for this: in metals, the electron gas tends to shield the positive nuclei. On the other hand, noble gases interact through weak van der Waals forces at long range, so it is reasonable to truncate them smoothly. Short-range interactions are also very practical, since the cost of doing MD simulations goes up like the number of neighbors about each atom. For massively parallel computations, we use domain decomposition [12], where space is broken up into cubic subcells whose side length is r_{max}. In that event, in order to compute the forces on a given atom, we need only look at neighboring atoms within its own subcell and in the immediately neighboring subcells. About 90% of an MD simulation is spent in computing forces, but domain decomposition keeps the cost linear in the total number of particles, N. If δ is the time step and t is the total length of physical time for the calculation (such that the number of time steps is $n = t/\delta$), the cost in computer time, or human time, or money, is given by

$$\$ = CNn \, ,$$

where C is a constant that represents the complexity of the interatomic interactions and is proportional to the cube of r_{max}. Since the EAM forces require one pass through the neighbors to obtain the local density information for each atom as well as the pairwise-additive part of the interaction, and then a final pass to compute the density-dependent contributions, EAM potentials typically require twice as much computer time as simple pairwise-additive potentials. Angular-dependent terms, explicit three-body or (God forbid!) four-body terms, and chemical reactivity pile on the complexity even further. (And add on the \$s—and in MD, as in many other worthwhile pursuits, time is money.)

The total number of atoms that appears in the above equation is proportional to the length L_x in the direction of shock-wave propagation (e.g., the x direction), as is the total time t for the NEMD shock simulation. Thus, the cost of simulation is quadratic in L_x, as well as in the cross-sectional dimension $L_y \approx L_z$ in 3-D.

4.3. NEMD Shock Simulation Results

The panorama of shock phenomena that has been studied to date includes waves propagating in a variety of crystal orientations (producing different kinds and degrees of plastic deformation) [1j], phase transitions [13], detonations [14,15], and spallation (the result of compressional waves produced in an impact propagating all the way to free surfaces, whereupon rarefaction waves are generated; when rarefaction waves collide, the material is put into tension, and voids are generated at the spall plane, where the material can come apart, provided that the original shock wave is sufficiently strong) [8,16]. Although spallation is an important phenomenon in shock physics, as is the chemistry of energetic molecules that leads to detonation, we will leave these very interesting topics to future discussions, and focus here on the initial unreactive shock waves.

As shock strength increases, so does the density ρ of the shocked material. Because the sound velocity goes roughly linearly with density, the vibrational frequency increases more rapidly than linearly with density (something like the power 4/3), so that the increase in the number of time steps grows roughly linearly with volumetric strain caused by the shock wave (V is the volume):

$$\varepsilon = 1 - \frac{V}{V_0} = 1 - \frac{\rho_0}{\rho} = \frac{u_p}{u_s}.$$

This makes very strong shock waves only somewhat more expensive to simulate, were it not for the significant speed-up of transformation mechanisms, as represented in part by a corresponding sharpening up of the shock front. In fact, a general rule of thumb appears to be that the shock thickness is inversely proportional to the shock strength, or strain ε.

The bottom line is that NEMD is limited most severely in its ability to reproduce the full set of mechanisms that come into play at low shock strengths. There, the thickness of the shock wave can become larger than the affordable system size, due to the scale dictated by the separation of pre-existing defects that serve as nucleation sites for phase transformation or plasticity. Moreover, real metals are not single crystals, but rather, polycrystalline, with grain sizes that range from nanometers (very fine-grained, hence, rare and expensive) to tens or hundreds of micrometers for typical engineering applications. As such, the distribution of defects is characterized (at least) by the grain size, which is usually enormous in comparison to the largest NEMD system that is presently possible.

In the case of weak shocks in perfect crystals, the response is entirely elastic, below the so-called Hugoniot elastic limit (HEL), above which plastic flow occurs (see Fig. 4.2). In polycrystalline materials, and even in single crystals that contain dislocations and other sources of heterogeneous nucleation sites, the HEL occurs at very low strains (a couple of percent). In perfect single crystals, the HEL is more extreme, on the order of 10–15%. Below this limit, Zybin and Zhakhovskii, as well as we at Los Alamos, have observed the propagation of steady elastic waves [1j, 17].

The mode of deformation in steady elastic shocks is the onset of transverse distortion, which makes the emerging solitary wave train "lock in" to a steady oscillatory profile, provided that the cross-sectional length is above a critical

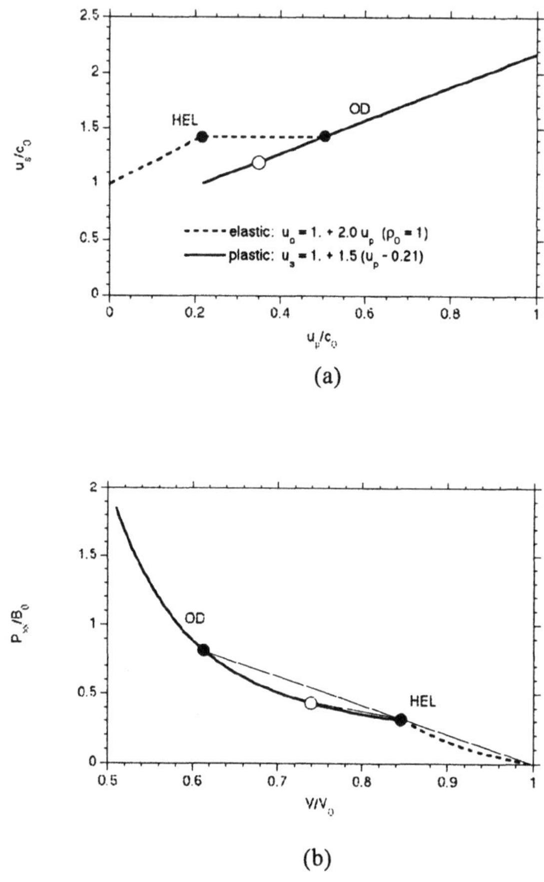

Figure 4.2. Model elastic-plastic Hugoniot (HEL = Hugoniot elastic limit, OD = overdriven limit): (a) normal pressure-tensor component P_{xx} vs. volume V (scaled by bulk modulus B_0 and normal volume V_0, respectively); (b) shock velocity u_s vs. particle velocity u_p (scaled by longitudinal sound velocity c_0). In (a), the Rayleigh line construction (light straight line) has slope proportional to the square of the wave speed. The open circle designates the amplitude of a typical elastic–plastic shock (fast elastic precursor, followed by slower plastic wave).

threshold. This characteristic length appears to be proportional to the longitudinal thickness of the oscillatory profile. One is tempted to conclude that the coupling in the transverse and longitudinal directions dictates the minimum scale of the distortion, and that when the cross-section is too small, the distortion is locked *out*, leading to an ever-growing, non-steady soliton profile.

In the case of elastic–plastic behavior above the HEL, the steady elastic precursor exhibits a transverse distortion that looks like a premonition of the shear deformation that comes later with the plastic wave. In the plastic wave, the deformation is irreversible—atoms slip into new positions that are a fraction of the lattice spacing (something of the order of one-half) away from their initial positions (see Fig. 4.3) .

Just as the elastic wave propagation speed depends on crystal orientation, so does the plasticity, both in kind and in degree [1j]. Different slip systems are available in different crystal directions; some are harder to activate than others, and the number available also differ, due to geometrical considerations in the crystal lattice. In the shock front, the pressure-tensor component in the shock propagation direction P_{xx} is larger than the transverse components, due to uniaxial compression. This leads to a buildup of the shear stress τ:

$$2\tau = P_{xx} - \tfrac{1}{2}(P_{yy} + P_{zz}).$$

Hydrostatic pressure is defined as $P = (P_{xx}+P_{yy}+P_{zz})/3$; hence,

$$P_{xx} = P + \tfrac{4}{3}\tau.$$

As plastic deformation occurs, the shear stress is relieved, though not necessarily all the way back to zero, as it would be in a fluid-like state where $P_{xx} = P$ (see shock profiles in Fig. 4.4).

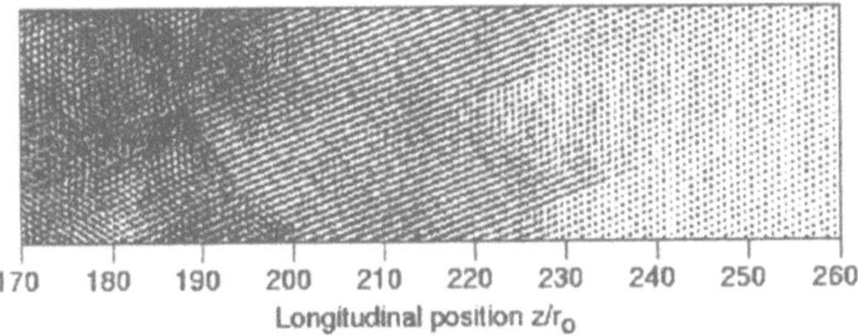

Figure 4.3. Side view of elastic–plastic shockwave propagation (to the right in the *z*-direction), revealing the faster, steady elastic precursor and its subsequent zone of elastic distortion (near $z = 230$), followed by the slower, steady plastic wave and its irreversible deformation ($z < 200$). (Note that atoms in the cold unshocked crystal, which are moving to the left toward the impact plane, line up and appear as simple dots in this orthographic projection onto a transverse plane, while elastic distortions appear as lines, and plastic deformation is a mess.)

160   Brad Lee Holian

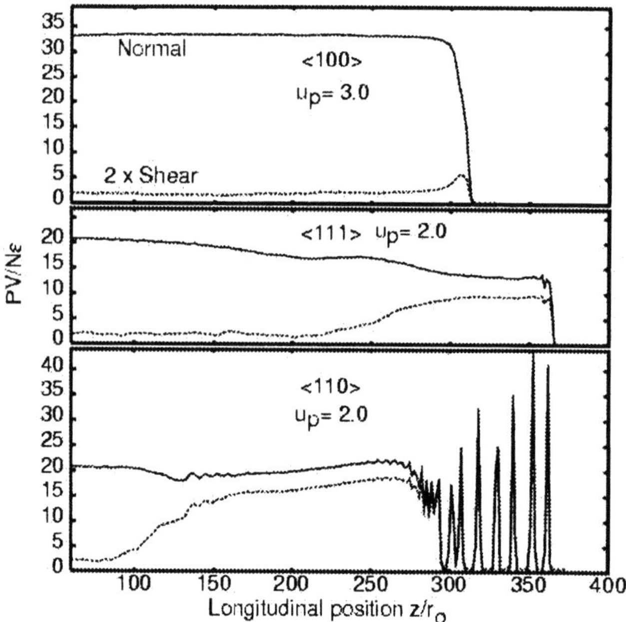

Figure 4.4. Spatial shock-wave profiles of the pressure-volume tensor as a function of propagation distance (the normal component is shown as a solid line, while twice the shear is a dotted line), for three fcc crystal directions: (a) $\langle 100 \rangle$ overdriven; (b) $\langle 111 \rangle$ elastic-plastic (with a steady oscillatory elastic precursor); (c) $\langle 110 \rangle$ elastic–plastic (note that *non-steady* soliton pulses are marching out in front of the steady, oscillatory elastic precursor near $z = 300$). The shock is in the z-direction and the piston velocity, u_p, is in units of $\sqrt{(\varepsilon_0/m)}$.

In the face-centered cubic (fcc) lattice, propagation along the usual unit cell edge, labeled $\langle 100 \rangle$, produces partial dislocations a few tens of lattice planes beyond the impact surface, which then propagate back toward the impact plane and forward along with the shock front, with no perceptible elastic precursor, at least for systems whose atoms interact via the Lennard-Jones (6–12) pair potential or a many-body EAM potential. Stacking faults are left behind as remnants of the shock-induced dislocation production [1c,1f,1h].

In very early work on static uniaxial compression of the $\langle 100 \rangle$ fcc lattice, Mogilevsky observed similar behavior [18], but at noticeably higher strains than the HEL in NEMD shock-wave simulations. (Homogeneous uniaxial compression is achieved instantaneously at time zero by shrinking all x-coordinates, including the periodic length of the computational volume L_x, by the uniaxial compression factor, $1 - \varepsilon$).

4.3.1. Plastic Deformation via the Uniaxial Hugoniostat

We have recently invented a new technique to elucidate the final damaged states behind the shock front [19]. It is based on Mogilevsky's instantaneous uniaxial strain for the initial condition, which favors the production of dislocations and other defects. But then the system is allowed to relax toward the final energy on the shock Hugoniot, guided by a feedback method related to Nosè–Hoover thermostatting [20]—hence, we call it the uniaxial "Hugoniostat". Behind a steady, planar shock wave, the final (Hugoniot) internal energy (kinetic energy relative to the center-of-mass motion plus potential energy Φ) in a sample volume $V = L_x L_y L_z$ is

$$E = E_0 + \tfrac{1}{2}(P_{xx} + P_0)(V_0 - V),$$

where subscript "0" indicates initial values.

In the Hugoniostat, the final volume V can be held fixed and chosen to represent the shock strength; but a more reliable way to get realistic final temperatures behind the shock is to allow L_x to fluctuate (so-called uniaxial barostatting based on the Nosè–Hoover method [20]), with P_{xx} fixed [10]. In principle, either variable, V or P_{xx}, should characterize the shock strength, but it turns out that fluctuations in L_x allow defects to form in a way that more closely resembles the processes in the NEMD shock front, especially in the two-wave, elastic–plastic regime.

The NEMD shock-wave simulation techniques described earlier are explicitly inhomogeneous, while the uniaxial Hugoniostat is completely homogeneous (PBCs in all directions); there is no wave propagation and no spatial gradients at the shock front to worry about. Hence, much shorter systems can be used, but not necessarily smaller cross-sections, since the deformation patterns have to fit into the computational cell, no matter what.

The results of the *uniaxial* Hugoniostat agree very well with the final states (energy and defective structures) of full NEMD simulations—note that we emphasize *uniaxial*, since isotropic compression does *not* lead to plastic deformation. Not all crystal directions are equal, however, as we learned quite quickly (to our chagrin and amazement). In particular, the $\langle 111 \rangle$ and (especially) the $\langle 110 \rangle$ directions are much more interesting than the $\langle 100 \rangle$ case. Both of these other directions exhibit elastic precursors, and have more complex patterns of deformation [10]. The uniaxially compressed $\langle 100 \rangle$ system produces octahedral-shaped stacking-fault zones on four equivalent slip systems; $\langle 111 \rangle$, which has three active available slip systems, produces tetrahedral-shaped zones; and $\langle 110 \rangle$, which is automatically anisotropic and has two available slip systems, quickly disintegrates into utter chaos as a rapid transformation occurs, resembling a martensitic phase transition [21]. This highlights the fact that the terms "plasticity" and "phase transformation" begin to blur into one another under shock-wave conditions. Consequently, polycrystalline samples subjected to shock waves—although they may appear to be more homogeneous than single crystals from the macroscopic point of view—become rich with complexity and heterogeneity at the microscopic level.

4.3.2. Shock-Induced Solid-Solid Phase Transformations

We have also begun to study a bona fide solid-solid phase transition under shockwave loading [13], namely, body-centered cubic (bcc) iron going to the hexagonal close-packed (hcp) phase. A shockwave of sufficient intensity in the bcc $\langle 100 \rangle$ direction shows the clear-cut formation of hcp twins (mirror-image crystal orientation), while the $\langle 110 \rangle$ bcc crystal produces a complex fine-grained polycrystalline pattern of phase transformation, curiously reminiscent of the $\langle 110 \rangle$ fcc mode of plasticity. Of course, the crystallites are nanoscale in dimension, but there is no reason to imagine that such a pattern could not be seen in careful, time-resolved, shock x-ray diffraction experiments.

4.3.3. Polycrystalline Complexity

In one very interesting and heroic calculation (involving *only* 24 million atoms), Germann and coworkers [22] studied a nanocrystalline sample of bcc Fe, and discovered that the profile spread out to the length of the computational cell, never achieving a steady state. The scattering of the initially sharp wave by the various crystallites, as well as the multiple sources of phase-transition nuclei at grain boundaries, leads to a very amorphous distribution of transformed and untransformed material, at a slightly lower shock strength (by about 10%) than the critical strength for inducing the transformation in the perfect bcc crystal (either $\langle 100 \rangle$ or $\langle 110 \rangle$).

4.3.4. The Outer Limits

This polycrystalline study demonstrates that NEMD has a lower limit on shock strength, namely, the shock thickness must be less than the grain size, and with present computers, that cannot be much larger than about 0.1 μm, i.e., at the nanoscale. An upper limit on shock strength, where NEMD is best suited, also exists: the shock thickness cannot ever fall below the interatomic spacing r_0. At that point, the shocked material is no longer solid, but rather a hot, dense fluid. Moreover, at even higher shock strengths, ionization begins to occur, sending hot electrons into cold, unshocked material, causing preheating and a dramatic increase in shock thickness—to hundreds of atomic spacings. At that point, standard NEMD is no longer appropriate: pressure- or temperature-induced ionization implies that the electrons are no longer in the ground state. An electronic phase transition takes place, and the short-ranged pair potential (or many-body EAM potential) changes to a long-range many-body potential appropriate to a plasma.

For very high strength shocks up to the point of ionization, and down to weak shocks, where mesoscale, defect-induced, heterogeneous nucleation begins to dominate the dissipative processes, one can say that NEMD describes shock phenomena very well. Over twenty years ago, we observed that the velocity distribution in the shock front, for solids, liquids, and gases, is significantly anisotropic and distinctly nonequilibrium, being well approximated by a bimodal distribution [1e,1i]. Moreover, for solids, the final state can be distinctly nonequilibrium, in that the stress tensor need not be isotropic, and the structures can be highly defective—metastable on very long time scales. Thus, the approximation of local thermodynamic equilibrium is never obeyed, particularly in the shock front, but neither, necessarily, far behind the front in solids.

Local thermodynamic equilibrium is sometimes not a bad approximation, however. For the case of fluids, the Navier–Stokes hydrodynamic description is remarkably good, as shown by Hoover [23], even for strong shocks [1e]. In principle, we should be equally well able to develop a plasticity model of comparable quality for solids, though the goal has not yet been achieved. Part of the difficulty is that NEMD results for solids have been restricted to perfect crystals, whose yield strength (the threshold for plastic flow, or HEL) is very high compared to real polycrystalline or even single crystalline samples containing pre-existing defects. We observe that perfect crystal yielding occurs at shock pressures approaching the shear modulus of the material. This corresponds to strains that are much larger than observed in experiments [24], by as much as an order of magnitude.

Nevertheless, NEMD can still be useful in elucidating unit processes, or mechanisms of the initiation of plastic flow and phase transformation in the heterogeneous nucleation regime for weak shock waves. There, grain boundaries, large-scale impurities or inclusions, and pre-existing dislocations can be critical to focusing the stress in a passing shock front, triggering the onset of dissipative transverse flow.

4.4. What the Hell is MD Good For, Anyway?

I will now confront a question on the minds of many people, but posed most eloquently by Jim Asay in an e-mail message sent to me in the course of preparing for a "What Is a Shock Wave" plenary session:

> "Do molecular-dynamics simulations of shock waves show
> any features that are *contrary* to 'accepted' notions of shock
> propagation?"
>
> —J.R. Asay, *17 February 2001*

Asay's Question addresses the most profound issues that we as molecular dynamicists have had to confront in this field of shock-wave phenomena, which, as I have said above, provides the greatest window of opportunity to learn about atomistic processes. At first blush, this question might appear to be a double-pronged challenge to NEMD: Are MD results real, or is it all computational dreaming devoid of any real physics? Or is everything already known from experiment and the classical picture of shock waves, and thus, there's no need to do MD? Alternatively, the question can be taken in its more positive form: Have we learned new insights, and can we hope to gain more as NEMD becomes even more powerful? What can these atomistic simulations teach us about new experimental probes and the future of continuum engineering modeling?

One could begin to approach the answer from a historical perspective. Early on, MD simulations, with the observation of ever-growing solitary wave trains, raised the question about the existence of steady shock waves in solids, and therefore, about the justification for the classical Hugoniot construction [1b]. The resolution came in just a couple of years, when solitons were shown to be rather fragile objects in 3D, particularly in the overdriven plastic-flow regime [1c]. Now, we know that even in the elastic regime in perfect crystals, solitons get locked into a steady oscillatory profile, because of transverse elastic distortion—provided that sufficient cross-sectional dimension is allowed in the simulations [17].

Overall, the answer to Asay's Question is mainly, No, there have not been any truly revolutionary surprises—at least not within the window of opportunity afforded by present-day NEMD simulations. As I said before, we do see some intriguing subtleties of shock propagation at the atomistic level: Local thermodynamic equilibrium is violated. Temperature and stress are anisotropic in the shock front, and residual shear stress can be left behind in solids, along with a fascinating zoo of different defect structures. The elastic precursor and the damaged plastic final state are both metastable. This suggests that there is difficulty imposing equilibrium thermodynamics, although the consequences may be difficult to quantify if sufficiently small. But, earthquakes destroying the temple of the classical view of shock-wave phenomena?—No.

Does that mean that we have learned nothing really new from atomistic simulations? In fact, we have learned a great deal. In particular, we have been able to visualize the kinds of phenomena that occur at the atomic scale, and therefore have a much better physical basis for the kinds of mesoscale (or even continuum) models that need to be developed.

It is important—not only for experimentalists, but also for practitioners of NEMD—to realize that molecular dynamics is not a "theory of everything." MD requires either direct input from quantum-mechanical treatment of electronic degrees of freedom for interatomic interactions (which, at the present time, can be prohibitively expensive), or else it requires more practical semiempirical fits to experimental data, or quantum calculations, or both. But even if we consider model potentials, such as Lennard-Jones, we can gain tremendous insight into generic physical mechanisms—even in 2-D! MD fits into our overall picture of micro-, meso-, and macroscopic physics by providing a physically reliable source of input into mesomechanical models, which average over the behavior of large ensembles of atoms, one way or another (see, for example, [25]). The thing to keep in mind from past history, however, is that continuum behavior happens on shorter time and distance scales than has been traditionally imagined. As Alder and Hoover showed us, the description of hydrodynamic behavior of fluids comes out as a direct consequence of the behavior of a remarkably small number of atoms. Solids are more challenging in that regard, but not entirely hopeless. In any event, we cannot hope to put fundamental physics modeling into continuum engineering simulations without mesoscale modeling; and mesoscale modeling cannot be done reliably without atomistic input.

4.5. Fulture Challenges and Opportunities

Having given at least a partially satisfactory answer to Asay's Question, namely, that no show-stoppers have been discovered by NEMD simulations of shock-wave phenomena, and that, nevertheless, deep insight has been gained into the way things work at the level of atoms, there are a number of interesting open questions and new frontiers to be explored in NEMD shock simulations in the next few years.

First and foremost is the issue of whether or not there are truly stable and steady planar shock waves, particularly in underdriven, two-wave regimes. By that I mean, can we conclusively demonstrate that shock waves are always planar, or do they ever exhibit galloping, runaway, non-planar fronts? It does no good to answer that, to the best of our ability, we see that everything appears to be "just fine". There are long-time thermally activated processes that may be just beyond our ability to detect. For example, if there are almost imperceptible heating mechanisms that go like the logarithm of time, or if there are subtle system-size effects that go like the logarithm of the number of particles or cross-sectional dimension, then we could be in fundamentally deep trouble. It may appear at the present level of resolution that there is "no problem", but I would be loathe to bet my retirement funds on that.

As to more mundane challenges, i.e., the ones that seem to have no obvious hidden pitfalls, the next decade will surely see an explosion in complexity of atomic and molecular interaction potentials, including the possibility of incorporating more sophisticated quantum-mechanical treatments into NEMD. Already, shock-induced chemistry is within reach, both at the semiempirical level [14,15] and at the level of tight-binding MD [26].

In the very near future, we can expect that ultra-fast dynamic x-ray diffraction, performed on-the-fly in shock-wave experiments (either gas gun- [24] or laser-driven [27]) will be able to verify the kinds of defect structures that we have seen in NEMD simulations. We should at least also be able to place limits on the relaxation times of plastic-flow and phase-transition mechanisms. Thus, the future of atomistic modeling has never been brighter, but the heady quest for ever-increasing complexity needs always to be sobered by facing Asay's Question squarely and honestly.

Acknowledgments

It is my distinct pleasure to thank Bill Hoover and Berni Alder for their inspiration to me through the years. I add to that distinguished pair the names of my friends and collaborators at Los Alamos and elsewhere, without whom this journey of discovery would have been not nearly so much fun: Tim Germann, Jim Hammerberg, Kai Kadau, Joel Kress, Peter Lomdahl, Jean-Bernard Maillet, Michel Mareschal, Ramon Ravelo, Carter White, and Sergey Zybin. Finally, I am indebted to experimental colleagues for their thought-provoking comments, occasional support for MD as a useful tool for mutual enlightenment, and their constant contributions and prods to theory over the last thirty years: Jim Asay, Lalit Chhabildas, Yogi Gupta, Dennis Hayes, and Dennis Grady.

References

[1]   (a) A. Paskin and G.J. Dienes, *J. Appl. Phys.* **43**, p. 1605 (1972);

(b) D.H. Tsai and R.A. MacDonald, in *Proceedings of the 1976 International Conference "Computer Simulations for Materials Applications," Gaithersburg, MD, 1976* (eds. R.J. Arsenault et al.) *Nuclear Metallurgy* **20**, p. 489 (1976);

(c) B.L. Holian and G.K. Straub, *Phys. Rev. Letters* **43**, p. 1598 (1979);

(d) V.Y. Klimenko and A.N. Dremin, in *Detonatsiya, Chernogolovka,* (eds. O.N. Breusov et al.) Akademiya Nauk, Moscow, p.79 (1978).

(e) B.L. Holian, W.G. Hoover, B. Moran, and G.K. Straub, *Phys. Rev. A* **22**, p. 2798 (1980);

(f) B.L. Holian, *Phys. Rev. A* **37**, p. 2562 (1988);

(g) B.L. Holian, *Shock Waves* **5**, p. 149 (1995);

(h) B.L. Holian and P.S. Lomdahl, *Science* **280**, p. 2085 (1998);

(i) V.V. Zhakhovskii, S.V. Zybin, K. Nishihara, and S.I. Anisimov, *Phys. Rev. Letters* **83**, p. 1175 (1999);

(j) T.C. Germann, B.L. Holian, P.S. Lomdahl, *Phys. Rev. Letters* **84**, p. 5351 (2000); B.L. Holian, T.C. Germann, P.S. Lomdahl, J.E. Hammerberg, and R. Ravelo, in *Shock Compression of Condensed Matter—1999* (eds. M.D. Furnish et al.), American Institute of Physics, New York, p.35 (2000).

[2]   B.J. Alder and T.E. Wainwright, in *International Symposium on Statistical Mechanical Theory of Transport Processes, Brussels, 1956* (ed. I. Prigogine) Interscience, New York, p.97 (1958); ibid., *J. Chem. Phys.* **27**, p. 1208 (1957).

[3]   B.J. Alder and T.E. Wainwright, *Phys. Rev. A* **1**, p. 18 (1970).

[4]   E.G.D. Cohen and W. Thirring, eds., *The Boltzmann Equation*, Springer-Verlag, Vienna and New York, (1973).

[5]   W.G. Hoover and W.T. Ashurst, *Adv. Theor. Chem.* **1**, p. 1 (1975).

[6]   O.B. Firsov, *Sov. Phys.–JETP* **36**, p. 1076 (1959).

[7]   B.L. Holian, A.F. Voter, N.J. Wagner, R.J. Ravelo, S.P. Chen, W.G. Hoover, C.G. Hoover, J.E. Hammerberg, and T.D. Dontje, *Phys. Rev. A* **43**, p. 2655 (1991); B.L. Holian and R. Ravelo, *Phys. Rev. B* **51**, p 11275 (1995); see original EAM reference M.S. Daw and M.I. Baskes, *Phys. Rev. B* **29**, p. 6443 (1984).

[8]   N.J. Wagner, B.L. Holian, and A.F. Voter, *Phys. Rev. A* **45**, p. 8457 (1992).

[9]   G.K. Straub, B.L. Holian, and R.G. Petschek, *Phys. Rev. B* **19**, p. 4049 (1979).

[10]   R. Ravelo, T.C. Germann, P.S. Lomdahl, and B.L. Holian, The Nature of Shock-Induced Plasticity: Comparison between NEMD and the Hugoniostat, in preparation.

[11]   Y. Mishin, M.J. Mehl, D.A. Papaconstantopoulos, A. F. Voter, and J.D. Kress, *Phys. Rev. B* **63** p. 224106 (2001).

[12]   D.M. Beazley and P.S. Lomdahl, *Par. Comput.* **20**, p. 173 (1994); ibid., *Par. Comput.* **11**, p. 230 (1997).

[13]   K. Kadau, T.C. Germann, P.S. Lomdahl, and B.L. Holian, Shock-Induced Phase Transition in Single-Crystal Iron, in preparation.

[14]   D.W. Brenner, D.H. Robertson, M.L. Elert, and C.T. White, *Phys. Rev. Letters* **70**, p. 2174 (1993).

[15]   T.C. Germann, J.-B. Maillet, B.L. Holian, and P.S. Lomdahl, Detonation Phenomena in a Model 2D and 3D Energetic Molecular Solid, in preparation.

[16]   A. Strachan, T. Cagin, and W.A. Goddard III, *Phys. Rev. B* **63**, p. 060103 (2001).

[17]   S.V. Zybin, T.C. Germann, P.S. Lomdahl, and B.L. Holian, Steady Elastic Shock Waves, in preparation.

[18]   M.A. Mogilevsky, in *Shock Waves and High Strain Rate Phenomena in Metals* (eds. L.E. Murr and M.A. Meyers) Plenum, New York, p.531 (1981).

[19]   J.-B. Maillet, M. Mareschal, L. Soulard, R. Ravelo, P.S. Lomdahl, T.C. Germann, and B.L. Holian, *Phys. Rev. E* **63**, p. 16121 (2001).

[20]   W.G. Hoover, *Phys. Rev. Letters* **42**, p. 1531 (1979).

[21]   J.P. Hirth, R.G. Hoagland, B.L. Holian, and T.C. Germann, *Acta mater.* **47**, p. 2409 (1999).

[22]   T.C. Germann, K. Kadau, and B.L. Holian, Shock-Induced Phase Transition in Polycrystalline Iron, in preparation.

[23]   W.G. Hoover, *Phys. Rev. Letters* **42**, p. 1531 (1979).

[24]   P.A. Rigg and Y.M. Gupta, *Phys. Rev. B* **63**, 094112 (2001).

[25]   P.V. Makarov, *Physical Mesomechanics* **1**, p. 57 (1998); K. Yano and Y. Horie, Phys. Rev. B **59**, p. 13672 (1999).

[26] J.D. Kress, S.R. Bickham, L.A. Collins, B.L. Holian, and S. Goedecker, *Phys. Rev. Letters* **83**, p. 3896 (1999).

[27] A. Loveridge-Smith et al., *Phys. Rev. Letters* **86**, p. 2349 (2001).

CHAPTER 5

Meso–Macro Energy Exchange in Shock Deformed and Fractured Solids

Yuri I. Mescheryakov

5.1. Introduction

One of primary problems in the dynamics of materials is developing an understanding of the coupling between microstructural features of a material and its macroscopic response on impact. Many theoretical models based on the microstructure dynamics, in particular on dislocation dynamics, have been developed to describe the macroscopic behavior of material under both uniaxial stress and uniaxial strain conditions. Nevertheless, this coupling is poorly understood both qualitatively and quantitatively. This is due to the incorrect, but commonly used, approach of linking the macroscopic response on impact with data describing the microstructural state that prevails *after* dynamic loading. In reality, adequate mathematical modeling of dynamic processes should be based on microstructural kinetics data obtained in real time, i.e., *during* the dynamic deformation and fracture processes. This requires that experimental technique provide not only measurements of macroscopic response such as the time-resolved free-surface-velocity profile for impacted specimens, but also kinetic characteristics of their internal structure. These characteristics provide information on the relative mobility of elementary carriers of deformation (ECD). Because the motion of ECDs in a heterogeneous medium is stochastic in nature, their kinetics must normally be described in the language of the particle velocity distribution function and its statistical moments. The term *mesoparticle kinetics*, as used herein, has the same meaning as in the physical kinetics of fluids and gases, i.e., it refers to the behavior of the particles as characterized by a distribution in velocity space. The width of that distribution, or the square root of the particle velocity dispersion, is defined in Section 5.2 as a *mean velocity fluctuation* of the mesostructure. This mean velocity fluctuation is a quantitative characteristic of mesoparticle kinetics.

It is generally now recognized that direct transition from dislocation dynamics to macroscopic plasticity is not possible because of collective interactions in dislocation ensembles and incorporation of large-scale carriers of deformation into plastic flow. In the case of shock wave propagation phenomena, that conclusion has been reached after numerous unsuccessful attempts to apply dislocation dynamics to the description of elastic precursor decay [1]. Collective interaction of dislocations results in deformation at some intermediate scale, the so-called mesoscopic level [2,3]. We think that all the events responsible for the observed change of the kinematical mechanism of deformation with changes in the external loading and boundary conditions are realized at the mesolevel.

5.2. Mesoscopic Scale Level

In considering the response of solids to dynamic loading, the primary interest concerns the dependence of macroscopic plasticity and strength of the material on the kinetics of internal structural modification and the scale of elementary carriers of deformation. Our understanding of the influence of mesostructural effects on quasistatic and dynamic deformation processes is incomplete. Historically, mesostructure effects were first introduced into models of quasistatic deformation processes in the form of a new kind of defect structure resulting from collectivization of dislocations and other defects of the crystalline lattice that occurs during large plastic deformation [2–4]. According to the early work, the most typical mesolevel feature is a spatial and charge heterogenization of dislocations on distances of the order of their free run. Mesovolumes ΔV having a polarized dislocation structure are considered as quasiparticles having an effective charge $q = \Delta\rho\Delta V$, where $\Delta\rho$ is the density of dislocation charge (see Fig. 5.1). It has been theoretically shown that these quasiparticles have a much larger radius of interaction than individual dislocations.

Further development of mesomechanics shows that nucleation of mesostructures during quasistatic loading of a solid occurs not only under large plastic deformation but also in the elastic region of deformation. In this latter case, the large-scale structures emerge as a result of nonlinear interaction of atoms. Modern nonlinear microdeformation theory grounded on precise solution of the nonlinear sine–Helmholtz equation [5] predicts instability of a crystalline lattice subjected to relatively large shear deformation in the nonlinear region of elastic loading. This leads to nucleation of such structures as elastics mesorota-

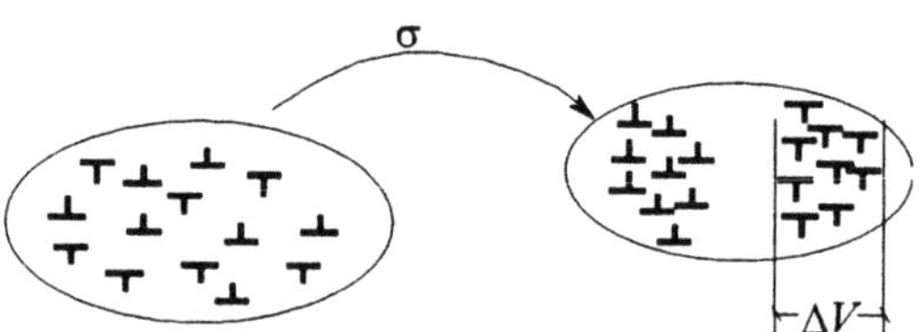

Figure 5.1 Polarization of dislocation structure at high deformations.

tions, shear bands, screw formations, and their combinations. These large-scale structures emerge long before the transition to macroscopic plasticity occurs. If the strain gradients do not exceed some critical value, these formations may disappear, so the process proves to be reversible at that stage. At higher strain gradients there exists a bifurcation transition from elastic to plastic deformation that results in irreversible nucleation of mesostructural features in the form of mesodefects—mesorotations, shear bands, dislocation cell structure, etc.

The mesoscopic scale level of deformation and fracture may often be subdivided into two or more sublevels. For example, microstructural investigations of spall zone for 30XH4M steel [6] shows that the process of dynamic fracture of this steel involves two scale levels: mesolevel-1 (10–20 μm) and mesolevel-2 (100–500 μm). In Fig. 5.2, the experimentally observed distributions of elementary steps of spall cracking are presented for horizontal (d and D) and vertical (h and H) pieces of the spall gap. These statistics show that spallation of this material proceeds at two scale levels simultaneously. Post-shock microstructural investigation specimens also shows that even if a process of dynamic straining occurs under one-dimensional conditions at the macrolevel, the motion of elementary carriers of deformation at the mesolevel is distinctly three-dimensional. Moreover, owing to continuity conditions at the mesolevel, the presence of the particle velocity distribution automatically implies the three-dimensional character of straining. As a result, such kinematical mechanisms of deformation and fracture as dynamic rotation can be realized at the mesolevel.

In considering the mesoscopic approach to dynamic deformation and fracture it should be noted that, to date, experimentally determinable kinetic characteristics are those belonging to the mesoscopic scale level. However, as distinct from the static and quasistatic situation, in dynamically loaded media the mesolevel does not represent a fully evolved structure. It is specifically a transient structure in which the scale of ECDs and the energy capacity at the mesoscopic scale level changes depending on the degree of non-steadiness of the deformation process. Energy capacity at the mesolevel characterizes a part of the kinetic energy initially given to the body from the external load that is transferred to the mesolevel in the form of velocity scattering of elementary carriers of deformation (mesoparticle velocity dispersion). In unsteady plastic waves, energy is exchanged between the mesolevel and the macrolevel. At the front of a compressive pulse the average motion (flux) of the medium is transformed into a velocity-distributed motion of mesoparticles. This occurs in the form of a decrease (loss) of mean particle velocity resulting from the so-called fluctuative decay of the shock wave. On the contrary, during the release stage of dynamic loading, a decrease of dispersion leads to an increase of the mean particle velocity.

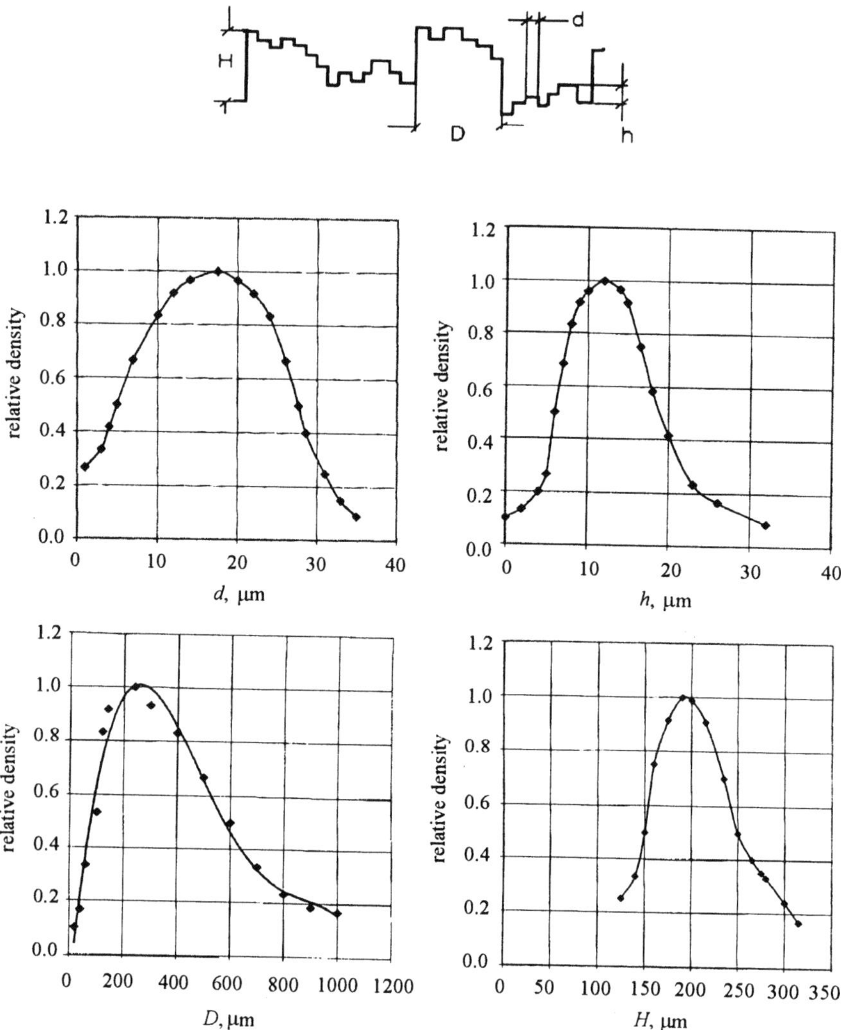

Figure 5.2 Spatial configuration of a two-scale spall crack and statistics of longitudinal and transverse components of spall-gap discontinuities.

5.3. Velocity Distribution Function and Its Statistical Moments

As with quasistatic straining, a set of experimental facts in dynamic plasticity provide evidence of mesostructure formation in the elastic region of dynamic loading. In particular, it has been found that particle velocity distribution at the mesolevel takes place not only at the plastic front of a compressive pulse but also

also at the elastic precursor. To prove that assertion two kinds of planar shock loading experiments conducted on D-16 (Russian: *Д-16*) aluminum alloy targets of different thickness have been performed [7]. For the first set of tests, the free-surface velocity was monitored with quartz gauges whereas, for the second set, it was measured using an optical interference technique. Results of these tests are presented in Fig. 5.3. Investigation of elastic precursor decay in uniaxial planar tests by using quartz gauges and an optical interference technique for measuring the free-surface velocity reveals a drastic difference in precursor behavior under identical loading conditions. The first set of targets shows that the elastic precursor attenuates from 11 to 2 kbar as the target thickness increases from 2 through 10 mm. In the second case, the the value of normal stress and free-surface velocity are related by the equation

$$\sigma = 0.5\rho c_l u_{\mathrm{fs}},$$

where c_l is the sound speed and u_{fs} is the free-surface velocity corresponding to the stress, σ. In contrast to the first set of targets, the amplitude of the elastic precursor measured in the second set of tests is invariable and equal to its steady-state value. The only explanation for this phenomenon is found in the destruction of the interference fringe pattern because of formation of mesostructure and initiation of the particle velocity distribution at the mesolevel. This means that formation of the mesostructure occurs long before the material transits into plastic state.

Unlike the case of quasistatic straining, the most important feature of high-strain-rate deformation of solids is the emergence of a space-time correlation among the elementary carriers of deformation. The collective effects cause the

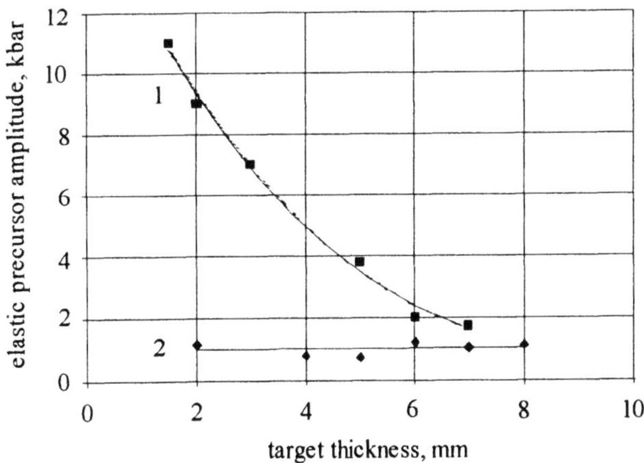

Figure 5.3 Elastic precursor amplitude versus target thickness in D-16 aluminum alloy measured with (1) quartz gauges and (2) with an interferometer.

formation of new structural elements having a larger scale than that of those elements initially present.

Because the experimental investigations presented here are based on a statistical approach to multiscale processes, we present definitions of variables that characterize the stochastic medium at the mesolevel. The description of non-equilibrium processes is grounded on the concept of physical kinetics for the particle velocity distribution function (PVDF), or probability density, commonly used in the mechanics of fluids and gases. According to the definition of the particle velocity distribution function, the quantity $f(\mathbf{r}, \mathbf{v}, t)\, d\mathbf{v}\, d\mathbf{r}$ represents the mathematical expectation of the number of particles in the volume $d\mathbf{r}$ at the position $\mathbf{r}$ that have velocities in the range from $\mathbf{v}$ to $\mathbf{v}+d\mathbf{v}$ at the time t. The normalization condition requires that integrating over the physical and particle-velocity spaces gives the total number of particles,

$$\int_{-\infty}^{\infty} f(\mathbf{r}, \mathbf{v}, t)d\mathbf{r}\, d\mathbf{v} = N(t) . \tag{5.1}$$

This assertion is valid for only one kind of elementary carrier of deformation at the mesolevel. In reality mesostructure includes several kinds of carriers of deformation and each type must be characterized by its own distribution function $f_a(\mathbf{r}, \mathbf{v}, t)$. The normalization condition for such a distribution is

$$\int f_a(\mathbf{r}, \mathbf{v}, t)d\mathbf{r}\, d\mathbf{v} = N_a(t) \tag{5.2}$$

where $N_a(t)$ is the number of mesoparticles of kind a present at time t [here, and henceforth, the integration is understood to be over the range $(-\infty, \infty)$ for each variable]. It follows from the normalization condition that the value $n_a(\mathbf{r}, t)$, which is determined by the relation

$$n_a(\mathbf{r}, t) = \int f_a(\mathbf{r}, \mathbf{v}, t)d\mathbf{v} , \tag{5.3}$$

is the number density of mesoparticles of type a. For the mass density one obtains

$$\rho(\mathbf{r}, t) = \sum m_a n_a = \sum m_a \int f_a(\mathbf{r}, \mathbf{v}, t)d\mathbf{v} , \tag{5.4}$$

where m_a is the mass of each of the particles of type a.

Hydrodynamic flow of a medium is characterized by the velocity of mass transport. We define the mean mass velocity at the position $\mathbf{r}$ and time t by relation

$$\mathbf{u}(\mathbf{r},t) = \frac{1}{\rho(\mathbf{r},t)} \sum_a m_a n_a(\mathbf{r},t)\langle \mathbf{v}_a \rangle, \tag{5.5}$$

where

$$\langle \mathbf{v}_a(\mathbf{r},t) \rangle = \frac{1}{n(r,t)} \int \mathbf{v}_a f_a(\mathbf{r},\mathbf{v}_a,t)d\mathbf{v}_a \tag{5.6}$$

is the mean value of the velocity, $\mathbf{v}_a$, of the particles of type a at the place $\mathbf{r}$ and the time t.

Besides the flow transportation of material there exists a relative fluctuative motion of particles of type a with the velocity

$$\mathbf{c}_a(\mathbf{r},\mathbf{v}_a,t) = \mathbf{v}_a - \mathbf{u}(\mathbf{r},t). \tag{5.7}$$

The average value of $c_a(\mathbf{r},\mathbf{v}_a,t)$,

$$\frac{1}{n(r,t)} \int f_a(\mathbf{r},\mathbf{v}_a,t)(\mathbf{v}_a - \mathbf{u})d\mathbf{v}_a = \langle \mathbf{c}_a(\mathbf{r},t) \rangle = \langle \mathbf{v}_a - \mathbf{u}(\mathbf{r},t) \rangle, \tag{5.8}$$

is called the *diffusion velocity*. It defines the flux density of mass for the a-component of the particles

$$\mathbf{J}_a = m_a n_a \langle \mathbf{c}_a \rangle = m_a \int f_a(\mathbf{r},\mathbf{v},t)\mathbf{c}_a\, d\mathbf{v}. \tag{5.9}$$

The sum of diffusion fluxes for all components of the mesostructure equals zero:

$$\sum_a n_a m_a \langle \mathbf{c}_a \rangle = \sum_a n_a m_a \langle \mathbf{v}_a \rangle - u \sum_a n_a m_a = 0. \tag{5.10}$$

Note that, in a one-component medium, the diffusion velocity concept is meaningless and the diffusion flux is absent. In the case of one kind of particle, Eqs. 5.4 and 5.5 can be written

$$\rho(\mathbf{r},t) = \int f(\mathbf{r},\mathbf{v},t)d\mathbf{v}; \quad \mathbf{u}(\mathbf{r},t) = \frac{1}{\rho(\mathbf{r},t)} \int f(\mathbf{r},\mathbf{v},t)\mathbf{v}\, d\mathbf{v}. \tag{5.11}$$

In this case, in order to characterize the fluctuative properties of dynamically deformed heterogeneous medium, one can use the velocity dispersion

$$D^2 = \int (\mathbf{v}-\mathbf{u})^2 f(\mathbf{r},\mathbf{v},t)d\mathbf{v}. \tag{5.12}$$

For our purposes, it seems to be convenient, together with dispersion, to use the square root of the mesoparticle velocity dispersion, D, which here and henceforth is called the *mean velocity fluctuation*.

One of the first steps in developing the kinetic theory, including meso-structure, of either fluids or solids, is definition of the equilibrium distribution function. In the gas and fluid kinetics the *local Maxwellian* distribution function

$$f^0{}_{LM}(\mathbf{r}, v, t) = n(\mathbf{r}, T)\left(\frac{m}{2\pi kT}\right)^{3/2} \exp\left[-\frac{m(v-u)^2}{2kT}\right] \qquad (5.13)$$

is commonly used for that purpose. In the case of solids, one can use an analogy with the kinetic theory of gases, replacing the coefficient $2kT/m$ by the meso-particle velocity dispersion D^2 [8]:

$$f^0(\mathbf{r}, \mathbf{v}, t) = n(\mathbf{r}, t)\frac{\pi^{-3/2}}{D_1 D_2 D_3}\exp\left[-\sum_{i=1}^{3}\frac{[v_i - u_i(r, t)]^2}{D_i^2}\right], \qquad (5.14)$$

where $n(\mathbf{r}, t)$ is the mean density of mesoparticles, $u_i(\mathbf{r}, t)$ is the i-component of the mass velocity, and $D_i^2(\mathbf{r}, t)$ is a mean dispersion of the i-component of the mass velocity. In other words, particle velocity dispersion plays the role of a mesolevel "temperature". Sometimes in the Western literature, in particular in molecular dynamics simulations, it is called a "granular temperature".

The first statistical moment of the PVDF, i.e., the average particle velocity $\mathbf{u}(\mathbf{r}, t)$, characterizes the macroscopic behavior of medium under dynamic strain conditions. Commonly used free-surface velocity profiles measured by means of gauges or a VISAR (a particular laser-interferometer configuration for measuring particle velocity) contain only information about the average behavior of material. Mesoscopic effects such as shear banding and mesorotations are determined by higher-order statistical moments, such as particle velocity dispersion and excess, i.e., asymmetry of the velocity distribution function. Their values provide a statistical characterization of the relative velocities of mesoparticles. This means that, when speaking about mesoparticle dispersion under conditions of dynamic deformation, one cannot determine the specific form (dislocation groups, shear bands, tilt boundaries, rotations, etc.) of elementary carriers of deformation at the mesolevel. Mesoparticles in a dynamically deforming body are interpreted as fluctuations of the mass velocity or local strain rate fields.

Nevertheless, it is not too hard to estimate their typical scale by using the principles of interference used to measure the particle velocity dispersion. Measurement of the particle velocity dispersion is based on the concept that different particles at the free surface of the target, and lying within the laser beam spot, give different Doppler frequency shifts. Effective interaction of individual particles and laser radiation is possible only when the dimension of the particle significantly exceeds the laser radiation wavelength (approximately ten times). On the other hand, uniform broadening of the radiation spectrum is achieved only if there is a sufficient number of similar particles within the diameter, d_b, of the laser beam spot on the free surface of the target. From this, the diameter, d_p, of the particle is estimated to fall in the range

$$\lambda \ll d_p \ll d_b . \tag{5.15}$$

As a rule, the laser beam has a transverse dimension of about of $70-100$ μm at its focal point on the free surface of a target, whereas the laser radiation wavelength is approximately $0.5-1.0$ μm. In this case, the size of the particle lies in the range $7-10$ μm. In accordance with a classification considered in [2–4], this size belongs to mesoscopic scale level.

The mesoparticle velocity distribution resulting in broadening the reflected laser radiation spectrum leads, in turn, to a decrease of the interference fringe contrast, which gives quantitative information about the particle velocity dispersion at the mesolevel [9,10]. When the mesolevel is subdivided into two sublevels, elementary carriers of deformation and fracture for the sublevels may suffer their own particle velocity distributions. In this case there may be three different situations depending on the degree of the velocity distribution at these levels:

1. Particle velocity dispersion at the mesolevel-1 is greater that at the mesolevel-2.

2. Particle velocity dispersion at the mesolevel-2 is greater than at the mesolevel-1.

3. The dispersions at these two levels are the same.

These cases are presented in Fig. 5.4 in the form of velocity–space configurations. The laser beam of the interferometer being focused on the free surface of the target just corresponds to the meso-2 scale level. In other words, when the laser interference technique is used, one deals with the velocity history for an individual element of mesolevel-2. This means that the macroscopic particle velocity u_{av} that results from averaging the particle velocity distribution at the mesolevel-2 coincides with the average particle velocity at the mesolevel-1 only when the particle velocity dispersion for the meso-1 and meso-2 scale levels are equal. In all the other cases, the macroscopic velocity can be obtained from averaging the velocity distribution at the mesolevel-2. For that purpose, simultaneous monitoring the free surface of the target at several points, as developed by Barker [11], can be used.

The previously described difference in elastic precursor response, as measured with quartz gauges and with an interferometer, is explained in terms of the two-scale mesoscopic kinetic approach to shock-wave propagation. This difference results from the distinction in the velocity distribution behavior at mesolevel-1 and mesolevel-2. The qualitative picture of the velocity distribution distribution at these scale levels is presented in Fig. 5.5. For mesolevel-1, the

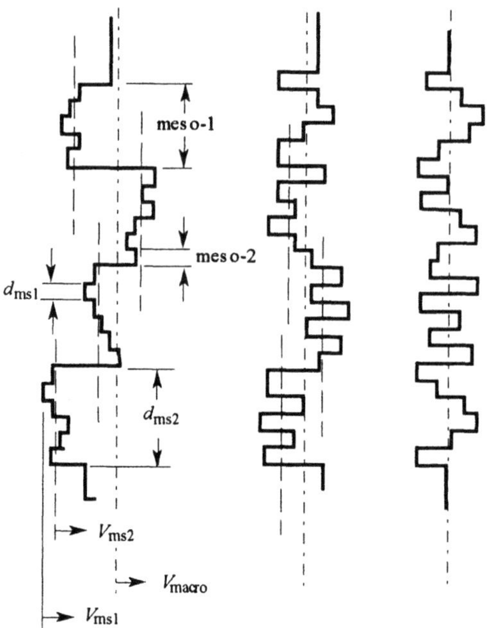

Figure 5.4 Space–velocity configurations of the shock-wave front for different relative values of the velocity dispersion at the mesolevel-1 and mesolevel-2.

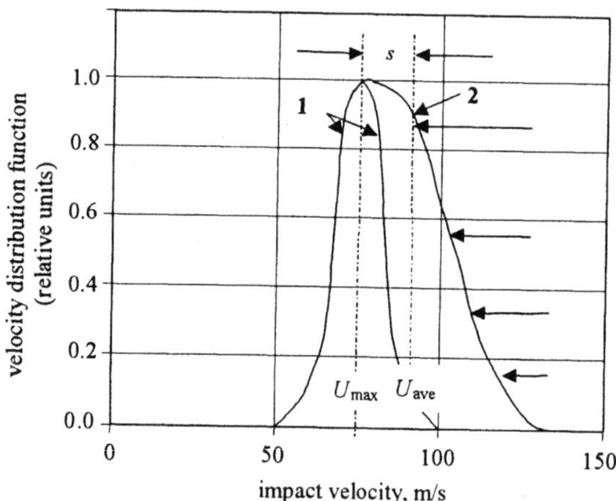

Figure 5.5 Qualitative picture of (1) the velocity distribution function at the mesolevel-1 and (2) mesolevel-2 for D-16 aluminum alloy.

particle velocity is in equilibrium. It has a symmetrical (Maxwellian) shape and is centered at the average velocity corresponding to the dynamic yield limit. For mesolevel-2, the velocity distribution at small target thicknesses is not in equilibrium. The particle velocity distribution function itself turns out to be non-symmetrical and has a positive excess. In this case, the average value of the particle velocity, u_{av}, is known to be different from the value u_{max} corresponding to the maximum of the distribution. As the elastic precursor propagates into target, the particle velocity distribution tends toward equilibrium, so values u_{av} and u_{max} become equal.

In accordance with the foregoing estimate, only the velocity distribution at mesolevel-1 is measured when the free-surface velocity of the target is measured with the velocity interferometer focused on a spot of dimension $d_b = 100\,\mu m$ (i.e., within one structural element of mesolevel-2). This distribution becomes equilibrated (symmetrical) over a very small distance from the loaded surface (≈ 1 mm). Therefore, the interferometer measures only the average particle velocity at mesolevel-1, which remains essentially unchanged as the elastic precursor propagates over large distances from the loaded surface.

When one uses a quartz gauge, the free-surface response is measured in the form of normal stress, which relates to the free-surface velocity as follows:

$$\sigma = 0.5\,\rho\,c_p\,u_{av},$$

where c_p is a plastic wave velocity and u_{av} is the free-surface velocity averaged over the non-equilibrium mesolevel-2 velocity distribution. When the velocity distribution at mesolevel-2 tends to equilibrium, $u_{av} \rightarrow u_{av}^{max}$, this results in a decrease of the elastic precursor amplitude from its initial value defined by the initial value of the average velocity $u_{av}^{i} = u_{av}^{max} + s$ to its final value of $u_{av}^{f} = u_{av}^{max}$ (see Fig. 5.5).

It has been shown experimentally that, for unsteady processes, the mesoparticle velocity distribution varies along the average velocity profile $u(t)$. The maximum dispersion occurs at the middle of the steady plastic wavefront. The dispersion behavior is not so evident for unsteady shock fronts. It may increase along the plastic front up to the compressive pulse plateau. Qualitative pictures of the behavior of the mesoparticle velocity distribution at the mesolevel-1 for a steady plastic front and an unsteady front are shown in Figs. 5.6 and 5.7, respectively. In each case, the rate of Maxwellization of the velocity distribution at the mesolevel-1 turns out to be much smaller than the rate of change of the mean velocity at the plastic front. Therefore, for the majority of materials, the velocity distribution at the mesolevel-1 in steady shock waves is in equilibrium; it has a symmetrical (Maxwellian) shape although its width changes along the plastic front.

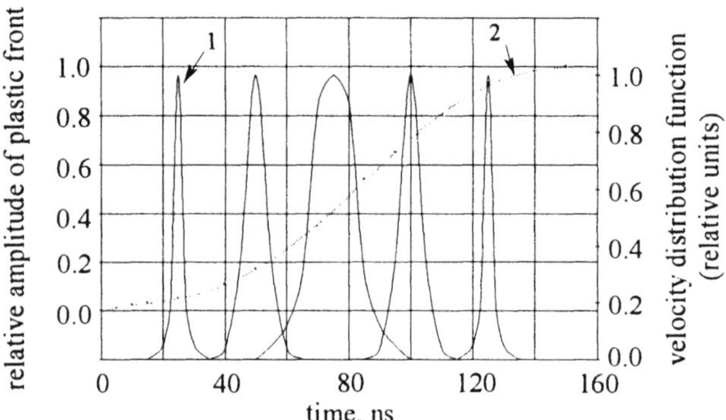

Figure 5.6 Qualitative picture of the behavior of (1) the particle velocity distribution function for (2) a steady shock wave.

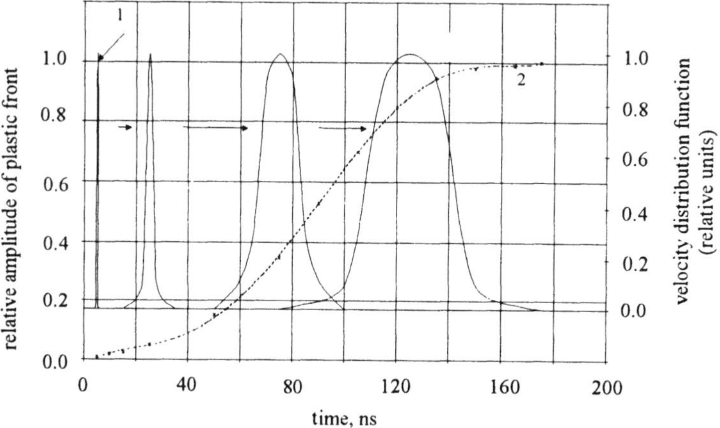

Figure 5.7 Qualitative picture of the behavior of (1) the particle velocity distribution function for (2) an unsteady shock wave.

Let us consider this approach in detail, since it is thought to be of great importance for understanding the nature of energy exchange between the macrolevel and the mesolevel. In accordance with the classification suggested by Duvall [12] there exist three kinds of wave decay, (i) geometrical decay, (ii) hydrodynamic decay and (iii) Maxwellian decay.

The first kind of decay relates to spatial geometry of the shock wave. For spherical waves the decay is proportional to $1/r^2$, for cylindrical waves the decay is proportional to $1/r$, and for plane waves decay does not depend on the propagation distance. Hydrodynamic decay results from the circumstance that a decompression wave propagating in shock-compressed material propagates faster than, and overtakes, the shock front. For triangular pulses, this results in attenuation of the shock. Lastly, Maxwellian decay results from dissipative processes occurring in a shock wave.

Besides the kinds of wave decay that have been enumerated, which can occur in processes that do not involve mesostructural fluctuations, these fluctuations can cause an additional kind of wave decay, called *fluctuative decay*. The estimate presented above shows that the spatial scale of structural nonuniformity prevailing during shock-wave propagation is attributed to effects at the mesolevel. The most typical feature of the mesolevel is a space and time correlation of the mesoparticle motion that occurs during dynamic straining of material. The nature of fluctuative decay relates to stochastic motion of mesoparticles. Because the space–time correlation of the motion of mesoparticles has a random character, the motion of a mesoparticle in the shock loaded solid is a stochastic process. It is closely reminiscent of the motion of charged particles in plasma where, owing to the long-range character of the interaction, the motions of particles are also correlated. Therefore, as the first step, we briefly review the nature of the fluctuative decay in plasma.

Fluctuative braking for an individual particle in plasma has been derived by J. Hubburd [13]. It has been shown that the trajectory of a probe particle in plasma is stochastic and the particle experiences retardation proportional to the gradient of the velocity dispersion in velocity space. One of the fundamental equations describing the stochastic motion of particles with non-local correlations is known to be the Fokker–Plank equation,

$$\frac{Df}{Dt} = -\frac{\partial}{\partial v}\left[F_1\, f(v,t)\right] + \frac{1}{2}\frac{\partial^2}{\partial v^2}\left[F_2\, f(v,t)\right]. \tag{5.16}$$

Here $f(v,t)$ is the particle velocity distribution function or probability density. Diffusion coefficients F_1 and F_2 characterize a change of the particle velocity distribution function due to random interactions of particles with each other. The first diffusion coefficient, or dynamic friction coefficient, $F_1 = d\langle \Delta v\rangle / dt$ characterizes a rate of change of mean velocity due to mutual interactions of the particles. Here $\langle \Delta v\rangle$ is the average change of the particle velocity owing to these interactions. Thus, the value F_1 has a sense of deceleration and, when multiplied by particle mass, defines the value of the fluctuative friction force directed opposite to the driving force. As a matter of fact, this value defines an additional kind of decay which can be classified as a *fluctuative braking* of particles or *fluctuative decay*. It disappears when the particle velocity dispersion becomes negligible.

The second diffusion coefficient, $F_2 = \langle d(\Delta v\,\Delta v)\rangle / dt$, in the Fokker–Plank equation defines the rate of change of the particle velocity dispersion. In the case of non-local correlation of particle motion, there exists the following relation between diffusion coefficients [13]:

$$F_1 = F_0 + \frac{1}{2}\frac{dF_2}{dv} \ . \tag{5.17}$$

It is seen from Eq. 5.17 that the dynamic friction coefficient, and hence the value of the fluctuative friction, depends on the rate of change of the second diffusion coefficient in the velocity space. This conclusion is very important for understanding the nature of fluctuative decay, not only for an individual particle having a random trajectory but also for flux motion of particles.

In the case of dynamic straining one deals with the totality of particles, which can be characterized by their average velocity and velocity dispersion. As in the case of an individual particle in plasma, the average particle velocity associated with dynamic straining also suffers fluctuative decay resulting from the stochastic motion of the particles around some average velocity. As a result, the mean particle velocity, u_{av}, decreases by some value Δu, which is also determined by the rate of change of the particle velocity dispersion, D^2. The dependence of the velocity loss, Δu, on particle velocity dispersion identical to Eq. 5.17 is obtained in [14]:

$$\Delta u = c_m \frac{d(D^2)}{du} \ , \tag{5.18}$$

where $c_m = dE/d(D^2)$ is the energy capacity of the mesolevel, and E is the internal energy. The mesolevel energy capacity is similar to the well-known thermal capacity $c_v = dE/dT$ for the atomic scale level. According to Eq. 5.18, the mean velocity loss is proportional to the rate of change of dispersion in the velocity space. Detailed explanation of the physical meaning of energy capacity of a medium at the mesolevel is given in [14].

Equation 5.18 defines the current value of the mean particle velocity attenuation. In order to obtain the total decrease of the mean particle velocity due to fluctuative decay it is necessary to integrate Eq. 5.18 over the duration of the deformation process. For example, the total decrease of mean particle velocity (velocity loss at the plateau of compressive pulse) for a shock front having rise-time τ is given by the equation

$$\Delta U = \int_0^\tau \Delta u \, dt \ . \tag{5.19}$$

5.4. On the Propagation of a Plane Shock Wave in a Heterogeneous Medium

Now, let us consider propagation of a one-dimensional plane wave in a medium having a particle velocity distribution at the mesolevel. The macroscopic motion of the medium is described by two balance equations, the Lagrangian form of which is

$$\rho \frac{\partial u}{\partial t} = \frac{\partial \sigma}{\partial x} \qquad (5.20)$$

for impulse conservation and

$$\frac{\partial u}{\partial x} = \frac{\partial \varepsilon}{\partial t} \qquad (5.21)$$

for continuity. Here $\partial \varepsilon / \partial t$ is the total (elastic plus plastic) strain rate. It may be represented in the form

$$\frac{\partial \varepsilon}{\partial t} = \frac{\partial \varepsilon^e}{\partial t} + \frac{\partial u^{pl}}{\partial x} .$$

Then the continuity equation can written

$$\frac{\partial u}{\partial x} = \frac{\partial \varepsilon^e}{\partial t} + \frac{\partial u^{pl}}{\partial x} . \qquad (5.22)$$

This reresentation is similar to that used in the continuum dislocation theory where the last term is to be expressed using the dislocation flow tensor [15,16]:

$$\rho_0 \frac{\partial u_m}{\partial t} = \frac{\partial \sigma_{mn}}{\partial x_n} ;$$

$$\frac{\partial u_m}{\partial x_n} = \frac{\partial w_{mn}}{\partial t} + J_{mn} , \qquad (5.23)$$

$$\sigma_{ik} = \lambda_{iklm} w_{lm} ; \qquad u_m = \frac{\partial U_m}{\partial t} .$$

Here J_{mn} is the dislocation flow tensor, w_{mn} is the elastic distortion tensor, and U_m is the total (elastic plus plastic) displacement.

In a one-dimensional case, the total stress component in the wave propagation direction, σ_1, consists of a spherical part, P, and a deviator part, S:

$$\sigma_1 = S_1 - P . \qquad (5.24)$$

In accordance with the approach developed in [8,14], the normal stress must be subdivided into two parts: $\sigma_1 = \sigma_1^e + \sigma_1^m$. The first part represents the elastic properties of medium connected with the cold compression and shear:

$$\sigma_1^e = P^e - S^e . \qquad (5.25)$$

For the one-dimensional propagation of an elastic–plastic wave, the elastic strain in Eq. 5.22 is

$$\varepsilon^e = \frac{\sigma_1^e}{\lambda + 2\mu},\tag{5.26}$$

where λ and μ are Lamé's constants.

Taking into account Eq. 5.26, the momentum and mass balance equations 5.20–5.22 can be reduced to the wave-type equation

$$\frac{\partial^2 u}{\partial x^2} - \frac{1}{c_l^2}\frac{\partial^2 u}{\partial t^2} = \frac{\partial^2 u^{\mathrm{pl}}}{\partial x^2} - \frac{1}{\lambda + 2\mu}\frac{\partial^2 \sigma_1^{\mathrm{m}}}{\partial x \partial t},\tag{5.27}$$

where

$$c_l = \left(\frac{\lambda + 2\mu}{\rho}\right)^{1/2}$$

is the longitudinal elastic velocity.

In accordance with [8,14], the second part of stress tensor $\sigma_1^{\mathrm{m}} = S_1^{\mathrm{m}} - P^{\mathrm{m}}$ is connected with the mesoscopic effects where S_1^{m} is responsible for the stress relaxation at the mesolevel and the spherical part is related to the mesoparticle velocity dispersion

$$p^{\mathrm{m}} = c_{\mathrm{m}}\,\rho\,D^2.\tag{5.28}$$

The plastic mass velocity is the first statistical moment of the particle velocity distribution function (see Eq. 5.9). In this paragraph we consider only the influence of the mesoparticle velocity dispersion on the wave propagation. Therefore, we accept that the stress deviator S_1^{m}, determined by the non-local model, equals zero. Then one obtains

$$\frac{\partial^2 u}{\partial x^2} - \frac{1}{c^2}\frac{\partial^2 u}{\partial t^2} = \frac{1}{\rho}\frac{\partial^2}{\partial x^2}\int v f\,dv + \frac{c_{\mathrm{m}}}{c_l^2}\frac{\partial(D^2)}{\partial x \partial t}.\tag{5.29}$$

The particle velocity distribution function, f, can be represented in a form consisting of two parts, an equilibrium part, f_0, and a non-equilibrium part, f_1:

$$f = f_0 + f_1.\tag{5.30}$$

where

$$f_0 = \frac{n}{D\sqrt{2\pi}}\exp\left[-\frac{(v-u)^2}{D^2}\right].\tag{5.31}$$

In the unsteady wave, the particle velocity at the plastic front is determined only by the non-equilibrium part of the particle velocity distribution function. To determine the non-equilibrium access f_1, one can use a relaxation form of kinetic equation:

$$\frac{\partial f}{\partial t} + v \frac{\partial f}{\partial r} + \langle \dot{v} \rangle \frac{\partial f}{\partial t} = -\frac{f - f_0}{\tau_r},$$
(5.32)

where τ_r is the relaxation time. For the sake of analytical simplicity in presentation we consider a spatially uniform case. Restricting ourselves to the first-order terms in Eq. 5.29, we obtain

$$\langle \dot{v} \rangle \frac{\partial f_0}{\partial v} = -\frac{f_1}{\tau_r}.$$
(5.33)

Here, the mean mesoparticle acceleration $\langle \dot{v} \rangle$ should be determined from the interaction potential Π for the mesoparticles as $\langle \dot{v} \rangle = (1/m) \mathrm{grad}\, \Pi$. Since the potential of interaction of the mesoparticles is unknown, we use an approximation according to which the total pressure in a dynamically loaded medium is a sum of cold pressure and "mesoparticle pressure". The latter emerges from introduction of the "mesoscopic temperature" defined by the mesoparticle velocity dispersion D^2. According to the statistical approach, the average mesoparticle velocity u_{av} coincides with the flow velocity u, so that the mean mesoparticle acceleration can be obtained from the equation of motion in which the right-hand side is the gradient of mesoparticle pressure:

$$\rho \langle \dot{v} \rangle = c_m \frac{\partial [\rho (D^2)]}{\partial x}$$
(5.34)

In this case, the non-equilibrium part of the distribution function is:

$$f_1 = -\tau_r c_m \frac{\partial D^2}{\partial x} \frac{\partial f_0}{\partial v}.$$
(5.35)

Then the integral in the right-hand side of Eq. 5.29 is

$$\int v f_1 \, dv = \tau_r c_m \frac{\partial |D^2|}{\partial x},$$

and Eq. 5.29 can be written

$$\frac{\partial^2 u}{\partial x^2} - \frac{1}{c_l^2} \frac{\partial^2 u}{\partial t^2} = c_m \tau_r \frac{\partial^2}{\partial x^2} \left(\frac{\partial D^2}{\partial x} \right) + \frac{c_m}{c_l^2} \frac{\partial^2 (D^2)}{\partial x \partial t}.$$
(5.36)

Write the following equations:

$$\frac{\partial}{\partial t} \left[\frac{\partial (D^2)}{\partial x} \right] = 2 \frac{\partial}{\partial t} \left[u \frac{D}{u} \frac{\partial D}{\partial x} \right] = 2 \frac{\partial u}{\partial t} \left[\frac{D}{u} \frac{\partial D}{\partial x} \right] + 2u \frac{\partial}{\partial t} \left[\frac{D}{u} \frac{\partial D}{\partial x} \right]$$

$$\frac{\partial}{\partial x} \left[\frac{\partial (D^2)}{\partial x} \right] = 2 \frac{\partial}{\partial x} \left[u \frac{D}{u} \frac{\partial D}{\partial x} \right] = 2 \frac{\partial u}{\partial x} \left[\frac{D}{u} \frac{\partial D}{\partial x} \right] + 2u \frac{\partial}{\partial x} \left[\frac{D}{u} \frac{\partial D}{\partial x} \right]$$

$$\frac{\partial^2}{\partial x^2}\left[\frac{\partial(D^2)}{\partial x}\right] = 2\frac{\partial^2}{\partial x^2}\left[u\frac{D}{u}\frac{\partial D}{\partial x}\right] + 4\frac{\partial u}{\partial x}\frac{\partial}{\partial x}\left[\frac{D}{u}\frac{\partial D}{\partial x}\right] + 2u\frac{\partial^2}{\partial x^2}\left[\frac{D}{u}\frac{\partial D}{\partial x}\right].$$

Then Eq. 5.29 can be written as follows:

$$\frac{\partial^2 u}{\partial x^2} - \frac{1}{c_l^2}\frac{\partial^2 u}{\partial t^2} = 2\frac{\partial^2 u}{\partial x^2}\left(c_m\tau_r\frac{D}{u}\frac{\partial D}{\partial x}\right) + 2\frac{\partial u}{\partial x}\frac{\partial}{\partial x}\left(2c_m\tau_r\frac{D}{u}\frac{\partial D}{\partial x}\right)$$
$$+ u\frac{\partial^2}{\partial x^2}\left(2c_m\tau_r\frac{D}{u}\frac{\partial D}{\partial x}\right) + 2\frac{c_m}{c_l^2}\frac{\partial u}{\partial t}\left(\frac{D}{u}\frac{\partial D}{\partial x}\right) + 2\frac{c_m}{c_l^2}u\frac{\partial}{\partial t}\left(\frac{D}{u}\frac{\partial D}{\partial x}\right) \tag{5.37}$$

or

$$\frac{\partial^2 u}{\partial x^2}\left[1 - 2\tau_r c_m\left(\frac{D}{u}\frac{\partial D}{\partial u}\right)\right] - \frac{1}{c_l^2}\frac{\partial^2 u}{\partial t^2} = 2\frac{c_m}{c_l^2}\frac{\partial u}{\partial t}\left(\frac{D}{u}\frac{\partial D}{\partial x}\right)$$
$$+ u\frac{\partial^2}{\partial x^2}\left(2\tau_r c_m\frac{D}{u}\frac{\partial D}{\partial x}\right)\frac{\partial u}{\partial x}\frac{\partial}{\partial x}\left(2\tau_r c_m\frac{D}{u}\frac{\partial D}{\partial x}\right) + 2\frac{c_m}{c_l^2}u\frac{\partial u}{\partial t}\left(\frac{D}{u}\frac{\partial D}{\partial x}\right). \tag{5.38}$$

Consider the simplest case when

$$2\tau_r c_m\left(\frac{D}{u}\frac{\partial D}{\partial x}\right) = 1. \tag{5.38}$$

If, furthermore,

$$\left(\frac{D}{u}\frac{\partial D}{\partial x}\right) = \frac{1}{2c_m\tau_r} = \text{const.}, \tag{5.39}$$

Eq. 5. reduces to

$$\frac{\partial^2 u}{\partial t^2} = -2c_m\frac{\partial u}{\partial t}\left(\frac{D}{u}\frac{\partial D}{\partial x}\right), \tag{5.40}$$

or

$$\frac{\partial^2 u}{\partial t^2} = -\frac{1}{\tau_r}\frac{\partial u}{\partial t}, \tag{5.41}$$

the solution of which is

$$u = u_0\exp\left(-\frac{t}{\tau_r}\right), \tag{5.42}$$

where

$$\tau_r = \left[2c_m\left(\frac{D}{u}\frac{\partial D}{\partial x}\right)\right]^{-1}. \tag{5.43}$$

Thus, Eq. 5.29 loses its hyperbolic character so the wave motion of medium disappears for the time of relaxation of the velocity distribution function for the mesostructure. This is the case when the plastic front transforms into a stress relaxation front [18].

The condition 5.38 can be written in the form

$$\tau_r c_m \left(\frac{D \dfrac{\partial D}{\partial t}}{u \dfrac{\partial x}{\partial t}} \right) = c_m \frac{D}{u} \frac{\dot{D}}{(c_p / \tau_r)} = 1.$$

(5.44)

The non-steady plastic front is determined by the non-equilibrium part of the velocity distribution function, which comes to equilibrium in the relaxation time τ_r during which the particles move a distance $u\tau_r$. For that time the plastic front propagates, on average, a distance $c_p\tau_f / 2$, where τ_f is the risetime of the compressive pulse. Then, we have the equality

$$u\tau_r = \frac{1}{2} c_p \tau_f .$$

(5.45)

By using Eq. 5.45, the condition 5.42 can be written in the form

$$c_m \frac{D}{u} \frac{\dot{D}}{\dot{u}} = 1,$$

(5.46)

where $\dot{u} = u / \tau_f$ is the mean particle acceleration over the plastic front. the condition 5.46 shows that decay of the plastic front is determined not only by the ratio of dispersion to mean particle velocity but also by their rates of change.

5.5. The Stochastic Approach to Change of Strain Kinematics

The process of shock-wave propagation in a random medium is a typical example of an interaction of external noise (fluctuative stress fields) and a dynamic variable (particle velocity). It is closely reminiscent of the well-known Uhlenbeck–Ornstein stochastic process for Brownian particles [19],

$$m\dot{u} = -\alpha u + F(t),$$

where the term $(-\alpha u)$ describes a frictional force and $F(t)$ is the fluctuative force due to random collisions.

The phenomenon of decrease of the flux velocity due to particle velocity distribution can be described in terms of a non-steady phase transition initiated by external noise, i.e, noise-induced changes in the macroscopic behavior of nonlinear systems. The situation appears to correspond closely to the well-known population genetic model that gives rise to the equation [20]

$$\dot{X}(t) = \lambda(t)\,X(t) - X^2(t) \tag{5.47}$$

in the presence of external noise. Equation 5.47 is a partial form of a wide class of phenomenological equations

$$\dot{X}(t) = f_\lambda(X),$$

The random parameter $\lambda(t)$ characterizes the influence of external conditions on the nonlinear system under consideration. It consists of two parts. One of these, λ_0, is regular while the other, ξ, is noise:

$$\lambda = \lambda_0 + D\xi.$$

In this equation, D is the intensity of noise. The detailed analysis in [20] shows that probability density curves for the dynamic variable $X(t)$ have two different positions (see Fig. 5.8), one of which corresponds to a zero value of the dynamic variable $X(t)$ and the other to some positive value, $X > 0$. The zero value corresponds to large values of noise intensity D.

The same can be seen for wave propagation in a non-uniform medium. Consider this situation in detail, regarding the mesoparticles as dislocations having a total Burgers vector B for the polarized dislocation structure shown in Fig. 5.2. In this case, one can use the Orowan equation for plastic deformation:

$$\varepsilon = nBl_\mathrm{d}. \tag{5.48}$$

In this equation, n is the mesoparticle number density and l_d is their run. Differentiating this equation with respect to time gives:

$$\frac{d\varepsilon}{dt} = Bl_\mathrm{d}\frac{dn}{dt} + Bn\frac{dl_\mathrm{d}}{dt}. \tag{5.49}$$

In accordance with dislocation dynamics, the time derivative of dislocation density, dn/dt, is given by the equation [21]

$$\frac{dn}{dt} = \alpha n - \beta n^2. \tag{5.50}$$

In this equation, α characterizes the rate of nucleation of mesoparticles in the process of polarization of the dislocation structure, and β is the coefficient that takes into account their locking. By definition, the strain can be expressed in the form

$$\varepsilon = \frac{dL}{dx} = \frac{\dfrac{dL}{dt}}{\dfrac{dx}{dt}} = \frac{u}{c_\mathrm{p}}, \tag{5.51}$$

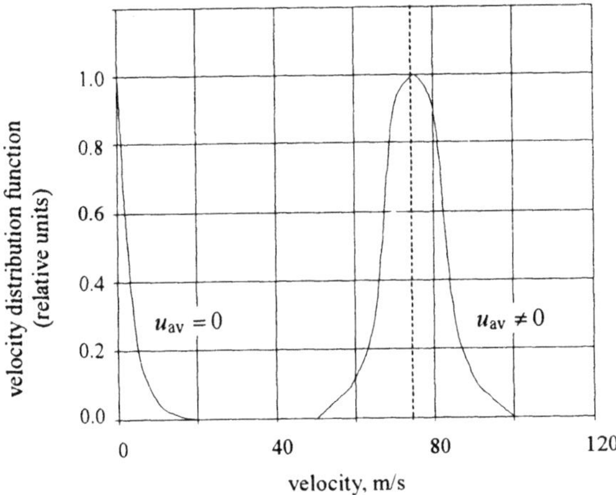

Figure 5.8. A qualitative picture of the positions of the velocity distribution function for different values of external noise.

where L is displacement, $u = dL/dt$ is the particle velocity, and c_p is the plastic wave speed. For steady plastic fronts, c_p = const. and the strain rate is given by

$$\frac{d\varepsilon}{dt} = \frac{1}{c_p}\frac{du}{dt}.$$
(5.52)

By using Eqs. 5.50–5.52 one obtains:

$$\frac{du}{dt} = c_p\left(Bl_d\frac{dn}{dt} + Bn\frac{dl_d}{dt}\right) = c_p\left[Bn\frac{dl_d}{dt} + Bl_d(\alpha n - \beta n^2)\right]$$

$$= c_p\left[Bn\frac{dl_d}{dt} + \alpha Bnl_d - \beta\frac{(Bnl_d)^2}{Bl_d}\right] = c_p\left[Bn\frac{dl_d}{dt} + \alpha\varepsilon - \frac{\beta}{Bl_d}\varepsilon^2\right]$$
(5.53)

$$= \left[c_p Bn\frac{dl_d}{dt} + \alpha u - \frac{\beta}{Bl_d}\frac{u^2}{c_p}\right].$$

Assuming that mesoparticle run is invariable, Eq. 5.53 can be transformed into an equation similar to Eq. 5.47 for the genetic model:

$$\frac{du}{dt} = \alpha u - \chi u^2,$$
(5.54)

where $\chi = \beta/Bl_d c_p$. In that equation, α^{-1} can be interpreted as the characteristic time of macroscopic evolution of the non-linear system and χ^{-1} is the distance over which the locking of mesoparticles occurs. In order to change over to a proper stochastic equation, one should express the parameter α as a sum of regular and noise terms: $\alpha = \alpha_0 + D\xi(t)$. The meaning of the noise intensity, D, will be made clear below. The stochastic equation corresponding to the deterministic equation 5.54 has the form

$$du = (\alpha u - \chi u^2)dt + D^2 u\, dW_t. \tag{5.55}$$

In this equation, W_t is the probability density for Wiener's random process [20]. A solution to this equation is sought in the form of a probability density function. Its time evolution is described by the proper Fokker–Plank equation corresponding to the stochastic equation 5.55. In the Sctratonovich interpretation that equation can be written [20]:

$$\frac{\partial f(u,t)}{\partial t} = -\frac{\partial}{\partial u}\left\{\left[\alpha u - \chi u^2 + \frac{D^2}{2}u\right]f(u,t)\right\} + \frac{D^2}{2}\frac{\partial^2}{\partial u^2}[u^2 f(u,t)]. \tag{5.56}$$

Extremes of probability density define the position of phase transitions on the velocity axis. They can be determined from the equation:

$$\left.\frac{\partial f(u,t)}{\partial u}\right|_{u=u_{\mathrm{m}}} = 0$$

or

$$\alpha u_{\mathrm{m}} - \chi u_{\mathrm{m}}^2 + \frac{D^2}{2}u_{\mathrm{m}} = 0,$$

which has the two roots:

$$u_{\mathrm{m}} = \begin{cases} 0 \\ \dfrac{1}{\chi}\left(\alpha - \dfrac{D^2}{2}\right) \end{cases}. \tag{5.57}$$

The first root means that probability density has an extreme at zero average velocity $u = 0$. The second root corresponds to a phase transition under the condition $\alpha = D^2/2$, which can be considered as a criterion for a noise-induced structural phase transition. When $\alpha > D^2/2$, the probability density maximum is shifted to the right on the velocity axis. We are now in a position to define the physical meaning of the noise intensity $D^2/2$. For this purpose, we remind ourselves that the first term on the right-hand side of the Fokker–Plank equation 5.56 includes the first diffusion coefficient F_1:

$$F_1 = \alpha u - \chi u^2 + \frac{D^2}{2}u. \tag{5.58}$$

and the second diffusion coefficient is

$$F_2 = D^2 u^2 . \tag{5.59}$$

The resonance condition $\alpha = \chi u$ means that the macroscopic spatial scale of evolution of the system considered, $l_{macro} = 1/\chi = Bl_d c_p /\beta = \tau_{macro} u$, equals the free run of mesoparticles before their locking. When $\alpha = \chi u$, Eq. 5.58 gives $F_1 = u D^2 /2$. On the other hand, the first diffusion coefficient in the Fokker–Plank equation, by definition, is $F_1 = d\langle \Delta v\rangle /dt$. When one applies the Fokker–Plank equation to description of stochastic motion of mesoparticles in the shock-deformed medium, the value $\langle \Delta v\rangle$ should be understood to be the mean velocity fluctuation, D, introduced above as the square root of the particle velocity dispersion, i.e., $F_1 = dD/dt$. Then one obtains

$$\frac{dD}{dt} = u\frac{D^2}{2}, \tag{5.60}$$

which can be rewritten in the form:

$$\frac{1}{u}\frac{dD}{dt} = \frac{d}{dt}\left(\frac{D}{u}\right) \tag{5.61}$$

under the condition that

$$\frac{dD}{du} \gg \frac{D}{u}. \tag{5.62}$$

Indeed,

$$\frac{d}{dt}\left(\frac{D}{u}\right) = \frac{1}{u}\frac{dD}{dt} + \frac{du}{dt}D = \frac{1}{u}\frac{du}{dt}\left(\frac{\dfrac{dD}{dt}}{\dfrac{du}{dt}} + \frac{D}{u}\right)$$

$$= \frac{1}{u}\frac{du}{dt}\left(\frac{dD}{du} + \frac{D}{u}\right) \cong \frac{1}{u}\frac{dD}{dt}.$$

In our case, the condition 5.62 means that the current (instant) rate of change of the mean velocity fluctuation is higher than its mean rate of change over the plastic wave front. Under the condition 5.62, Eq. 5.59 can be written

$$D^2 = \frac{F_2}{u^2} \cong \left\langle \frac{d}{dt}\left(\frac{D^2}{u^2}\right)\right\rangle . \tag{5.63}$$

From Eqs. 5.60 and 5.62 one obtains:

$$\frac{d}{dt}\left(\frac{D}{u}\right) = \frac{1}{2}\left\langle \frac{d}{dt}\left(\frac{D^2}{u^2}\right)\right\rangle . \tag{5.64}$$

This means that, under the resonance condition $\alpha = \chi u$, the rate of change of the mean velocity fluctuation, D, normalized by the current value, u, of the mesoparticle velocity, equals one-half of the rate of change of the velocity dispersion normalized by the current squared mean velocity, u^2.

The condition for occurrence of a phase transition, $\alpha = D^2/2$, is

$$\alpha = \frac{d\zeta^2}{dt} \tag{5.65}$$

or

$$\frac{\alpha}{\zeta} = 2\frac{d\zeta}{dt}, \tag{5.66}$$

where $\zeta = D/u$. At strain rates where $\alpha \ll d\zeta^2/dt$, the probability density curve lies near the particle velocity $u = 0$. If, however, the particle velocity dispersion changes slowly, one has the opposite situation in which $\alpha \gg d\zeta^2/dt$. In this case, a flux motion of mesoparticles takes place with a nonzero average velocity in the wave propagation direction.

The analysis presented shows that, in dynamic processes related to structural phase transitions, the most important characteristic becomes not only the mesoparticle velocity dispersion but also its rate of change. The foregoing analysis also shows that a dynamically deformed solid can display different types of behavior of elementary carriers of deformation at the mesolevel. The first type of behavior corresponds to the situation in which the mean particle velocity at the plastic front is zero. The second type of behavior corresponds to the opposite situation when all the mesoparticles move with the same velocity in the wave propagation direction. The transition between these states can be treated as a structural phase transition, the realization of which is determined by the rate of change of the particle velocity dispersion.

Equation 5.66 can be written in the form

$$\frac{D}{u}\frac{\partial}{\partial t}\left(\frac{D}{u}\right) = \frac{1}{\tau}, \tag{5.67}$$

where $\tau = 1/\alpha$ is the time of macroscopic evolution of the system considered. This equation can be written in the form

$$\frac{1}{\tau}\frac{u}{D} = \frac{d}{dt}\left(\frac{D}{u}\right). \tag{5.68}$$

In this case the inequality

$$\frac{1}{\tau}\frac{u}{D} \ll \frac{d}{dt}\left(\frac{D}{u}\right) \tag{5.69}$$

means that the rate of growth of the relative mean particle velocity (i.e. normalized by the mean velocity fluctuation) for the full duration of the process (in our case risetime of plastic front) is less than the current rate of growth of the mean velocity fluctuation normalized by the average particle velocity. In other words, the change of the particle velocity dispersion happens more quickly than the change of the average particle velocity at each and every moment. In the opposite case, when

$$\frac{1}{\tau}\frac{u}{D} \gg \frac{d}{dt}\left(\frac{D}{u}\right), \tag{5.70}$$

the change of the velocity dispersion occurs more slowly than of the average particle velocity. The first case corresponds to the situation when all additional kinetic energy supplied to medium immediately transforms into mesostructure formation and then on to chaotic motion of the mesostructure, so the average particle velocity does not change. In the second case, on the contrary, additional energy supplied to medium results in an increase of the flux motion of the mesostructure whereas the mesostructure formation turns out to be frozen. Thus, fluctuative decay of mean velocity depends sensitively on the relation between rate of change of the mean velocity fluctuation dD/dt and rate of change of the mean velocity (acceleration), du/dt.

5.6. Experiments on Dynamic Plasticity

Now we present a few typical examples of influence of the fluctuative properties of a medium on shock-wave propagation and attenuation. We compare the shock-wave behavior for three kinds of aluminum alloys having different relations between dD/dt and du/dt. Figure 5.9 shows an interferogram together with the deciphered free-surface-velocity profile, $u_{fs}(t)$, and the velocity dispersion history, $D(t)$, for a 2-mm thick AMg-6 (Russian *АМг-6*) aluminum alloy target loaded in a symmetrical planar test at the impact velocity of 160 m/s. This is a typical case when velocity dispersion is a maximum in the middle of the plastic front ($D = 13.2$ m/s) and the total velocity loss at the plateau of the compressive pulse is zero.

Another situation is provided in Fig. 5.10, which shows the analogous dependencies for a D-16 aluminum alloy target loaded at the impact velocity of 154 m/s. One can see that mean velocity grows to $u_{fs} = 92$ m/s after which a break occurs in the interference fringe signal (point B). A simple calculation shows that, over the piece AB of the interferogram, $du/dt = 0.33 \times 10^9$ m/s^2 and $dD/dt = 0.57 \times 10^9$ m/s^2, that is $dD/dt > du/dt$. A similar relation between the rate of change of the mean velocity fluctuation dD/dt and the mean acceleration du/dt leads to the noise-induced structural transition, when velocity dispersion grows catastrophically whereas the mean mass velocity decreases,

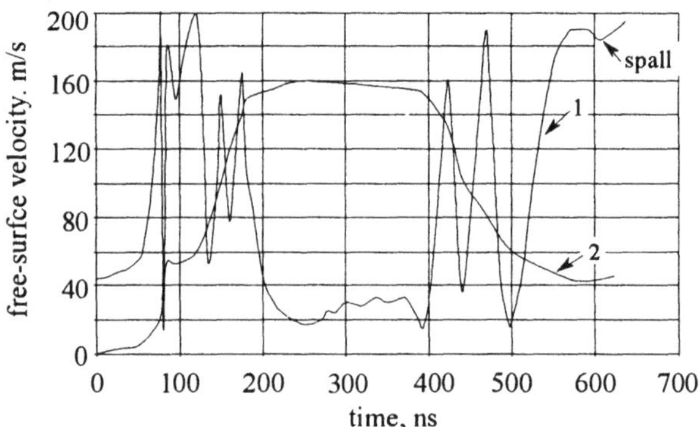

Figure 5.9. (1) Interference fringe signal and (2) free surface velocity profile for AMg-6 Al alloy target loaded at the impact velocity of 160 m/s. Here $du/dt > dD/dt$ along the overall plastic front.

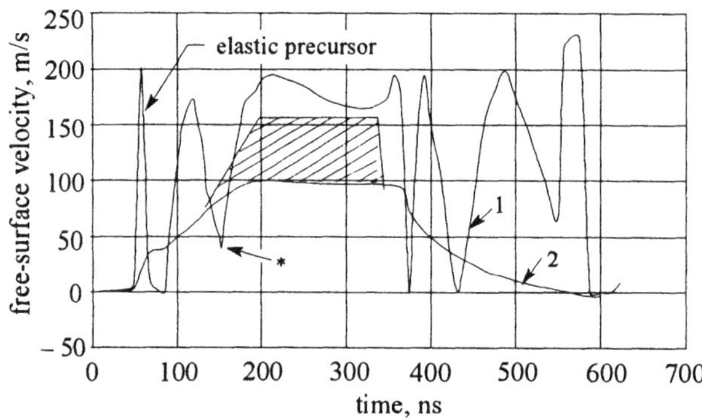

Figure 5.10. (1) Interference fringe signal and (2) free surface velocity profile for a D-16 Al alloy target loaded at the impact velocity of 154 m/s. $dD/dt = du/dt$ at the point indicated by *.

resulting in the velocity loss ΔU. In this concrete case, the velocity loss equals 62 m/s. At the release front of the compressive pulse the opposite relation is true. For the piece CD of the interferogram, $dD/dt = 0.43 \times 10^9$ m/s^2 and $du/dt = 0.55 \times 10^9$ m/s^2, so $dD/dt < du/dt$. Thus, the mean mass velocity decreases during the load front of compressive pulse in a D-16 aluminum alloy target, whereas it increases in the release front.

Now, let us consider the analogous fringe signal and free-surface velocity profile for VT-95 aluminum alloy presented in Fig. 5.11 ($u_{imp} = 160$ m/s). The break in the fringe signal occurs at a free-surface velocity of 88 m/s (point A), where $dD/dt = 1.16 \times 10^9$ m/s^2 and $du/dt = 0.8 \times 10^9$ m/s^2, so $dD/dt > du/dt$.

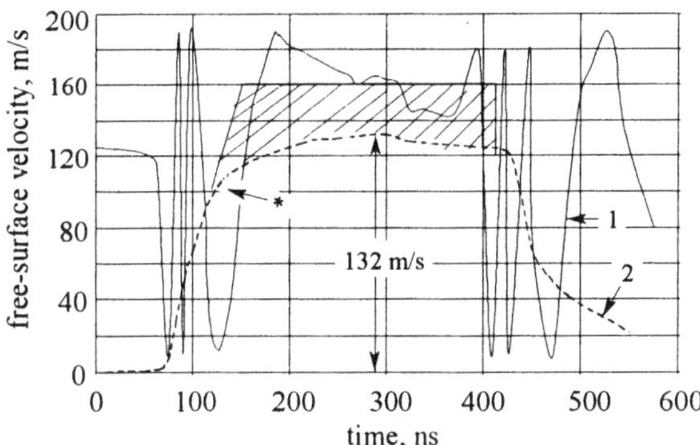

Figure 5.11. (1) Interference fringe signal and (2) free surface velocity profile for VT-95 Al alloy target loaded at the impact velocity of 160 m/s. $dD/dt = du/dt$ at the point indicated by *.

Thus, the noise-induced phase transition occurs almost immediately behind the elastic precursor. At the end of the plateau of the compressive pulse (point B), the rate of change of the mean velocity fluctuation becomes smaller than the deceleration of the mean mass velocity ($dD/dt < du/dt$), owing to relaxation processes in the plateau. In this case, instead of the velocity loss, we have an increase of the velocity due to pumping kinetic energy from the mesolevel to the macrolevel. As a result, the total amplitude of the compressive pulse is recovered at the release front. This is thought to be the reason why the number of fringes at the release front is greater than at the load front.

One more example of trigger-like behavior of the strain-rate-dependent macroscopic response of a material is provided in Fig. 5.12. We present two interferograms obtained for identical AMg-6 Al alloy targets loaded at different impact velocities. These interferograms are plotted on the same time scale. Both interference signals show a monotonic decrease of the fringe amplitude up to the onset of the plateau of the pulse. The fringe signal obtained at the impact velocity of 178 m/s shows a gradual transition of the plastic front into the plateau without velocity loss ($\Delta U = 0$). The second fringe signal, obtained at an impact velocity of 187 m/s, shows a break of the plastic front at the free-surface velocity of approximately 113 m/s where $dD/dt \cong du/dt$. In Fig. 5.13 the interference signal is presented separately for clarity. Thus, increase of impact velocity from 178 m/s to 187 m/s provides a widening of the velocity distribution sufficient for the criterion for structural phase transition to be fulfilled. Whereas the examples presented in Figs. 5.9 –5.11 show the influence of metallurgical properties on the dynamic plasticity of the material, the results presented in Fig. 5.12 demonstrate the influence of strain-rate and non-linear features.

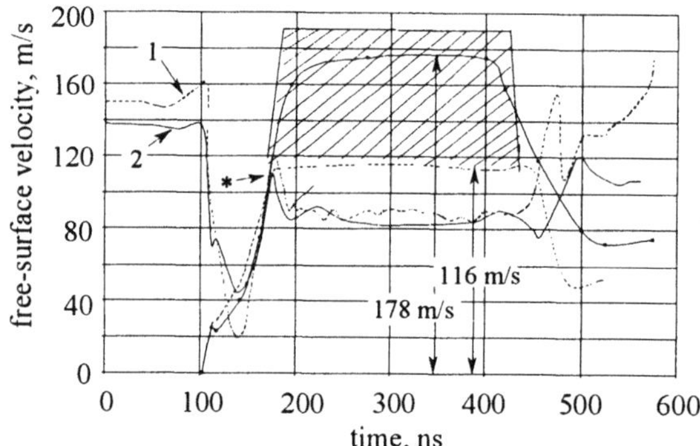

Figure 5.12. Interference fringe signals and free-surface-velocity profiles for Amg-6 Al alloy targets loaded at the impact velocity of (1, broken lines) 187 m/s and (2, solid lines) 178 m/s. At the point indicated by *, $dD/dt = 1.5 \times 10^9\,\text{m/s}^2$, and $du/dt = 1.34 \times 10^9$ m/s.

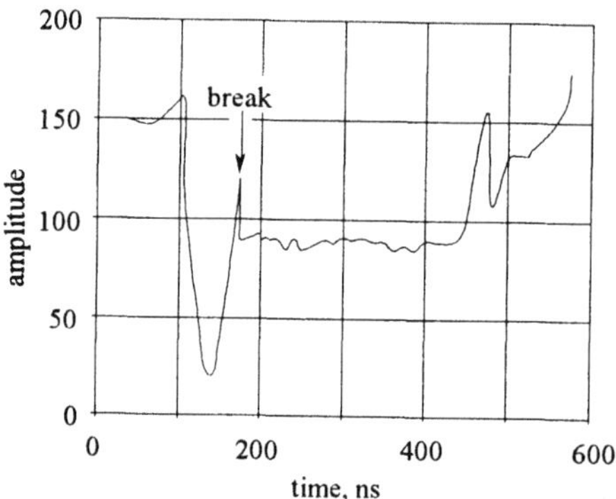

Figure 5.13 Interference fringe signal for the impact loading of a 2-mm thick AMg-6 aluminum alloy target loaded at the impact velocity of 187 m/s.

Finally, consider two typical examples of the dispersion behavior that occur during dynamic deformation of plane steel targets. In Fig. 5.14, time-resolved free-surface-velocity profiles and dispersion histories for viscous, high-strength 30XH4M steel and 16X11H2MBФA steel are presented. The first free-surface profile has been obtained at the impact velocity of 369 m/s. It is seen from Fig 5.13a that the measured value of the free-surface velocity is ≈300 m/s, i.e. the free-surface velocity loss equals 69 m/s. In Fig. 5.13b an analogous pair of dependencies is provided for the 16X11H2MBФA steel target loaded at the impact velocity of 365.5 m/s. The measured free-surface velocity in this case turns out to be 365 m/s, which is very close to the impact velocity. This means that fluctuative decay of the mean velocity for that material is very small or completely negligible. The velocity fluctuation's, behavior also differs from the

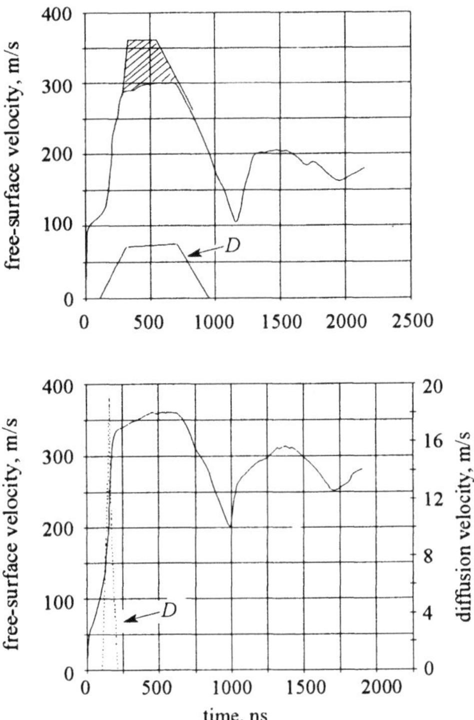

Figure 5.14 Temporal profiles of free-surface velocity and diffusion velocity of (a) 30XH4M steel (u_{imp} = 369 m/s, ΔU = 69 m/s), and (b) 16X11H2MBΦA steel (u_{imp} = 365.5 m/s, $\Delta U \cong 0$).

previous case. The maximum value of the velocity fluctuation for that test is 20 m/s, a value achieved in the middle of the plastic front of the compressive pulse. Use of Eqs. 5.16 and 5.17 for that case results in a zero value of the total velocity loss. This is because the dynamic friction coefficient has time to change its sign in the middle of plastic front, so that time integration over the plastic front in Eq. 5.17 gives a zero value for the velocity loss.

The dependence of macroscopic particle velocity on the particle velocity distribution is the fundamental feature of unsteady processes in the structured media. The specifics of structural rearrangement of dynamically deformed material are revealed by the decrease (loss) of the average (macroscopic) particle velocity. The value of the velocity loss characterizes the intensity of energy exchange between macrolevel and mesolevel. Physically, the particle velocity loss means that a certain part of the kinetic energy of macroscopic motion of the medium is transferred to individual particles at the mesolevel. As a result, the particle velocity dispersion at the mesolevel increases while the average velocity decreases.

The values of the velocity loss defined by Eqs. 5.16 and 5.17 can be checked experimentally. The experimental determination of the mean velocity loss is based on independent measurement of three dynamic characteristics: i) mean particle velocity, $u_p(t)$, ii) mean velocity fluctuation D (the square root of the particle velocity dispersion D^2), iii) loss of the mean velocity at the plateau of compressive pulse ΔU.

The mean particle velocity is determined by applying the equation $u_p = 0.5u_{fs}$ to the free-surface-velocity profile recorded with velocity interferometer or other velocity gauge. The experimental approach developed in [9,10] allows the particle velocity dispersion to be measured as a quantitative characteristic of the particle velocity heterogeneity during the shock-wave propagation. Independent measurement of the mean velocity loss, in accordance with [10], can be made by comparing a maximum value of the free-surface velocity at the plateau of compressive pulse, u_{fs}, and the impact velocity measured in a symmetrical collision, u_{imp}.

Let us now turn to three cases in which free-surface-velocity histories were obtained for Cr–Ni–Mo steel targets that were prepared using different thermal treatments and were loaded in the range 150–320 m/s. All of the data for these experiments are provided in Table 5.1 and the time-resolved profiles of the free-surface velocity, $u_{fs}(t)$ are presented in Figs. 5.15–5.17.

Case I. Symmetrical profile of mean velocity fluctuation (Fig. 5.15). The velocity at the plateau of the compressive pulse exactly equals the impact velocity in a symmetrical collision, i.e., $u_{fs} = u_{imp}$. The particle velocity loss is zero. The particle velocity dispersion achieves its maximum value at the middle of plastic front after which it decreases to zero long before the plastic front transits into the plateau.

In this case, we face the situation when the well-known free-surface approximation conditions are satisfied: $2u_p = u_{fs}^{max} = u_{imp}$. This means that decay

Table 5.1. Free-surface-velocity data

case number	impact velocity u_{imp}, m/s	maximum free-surface velocity u_{fs}^{max}, m/s	mean velocity fluctuation D, m/s	velocity loss ΔU, m/s
I*	150	148	28	0
II**	237	200	32	37
III**	320	140	—	180

* 28X3CHMBΦA steel   ** 30XH4M steel

of the mean particle velocity due to macro–meso energy exchange is absent. The same result follows from the application of Eqs. 5.16 and 5.17 to the mean velocity fluctuation profile presented in Fig. 5.15. In this case, the mean velocity fluctuation profile must be subdivided into two parts so the velocity loss ΔU must be calculated separately for the increasing and decreasing branches of the velocity dispersion profile.

1. Increasing branch: $D^2 = 544\,(\text{m/s})^2$; $dD/dt = 18 \times 10^7\,\text{m/s}^2$; $\Delta U^+ = 32.8\,\text{m/s}$.

2. The decreasing branch has the identical value of opposite sign since the dispersion profile for that shot is symmetrical, i.e., $\Delta U^- = -32.8\,\text{m/s}$.

Thus, the total mean velocity loss at the plateau of the compressive pulse is zero.

Case II. Non-symmetrical dispersion profile (Fig. 5.16). The value of the free-surface velocity is smaller than the impact velocity in a symmetrical collision, $u_{fs} \leq u_{imp}$. The particle velocity dispersion achieves its maximum value at the end of the plastic front, after which it remains invariable.

1. For the increasing branch: $D^2 = 723\,(\text{m/s})^2$; $dD/dt = 12 \times 10^7\,\text{m/s}^2$, $\Delta U^+ = 36\,\text{m/s}$,

2. The decreasing branch at the plastic front of the compressive pulse is absent (it appears only at the edge of the pulse), so $\Delta U^- = 0$.

The total mean velocity loss $\Delta U = \Delta U^+ + \Delta U^- = 36\,\text{m/s}$.

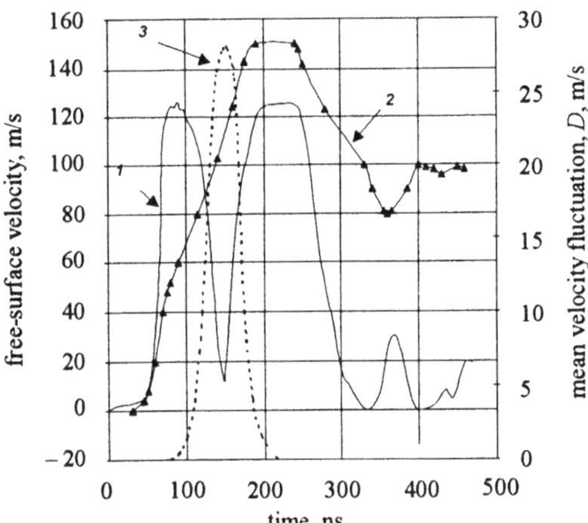

Figure 5.15. (1) Interference signal, (2) free surface velocity profile and (3) mean velocity fluctuation for annealed 28X3HCMBФA steel target loaded at the impact velocity of 150 m/s.

Case III. Non-symmetrical dispersion profile (Fig. 5.17). The value of the free-surface velocity is much smaller than the impact velocity in a symmetrical collision, i.e., $u_{\text{fs}}^{\max} \ll u_{\text{imp}}$.

The example in Fig. 5.17 reflects the situation in which the dispersion changes slowly along the plastic front. A more drastic case is provided in Fig. 5.15 in which the plastic front disappears entirely. This means that the mean particle velocity at the first plastic front disappears entirely because of fluctuative motion of medium. This case corresponds to the noise-induced structural phase transition described above and illustrated in Fig. 5.8. Owing to the large value of the mean velocity fluctuation, D, the free-surface velocity, u_{fs}, decreases from 320 m/s to 140 m/s, i.e., the velocity loss equals 180 m/s. However, at the back edge of the pulse, the mean particle velocity is restored and we can determine the pull-back velocity. It turns out to be approximately 150 m/s. This value is higher than the amplitude of the first plastic front. That is why the formal deciphering of the interference signal leads to negative pull-back velocity at the back edge of the pulse.

5.7. Kinetics of Mesostructure and Dynamic Strength

Now consider the dependence of the dynamic strength of solids on the mutual relation between mean velocity fluctuation, D, and mean velocity loss, ΔU. To begin, we present experimental observations of shock-wave behavior for several kinds of construction materials such as steels, aluminum alloys, copper, and beryllium in the impact velocity range 100–600 m/s. The experiments on shock loading of plane targets are performed using a light gas gun facility of 37-mm

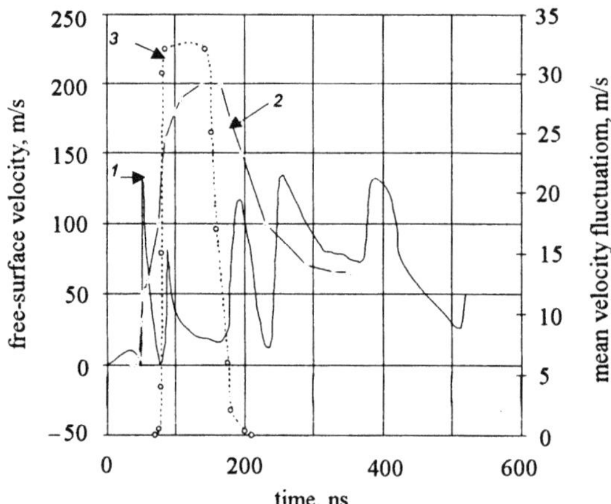

Figure 5.16. Interference signal and free-surface-velocity profile for an 30XH4M steel target loaded at the impact velocity of 237 m/s.

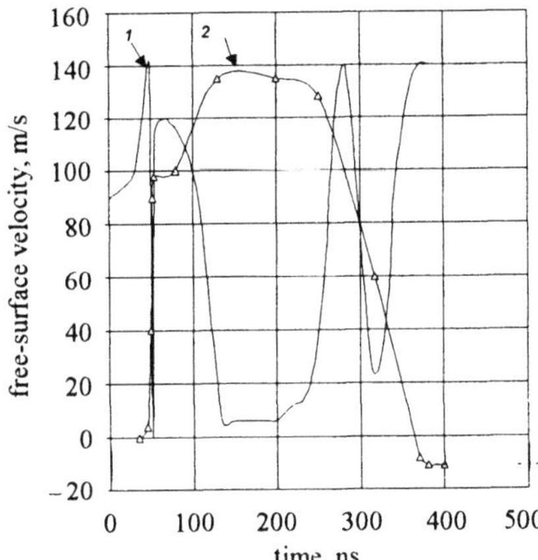

Figure 5.17 (1) Interference signal and (2) free-surface-velocity profile for an 30XH4M steel target at the impact velocity of 320 m/s.

bore diameter. All the targets are 52 mm in diameter and 5–10 mm thick. Two-millimeter-thick impactor plates of the same material as the target are mounted at the front of the projectile. The thicknesses of targets and impactors were adjusted to produce spallation. Time-resolved free-surface-velocity profiles were recorded using a two-channel velocity interferometer [10].

Case I. 40XCHMA steel. This is complex alloyed steel of sufficiently high hardness (HRc 55). Spall-strength characteristics for this steel were studied for two sets of targets that have been subjected to different thermal treatments.

For the first set of targets, dependencies of the pull-back velocity, W, the free-surface-velocity loss, ΔU, and the mean velocity fluctuation at the meso-level-1, D, on the impact velocity are presented in Fig. 5.18. This figure shows that W, D, and ΔU, are monotonically increasing functions of the impact velocity. It is seen that, for 40XCHMA steel within the impact velocity range 211–437 m/s, the maximum value of mean velocity fluctuation remains smaller than the mass velocity loss, i.e.,

$$D < \Delta U. \tag{5.72}$$

The dependencies of ΔU and D on u_{imp} have different slopes. This means that these curves intersect at a higher impact velocity. In order to find their intersection these curves are plotted in different scale in Fig. 5.19. It can be seen that the curves intersect at an impact velocity of ~700 m/s, which would correspond to a pull-back velocity of 250 m/s. Thus, within the impact velocity range 211–437 m/s this steel does not posses optimum spall-strength characteristics. Its thermal treatment corresponds to maximum dynamic strength at the impact velocity of ~700 m/s and higher.

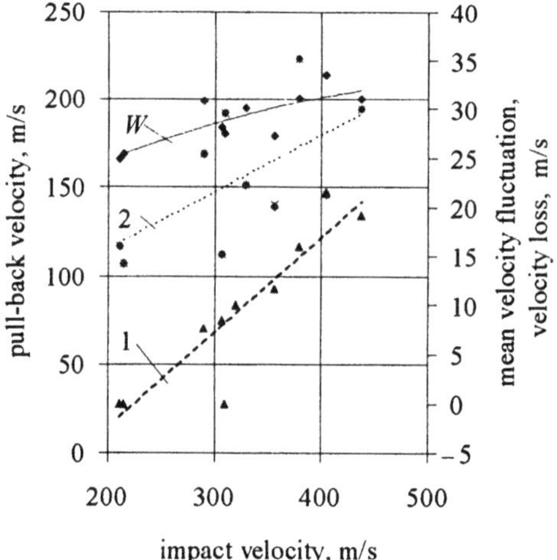

Figure 5.18 Pull-back velocity, W, mean velocity fluctuation, D, (1) and velocity loss ΔU, (2) versus impact velocity for the first set of 40XCHMA steel targets.

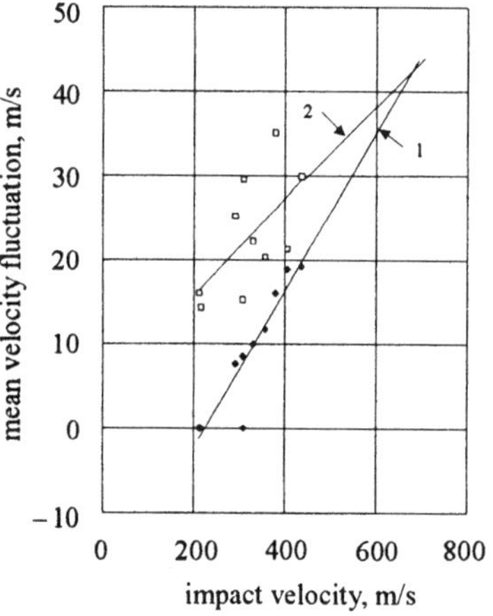

Figure 5.19 (1) Mean velocity fluctuation, D, and (2) velocity loss, ΔU, versus impact velocity for the first set of 40XCHMA steel targets.

For the second set of targets, we present the dependencies of W, D, and ΔU on u_{imp} in Fig. 5.20. One observes the following relationship between mean velocity fluctuation, D, and mass velocity loss at the plateau of compressive pulse, ΔU:

1. $D > \Delta U$ below an impact velocity of 495 m/s,

2. $D = \Delta U$ at the impact velocity of 495 m/s,

3. $D < \Delta U$ at impact velocities exceeding 495 m/s.

The maximum spall-strength corresponds to the impact velocity of 495 m/s. Its value coincides with the value of spall strength (~250 m/s) that is expected for the first set of targets if the optimum kinetic characteristics are provided.

Thus, for both sets of 40XCHMA steel targets, the maximum spall strength corresponds to the strain rate at which $D \cong \Delta U$. On the basis of spall tests of two sets of 40XCHMA steel targets, one can preliminarily conclude that the behavior of such kinetic characteristics as mean velocity fluctuation at the mesolevel-1 and mass velocity loss provide information on whether or not the thermal treatment of the steel corresponds to that maximizing its dynamic strength.

Case II. 38XH3MΦA steel (HRC 39). This is a complex Cr–Ni–Mo alloy steel. Under quasistatic conditions, this steel has high ductility and strength. Two sets of targets have been tested. The first set was made from the material as supplied and the second was made from the same steel after thermal treatment providing a homogenization of structure. As a result, the first kind of steel has obvious textural features, whereas the second steel has a homogeneous, equiaxed grain structure.

For the first set of 38XH3MΦA steel targets, dependencies of pull-back velocity, velocity loss, ΔU, and mean velocity fluctuation at the mesolevel-1, D, on the impact velocity are presented in Fig. 5.21. This kind of steel exhibits a non-monotonic dependence of spall strength on the strain rate. Dependencies of

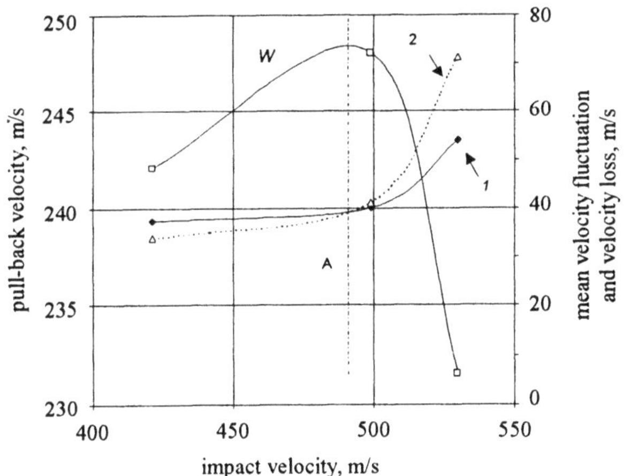

Figure 5.20 Pull-back velocity, W, (1) mean velocity fluctuation, D, and (2) velocity loss, ΔU, versus impact velocity for the second set of 40XCHMA steel targets.

the velocity loss, ΔU, and mean velocity fluctuation, D, at the mesolevel-1 have a point of contact that corresponds to the criterion :

$$D = \Delta U .\tag{5.73}$$

For the second set of 38XH3MФA steel targets, dependencies on the impact velocity of pull-back velocity, W, velocity loss, ΔU, and mean velocity fluctuation, D, at the mesolevel-1 are provided in Fig. 5.21. One can see that the pull-back velocity is essentially constant over the impact velocity range under investigation.

Case III. Beryllium. Beryllium is taken as an example of a dynamically brittle material, so influence of mesostructure kinetics on the dynamic strength seems to be interesting to compare with such ductile materials such as steels, aluminum alloys, and copper.

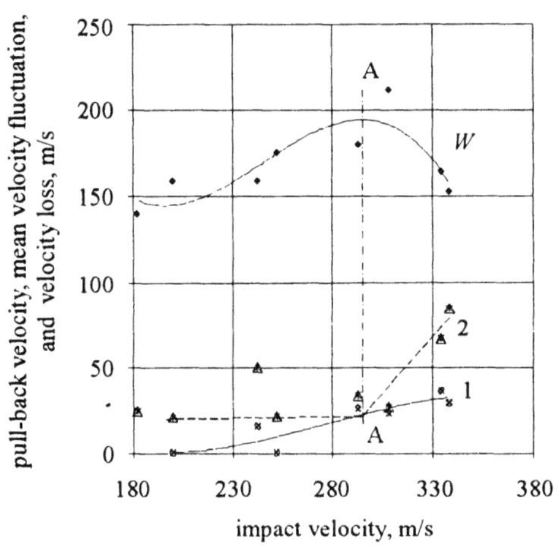

Figure 5.21. Pull-back velocity, W, (1) mean velocity fluctuation, D, and (2) velocity loss, ΔU, versus impact velocity for the first set of 38XH3MФA steel targets.

Experimental data on spallation for both ductile and brittle materials show that spall strength of a material depends sensitively on the relation between mean velocity fluctuation at the mesolevel-1, D, and velocity loss, ΔU. The relationship between these kinetic characteristics reflects the direction and intensity of shock-induced energy exchange between the macrolevel and the mesolevel. This energy exchange can flow in two ways. The first way corresponds to nucleation of mesostructure as a result of collective interactions of elementary carriers of deformation due to non-local correlation of their motion in non-equilibrium processes of dynamic deformation [14]. At this stage, the meso-

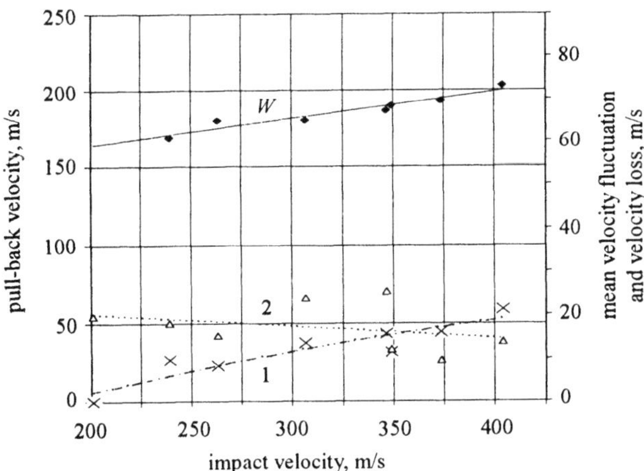

Figure 5.22. Pull-back velocity, W, (1) mean velocity fluctuation, D, and (2) velocity loss, ΔU, versus impact velocity for the second set of 38ХН3МФА steel targets.

particle velocity distribution emerges as a result of anharmonic oscillations of the lattice. It reflects the stochastic elastic response of the lattice to shock loading. The velocity fluctuations grow with increasing strain rate. Relaxation of local stresses begins together with growth of the amplitude of the fluctuations. In the steady plastic front, maximum velocity fluctuations occur in the middle of the front, after which the amplitude of fluctuation decreases and becomes zero at the end of the plastic front. From this instant, only the relaxation mechanism is active, which results in a change in the slope of the plastic front (point A in Fig. 5.23).

The difference between the impact velocity under symmetrical collision and the peak value of the free surface velocity, $\Delta U = U_{\text{imp}} - U_{\text{fs}}^{\text{max}}$, is presented in Fig. 5.24. One can see that, for thin beryllium targets, the peak free-surface velocity exceeds the impact velocity: $U_{\text{imp}} - U_{\text{fs}}^{\text{max}} < 0$. As the plastic front goes to the steady shape corresponding to a target thickness of 7 mm, the difference $\Delta U = U_{\text{imp}} - U_{\text{fs}}^{\text{max}}$ changes sign, i.e., $U_{\text{imp}} - U_{\text{fs}}^{\text{max}} > 0$. Thus, in the case of beryllium, there exist two directions of pumping the energy, depending on the strain rate:

- In unsteady shock waves the energy is pumped from mesolevel-1 to mesolevel-2 (a positive velocity loss at the plateau of the pulse).

- In steady shock waves the energy is pumped from mesolevel-2 to mesolevel-1 (a negative velocity loss at the plateau of the pulse).

The critical strain rate at which a structural phase transition occurs can be considered a very important characteristic of a dynamically loaded material. Taking the foregoing considerations into account, we accept, as an additional strength characteristic, the value of the particle velocity at the plastic front of the

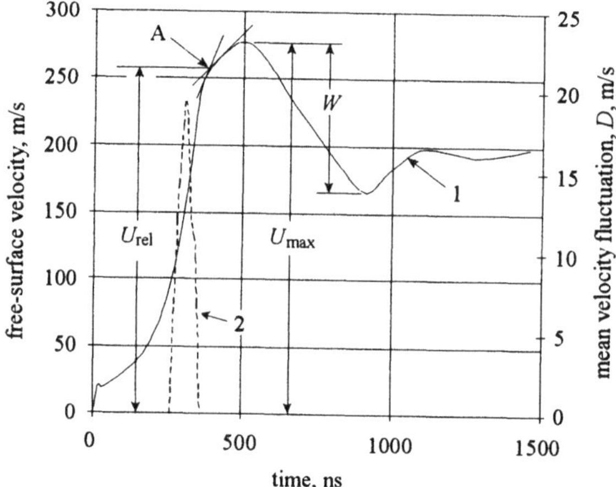

Figure 5.23 Free-surface velocity (1) and mean velocity fluctuation (2) profiles for a beryllium target at the impact velocity of 285 m/s. The point at which the slope of the plastic front changes is indicated by the letter A.

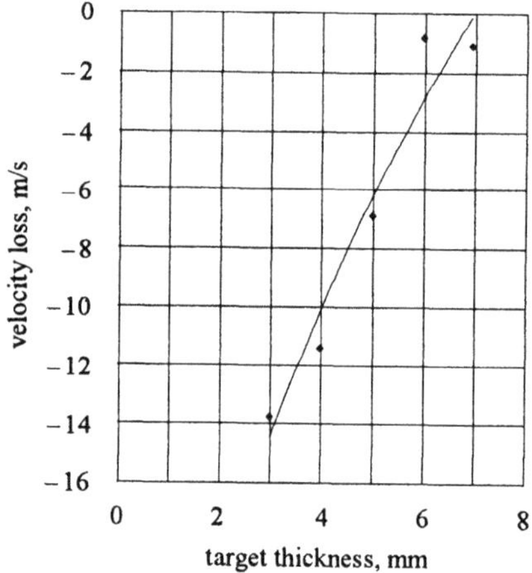

Figure 5.24 Velocity loss vs. target thickness for beryllium.

compressive pulse where the latter suffers a sharp change of its slope. We call this velocity the *relaxation threshold*, u_r. The dependence of u_r on the impact velocity is presented in Fig. 5.25. In addition, a symmetrical curve $u_{fs} = u_{imp}$ is plotted in this figure. One can see that, at small impact velocities (below $u_{imp} = 122$ m/s), the dependence of u_r on u_{imp} practically coincides with the line $u_{fs} = u_{imp}$, whereas at the higher value, $u_{imp} = 122$ m/s, it deviates downward. We call the point of the break in the dependence of u_r on u_{imp} the *threshold of dynamic*

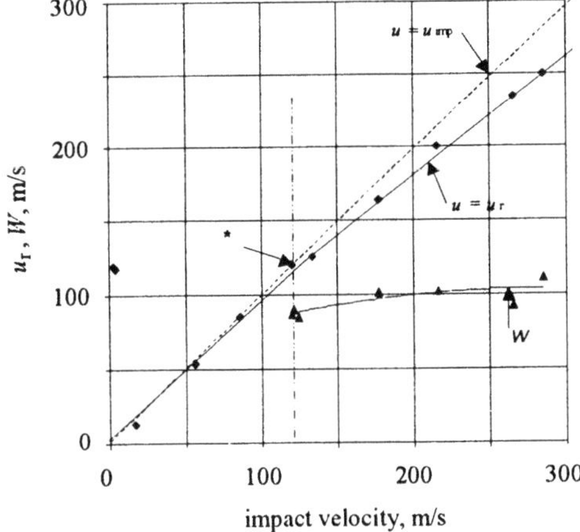

Figure 5.25 Dependence of the relaxation threshold, u_r, and pull-back velocity, W, on the impact velocity for beryllium.

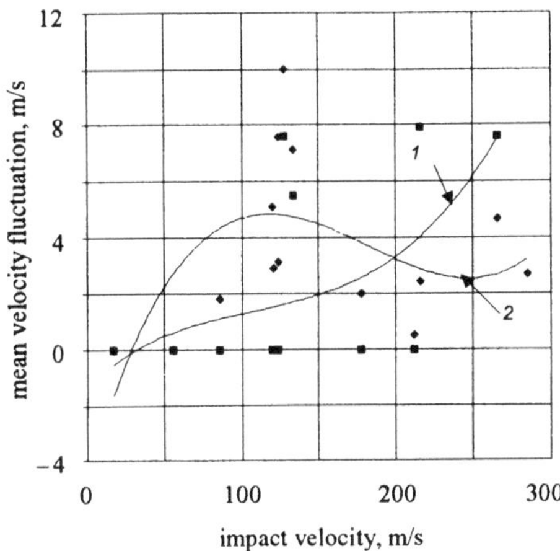

Figure 5.26 Mean velocity fluctuation versus impact velocity at (1) the mesolevel-1 and (2) mesolevel-2 for beryllium.

instability for compression, u_{ins}. In Fig. 5.25 we also present the dependence of the pull-back velocity. One can see that the onset of spallation coincides with the instability threshold, which means that preliminary fracture of the material occurs at the plastic front.

As a matter of fact, the instability threshold in compression corresponds to nucleation of structure at the mesolevel-2, which is responsible for fragmentation of material. The latter happens when the velocity dispersion at the mesolevel-2 becomes greater than the velocity dispersion at the mesolevel-1. In Fig. 5.26 we present the relative position and values of the velocity dispersion at the mesolevel-1 and mesolevel-2 for beryllium. At low and high impact velocities, $D_{m2} < D_{m1}$. In the middle of the impact velocity range, on the contrary, $D_{m2} > D_{m1}$. Accordingly, at low strain rates the free-surface profiles for beryllium have a relaxation drop of the velocity behind the elastic precursor similar to that of a typical viscous material. At high strain rates it also reveals ductile properties since its dynamic lies close to the hydrostatic curve (see stress–strain diagram in Fig. 5.27).

Analogous behavior of the relaxation threshold value, u_r, on the impact velocity is seen for M-2 copper (see Fig. 5.28). Instability of the plastic front for this material occurs at an impact velocity of 135 m/s. Pull-back velocity dependence for M-2 copper suffers a break at the impact velocity corresponding to the instability threshold. This means that the character of dynamic fracture after that strain rate changes very strongly. It may be assumed that compression instability threshold changes the kind and scale of kinematical mechanism of deformation at the mesolevel. Detailed analysis of that phenomenon from the point of view of thermodynamics of the processes at the shock wave front is given in [14].

5.8. Multiscale Energy Exchange and Dynamic Strength

In order to account for the energy exchange in the processes of dynamic plasticity and fracture we should write an internal energy balance in dynamically deformed solid where the micro-, meso and macro components of internal energy are separated. The energy balance has the form [14]:

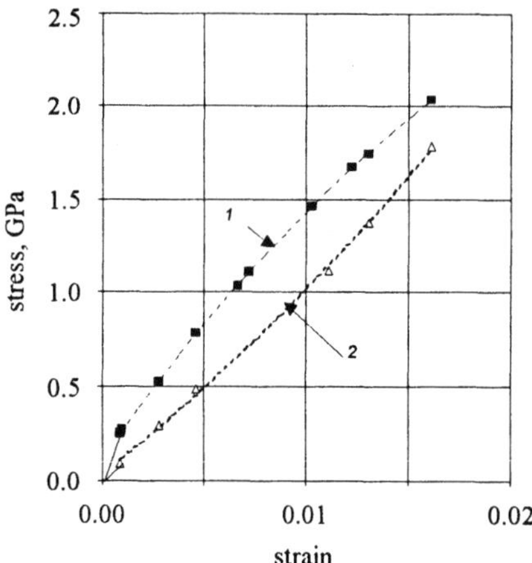

Figure 5.27. (1) Dynamic stress-strain diagram and (2) hydrostatic compression curve for beryllium.

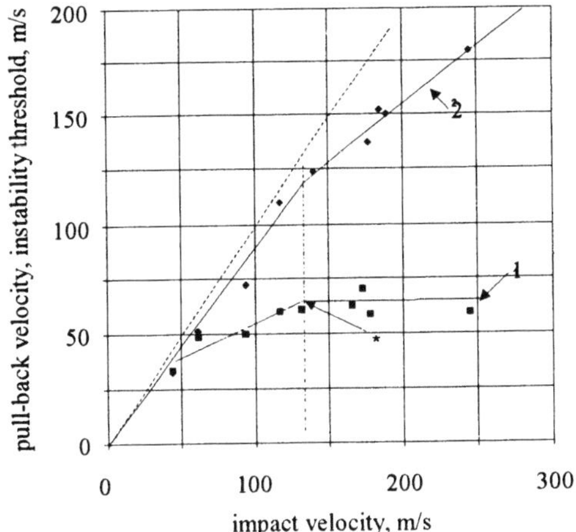

Figure 5.28. (1) Pull-back velocity, W, and (2) relaxation threshold, u_r, versus impact velocity for M-2 copper. The instability threshold is indicated by the symbol:*.

$$\frac{\partial E^e}{\partial t} + \sigma^e \frac{\partial \varepsilon^e}{\partial t} + \frac{\partial E^{ms}}{\partial t} + \sigma^{ms} \frac{\partial \varepsilon^{ms}}{\partial t} + \frac{\partial E^{mc}}{\partial t} + \sigma^{mc} \frac{\partial \varepsilon^{mc}}{\partial t} = 0. \qquad (5.74)$$

The sum of the first and the second terms characterizes the balance of elastic energy; $\partial E^e / \partial t$ is the change elastic energy supplied to medium and $\sigma^e \partial \varepsilon^e / \partial t$ is the change of the work of elastic forces. There may two different situations in the elastic range of dynamic straining:

$$1. \quad \frac{\partial E^e}{\partial t} - \sigma^e \frac{\partial \varepsilon^e}{\partial t} = 0, \qquad (5.75)$$

$$2. \quad \frac{\partial E^e}{\partial t} - \sigma^e \frac{\partial \varepsilon^e}{\partial t} > 0. \qquad (5.76)$$

The first case corresponds to the balance of internal elastic energy. If the balance of elastic energy is not fulfilled, the excess of internal energy is transferred to the mesolevel. Here, $\partial E^{ms} / \partial t$ is the change of the internal energy at the mesolevel. The internal energy at the mesolevel consists of two terms:

$$E^{ms} = c_m \rho D^2 + E^{ms}_{pot}. \qquad (5.77)$$

The first term on the right-hand side of Eq. 5.77 characterizes elastic fluctuation energy of mesostructure, which is determined by the mesoparticle velocity dispersion, D^2, and by the energy capacity of mesoscopic fluctuations, $c_m = \partial E^{ms} / \partial D$. The latter can be considered as the mesoscopic analog of the thermal capacity in thermodynamics. The second term E^{ms}_{pot} is the potential energy of the mesolevel resulting from elastic interactions of mesoparticles due to couple stresses. The change of internal energy at the mesolevel is associated with the processes of mesostructure formation, which are characterized by the term $\sigma_m \, \partial \varepsilon^{ms} / \partial t$. When

$$\frac{\partial E^{ms}}{\partial t} + \sigma_m \frac{\partial \varepsilon^{ms}}{\partial t} = 0 \, , \tag{5.78}$$

there is no energy exchange between the mesolevel and the macrolevel.

In the case, when neither of Eqs. 5.75 and 5.77 is fulfilled, the internal energy changes by means of macroscopic processes:

$$\frac{\partial E^{mc}}{\partial t} + \sigma_m \frac{\partial \varepsilon^{mc}}{\partial t} \, . \tag{5.79}$$

The first term characterizes the change of the internal energy by means of thermal conductivity and the second term is the change of the energy due to plastic work in case of a viscous solid and/or crack formation for brittle materials.

Now, we apply the foregoing energy balance to derive a criterion for dynamic fracture by using the representations introduced for the particle velocity dispersion, D^2, and the velocity loss, Δu. According to these representations there is current energy exchange between the macroscopic scale level and the mesolevel. This exchange is realized in the form of drain of energy from macrolevel through the velocity loss, Δu. This energy initiates fluctuations at the mesolevel. The quantitative measure of mesofluctuation intensity is the mesoparticle velocity dispersion, D^2. How much energy the mesolevel can absorb depends on (a) the value of the velocity dispersion, D^2, and (b) on the specific mesolevel energy capacity, c_m. In the dynamic equilibrium regime the energy of mesofluctuations equals the energy drain from the macrolevel. However, if the energy transferred from the macrolevel to the mesolevel is greater than the latter can absorb in the form of mesofluctuations, the excess energy is dissipated. The process of dissipation is described by the last two terms in Eq. 5.74. If we neglect the thermal conductivity in case of high-velocity processes, the only mechanisms of dissipation are viscousity and/or brittle fracture. This is the condition for onset of fragmentation of the material.

Suppose, for distinctness, that the energy excess goes into viscous fracture in uniaxial experiments. In other words, we accept that the only mechanism dynamic fracture is the viscous work while the thermal conductivity is negligible ($\partial E^{mc} / \partial t = 0$). Because we restricted ourselves energy exchange between the mesolevel and the macrolevel, the balance equation takes the form

$$\frac{\partial E^{ms}}{\partial t} + \sigma_m \frac{\partial \varepsilon^{ms}}{\partial t} = \sigma_m \frac{\partial \varepsilon^{mc}}{\partial t} \, . \tag{5.80}$$

Consider each term in this equation separately. The first term on the left-hand side characterizes the rate of change of the reversible energy of mesofluctuations. If we accept in Eq. 5.77 that the potential energy of mesorotation equals zero ($E_{\text{pot}}^{\text{ms}} = 0$), then

$$\frac{\partial E^{\text{ms}}}{\partial t} = \rho c_{\text{m}} \frac{\partial D^2}{\partial t}, \tag{5.81}$$

which, after integrating over the time, gives:

$$E^{\text{ms}} = \rho c_{\text{m}} D^2. \tag{5.82}$$

The second term in the left-hand side of Eq.5.80 characterizes the energy loss due to mesostructure formation. We accept that mesostructure formation is due to the energy loss, i.e. is determined by the velocity loss, Δu. In this case, $\sigma_{\text{m}} = \rho c_{\text{p}} \Delta u$ and we have

$$\sigma_{\text{m}} \frac{\partial \varepsilon^{\text{ms}}}{\partial t} = \rho \Delta u\, c_{\text{p}} \frac{\partial u}{\partial x}. \tag{5.83}$$

Integrating over the time gives [14]:

$$\int \sigma^{\text{ms}} \frac{\partial \varepsilon^{\text{ms}}}{\partial t} = -\rho \int \Delta u\, c_{\text{p}} \frac{\partial u}{\partial x} dt \approx \rho \Delta u^2. \tag{5.84}$$

The term on the right-hand side of Eq. 5.80 characterizes the viscous work. In our consideration, we assumed that the difference between energy loss and mesofluctuation goes into viscous work. The rate of change of energy due to viscous work is

$$\sigma_{\text{m}} \frac{\partial \varepsilon^{\text{mc}}}{\partial t} = \eta \frac{\partial u}{\partial x} \frac{\partial \varepsilon^{\text{mc}}}{\partial t}. \tag{5.85}$$

Here η is the kinematical viscosity and $\sigma^{\text{vs}} = \eta\, \partial u / \partial x$ is the viscous stress. After integrating over the time one obtains

$$\eta \int \frac{\partial u}{\partial x} \frac{\partial \varepsilon^{\text{vs}}}{\partial t} dt = \mu \Delta u \varepsilon^{\text{vs}}, \tag{5.86}$$

where $\mu = \eta / \rho$ is a dynamic viscosity. Substitution of Eqs. 5.82, 5.84 and 5.86 into Eq. 5.80 results in the balance equation:

$$\rho c_{\text{m}} D^2 - \rho c_{\text{m}} D^2 = \mu \Delta u\, \varepsilon^{\text{v}}. \tag{5.87}$$

This equation can be written in the form:

$$\left(1 - c_{\text{m}} \frac{D^2}{\Delta u^2}\right) = \mu \Delta u\, \varepsilon^{\text{vs}}. \tag{5.88}$$

If $\varepsilon^{vs} = \varepsilon_{cr}$ is interpreted as the critical strain for viscous fracture, this expression can be considered as criterion of viscous fracture of material. In this case, the mesoparticle velocity fluctuation, D_{cr} and velocity loss, Δu_{cr}, are also interpretable as critical values. As distinct from the other criterions of dynamic fracture, the criterion 5.88 supposes that fracture happens owing to a critical relation between energy losses and energy of mesofluctuations. Note that both values can be measured during the shock loading of a plane target under uniaxial strain conditions.

5.9. Conclusions

The results presented in this study may be considered as one of steps in developing our representations concerning a coupling between macro and micro characteristics of shock deformed medium. The main result is thought to be the discovery and experimental verification of meso–macro energy exchange that is realized in form of mutual change of the mean particle velocity loss and the particle velocity dispersion.

Depending on the strain rate, the energy exchange over the plastic front of shock wave may flow either gradually or as a trigger process in the form of the so-called noise induced structural phase transition. There is a rigid coupling between mean particle velocity loss over the shock wave and rate of change of the particle velocity dispersion.

Formation of two-scale mesostructure begins at the elastic precursor of elastic–plastic wave. Time of structure relaxation for the mesolevel-1 turns out to be much smaller than that for the mesolevel-2.

References

[1]   J.N. Johnson., O.E. Jones and. T.E. Michaels, "Dislocation dynamics and single-crystal constitutive relation." *J. Appl. Phys.* **41**, pp 2770–2279 (1970).

[2]   V.E. Panin, V.Yu. Grinjaev, T.F. Elsukova and A.G. Ivanchin, "Structure levels of deformation of solids". *Isvestja Vuzov. Fizika* **6**, pp. 5–22 (1982).

[3]   V.I. Vladimirov, V.N. Nikolaev, and N.M. Priemski, "Mesoscopic level of plastic deformation," In: *Physics of strength and plasticity* (ed. S.I. Zhurkov).Nauka. Leningrad, pp. 69–80, (1986).

[4]   V.V. Rybin, "Large plastic deformations and fracture of metals," *M. Mettalurgy.* 224 p. (1986) (in Russian).

[5]   E.L.Aero,"Microscale deformation in two-dimentional lattice:structural transitions and bifurcations at critical shear," *Physics Solid State* **42**, pp. 1147–1153 (2000). (Translated from *Fizika Tverdogo Tela.* **42**, pp. 1113–1119 (2000)).

[6]   Yu.I. Mescheryakov, N.A. Makhutov, and S.A. Atroshenko, "Micromechanisms of dynamic fracture of ductile high-strength steels," *J. Mech. Phys. Solids* **42**, pp. 1435–1450 (1994).

[7]   Yu.I. Mescheryakov., A.K. Divakov, and L.P. Fadienko, "On the particle velocity distribution at the elastic precursor of compression wave in Aluminum," *J. Tech. Phys.* **53**, p. 2050 (1983).

[8]   T.A. Khantuleva and Yu.I. Mescheryakov, "Kinetics and non-local hydrodynamics of mesostructure formation in dynamically deformed media," *Int. J. Phys. Mesomechanics* **2**, pp. 5–17 (1999).

[9]   J.R. Asay and L.M. Barker, "Interferometric measurements of shock-induced internal particle velocity and spatial variation of particle velocity," *J. Appl. Phys.* **45**, pp. 2540–2546 (1974).

[10] Yu. I. Mescheryakov, and A.K. Divakov, "Multiscale kinetics of microstructure and strain-rate dependence of materials," *Dymat Journal* **1**, pp. 271–287 (1994).

[11] L.M. Barker, "Multi-beam VISAR for simultaneous velocity vs. time measurements" In: *Shock Compression of Condensed Matter—1999* (ed. M.D. Furnish, L.C. Chhabildas, and R.S.Hixson., American Institute of Physics, Melville, New York, pp 999–1002 (2000).

[12] G.E. Duvall, " Maxwell-like relations in condensed matter. Decay of shock waves," *Irish J. Phys. Tech.* **7**, pp.57–69 (1978).

[13] J. Hubburd, "The friction and diffusion coefficients of the Fokker–Plank equation in a plasma," *Proc. Roy. Soc. A* **260**, pp 114–126 (1961).

[14] T.A. Khantuleva, this volume, Chapter 6.

[15] A.M. Kosevich, *Dislocation in the theory of elasticity*, Moscow. Nauka. (1978).

[16] T. Mura, "Continuous distribution of moving dislocations," *Phil. Mag.* **8**, pp. 843–853 (1963).

[17] J.J. Gilman. "The plastic wave myth" In: *Shock Compression of Condensed Matter—1991* (eds. S.C.Schmidt, R.D. Dick, J.W. Forbes, and D.G. Tasker), North Holland, Amsterdam, pp 387–389 (1992).

[18] T.Kihara and O.Aono, "Unified thiory of relaxation in plasma. Basic theorem", *J. Phys. Soc. Japan* **18**, pp. 837–851 (1963).

[19] G.E. Uhlenbeck, L.S. Ornstein, "On the theory of Brownian motion," *Phys. Rev.* **36**, p. 823 (1930).

[20] V.Horsthemke and R. Lefever, *Noise-induced transitions*, Springer-Verlag, New York, p. 297 (1984).

[21] J.J. Gilman, "Dislocation dynamics and the response of materials to impact," *Appl. Mech. Rev.* **21**, p. 767 (1968).

CHAPTER 6

The Shock Wave as a Nonequilibrium Transport Process

T.A. Khantuleva

6.1. Introduction

Shock-wave propagation involves mass, momentum, and energy transport in a medium. The state of a medium is determined by the exchange processes with surroundings through the interphase boundaries and the transport processes occurring within it. The transport processes in a medium are characterized by their rates, intensity, and the initial state of the medium. There a problem arises: how does one determine the state of the medium under given loading conditions? Following Maxwell, the medium state can be determined only with respect to the applied loading conditions. Under a short-time loading a medium demonstrates its solid properties and, when subject to a long-time force, the same medium behaves more as a liquid. There have been many attempts to describe plastic properties of deformed solids by methods similar to those used for hydrodynamic flow. Classical hydrodynamics, however, is valid only for fluids near thermodynamic equilibrium. From the point of view of nonequilibrium statistical mechanics, the deviation of the state of the system from thermodynamic equilibrium is closely connected with the correlation concept. A solid body subject to a very low velocity impact moves without deformation owing to the rigid correlation among the particles of the medium. All particles are linked together and their typical correlation length corresponds to the size of the body. In this case, we have mechanical transport only at the macroscopic scale. If we have a fluid in thermodynamic equilibrium, and if the applied loading isn't high-rate and intense, its behavior is that of an ideal fluid without relaxation. All particles are moving chaotically without any correlation. The typical correlation length is zero. More intense loading produces nonequilibrium transport characterized by finite space–time correlation scales. In all intermediate cases, we have a finite correlation length within which collective motion of particles of the medium occurs. This means that the medium does not behave as a simple fluid when subjected to load. Rather than being a structureless continuum, it is a medium having a complicated internal structure.

During shock-wave propagation, the effect of application of a time-dependent load propagates at a finite velocity and the state of the medium varies rapidly from the initial state to a nonequilibrium state determined by the maximal loading intensity. In this state, the medium cannot be considered as either a rigid body or an ideal fluid. The complicated nonequilibrium transport passes through a range of regimes at intermediate scale levels.

To address this problem, we need a model of a structured medium that can describe the nonequilibrium transport processes that occur at intermediate scale levels and embrace all limiting situations from ideal solids to ideal fluids. Such an approach is proposed in this chapter to describe nonequilibrium transport associated with shock-wave propagation in metals. This approach is based on the self-consistent non-local theory of nonequilibrium transport processes initially developed by the author for nonequilibrium flows of structured media.

6.2. Generalization of Classical Hydrodynamics

6.2.1. Deficiencies of the Classical Continuum Theories

The classical continuum model allows satisfactory description of elastic solids and of flows of real gases and liquids on sufficiently large space–time scales at moderate temperatures and densities and far from thermodynamic critical points and phase transitions. According to this model, the state of a system is characterized by mean-field quantities governed by the differential balance equations. In such a treatment, one considers the conservation laws for an element of the medium that has a linear size negligible in comparison with the typical flow length but much exceeding the scale of the internal structure of the medium. In the general case, the balance equations are incomplete insofar as the fluxes and forces cannot be uniquely defined at the macroscopic level without additional assumptions.

The fluxes can be determined within the framework of linear, irreversible transport thermodynamics [1]. This theory is based on the assumption that the thermodynamic state of a macroscopic system is near the local equilibrium state for which a temperature and entropy can be defined in the vicinity of every point and all equilibrium thermodynamic relations are satisfied. The hypothesis of linear relations between thermodynamic forces and dissipative fluxes is the simplest assumption permitting completion of the hydrodynamic balance equations. The Navier–Stokes equations, which are the main equations of theoretical hydrodynamics, have been obtained in this manner. The restrictions on their applicability are well known. They are not valid for flows with large hydrodynamic gradients, in Knudsen layers of thickness comparable to the mean free path scale that are near rigid boundaries, and for times of the order of the mean free molecular time following a change in the imposed boundary conditions. This last circumstance makes it difficult to state initial conditions for the

Navier–Stokes equations because, in this situation, it becomes necessary to establish dummy boundary conditions that differ from the real ones by amounts of the order $O(Kn)$ [2]. The problem is that the usual transport coefficients (e.g., viscosity or thermal conductivity) are not sufficient to describe transport processes at high deformation rates and under conditions in which there are spatial inhomogeneities having a scale comparable with the internal structural scales. Therefore, we face the problem of extending the transport coefficients in gases and liquids to the case of essentially nonequilibrium conditions.

6.2.2. Review of Approaches to Extension of Hydrodynamic Theories

There have been many attempts extend the classical hydrodynamic theory [5–20]. They may all be classified into one of three main types.

The first type of approach involves models using asymptotic expansions in terms of a small parameter. Because asymptotic series do not converge uniformly, attempts to extend them to the larger parameter values by using more terms of the series are not usually successful. For instance, for rarefied gases there exists a Knudsen number Kn^* such that, for all $Kn > Kn^*$, the equations obtained using the Chapman–Enskog procedure and truncated after a finite number of terms yield no solution for some problems [3]. Gradient models and moment theories are also of this first type, because it is implicitly assumed that there is a small parameter, which makes it possible to neglect the higher-order moments and gradients and, thus, complete the model [4]. It is worth remarking that the higher-order equations are too cumbersome and complicated to be of practical use.

Understanding the deficiencies of approaches of this first type has lead to the development of integral models, considering the integrals as a convolution of infinite series in terms of a small parameter. The non-local hydrodynamic models must be constructed on the basis of kinetic-level models. However, the transition from the microscopic description to the hydrodynamic level brings with it, in general, a fundamental problem that cannot be overcome without using asymptotic approximations or some additional simplifying assumptions.

The second type of approach to extending the classical hydrodynamic theory comprises non-local hydrodynamic models employing additional simplifications on the basic kinetic level to yield a hydrodynamic description. As a rule, the simplifications adopted are rather serious and strongly circumscribe the region of validity of the theory obtained (see, for example, [5]). Therefore, these approaches do not provide a theoretical basis for construction of macroscopic models of sufficiently broad applicability.

The approaches of the third type also yield non-local hydrodynamic model equations based on microscopic models or even first principles. However, in contrast to models of the second type, models of the third type are realized directly at the macroscopic level and are based on use of explicit space–time dependencies for the kernels of non-local integral equations. In the general case, macroscopic expressions for the integral kernels are unknown. Moreover, non-trivial approximations for them have not yet been obtained and guessing their space–time behavior is simply impossible. Hence, in practically all cases, it is necessary to invoke some model or empirical assumptions in order to get numbers.

Generalized hydrodynamic modeling of relaxation transport kernels for non-stationary processes must include both non-locality and memory effects. If the spatial and temporal relaxation scales differ by an order of magnitude, or if both are small, it is possible to separate the non-locality and memory effects and to construct different models in space and time. If the non-locality and memory effects cannot be separated, modeling for the relaxation kernels is essentially complicated because of the need to take into account correlations between these effects. Provided the correlations are small, these effects may be included in the relaxation kernels as simple space–time dependencies.

Relatively simple exponential and Gaussian models of the relaxation kernels at the fixed model parameters can be used for the description of media with sufficiently small relaxation times. In the general case, more complicated models are required. One should not hope to obtain models equally valid for different media without introducing the inverse influence of the medium properties on the transport kernels themselves. Moreover, it is necessary to use different model parameters for modeling different flow regimes of the same medium. Eventually, the validity of one or another model must be examined using a set of test problems. The use of oversimplified space–time dependencies for the transport kernels can lead to qualitatively incorrect results. Therefore, one prefers gradient theories of the first type in order to set the problem in terms of differential equations instead of integrodifferential equations. Some examples that permit comparing the two types of theories were presented in [8,9]. It was shown that the simplest exponential dependencies for the integral kernels lead to the equivalence of such non-local and gradient theories. In the general case the two type theories give quite different results.

So, of prime interest to the non-local theories appears a construction of the self-consistent relaxation models taking into account an inverse influence of the relaxing medium on the relaxation model, multiscale internal structure effects, and a wide range of loading conditions.

6.2.3. Rigorous Statistical Approaches

Concurrently with attempts to model nonequilibrium flows of real media, it seemed appropriate to validate the hydrodynamic theory using a molecular model of fluid behavior. Naturally, the validation problem is closely connected with another one of establishing validity limits for classical hydrodynamics and exploring possibilities for its generalization outside these limits. Therefore, statistical physics have lead to the same problem of modeling nonequilibrium flows of real media but proceeding from the first principles.

The main statistical mechanics problem is derivation of equations describing irreversible processes on the basis of multiparticle dynamics. The complete dynamic description of a multiparticle system is expressed in terms of the Liouville equation for the N-particle distribution function in phase space. To derive equations for irreversible processes it is necessary to adopt a less detailed and superfluous description. An idea proposed by Bogoliubov regarding the relaxation time hierarchy during system evolution occupies an important place in nonequilibrium statistical theory [10]. Establishing arbitrary initial conditions requires specification of many rapidly varying multiparticle distribution functions. With the passage of time (for gases, the time of particle interaction), the distribution functions become synchronized. Once synchronization is achieved, evolution of a system is characterized by a one-particle distribution, no matter what the initial distribution was. At this stage of evolution, the number of parameters describing the state of the system is reduced.

Past the next characteristic time (for gases, the mean time of a free molecular path), a further reduction of the description occurs. Then the evolution of the system is described by some macroscopic parameters (averaged molecular velocities). This is already the hydrodynamic stage of evolution. Each subsequent stage is characterized by a greater degree of chaotization of the system and a less detailed description of its evolutionary process.

In the middle of the 1950s a correlation function method (the Green–Kubo formulæ) was proposed for expressing transport coefficients through the equilibrium time correlation functions of irreversible fluxes [11–13]. These results are valid near a state of thermodynamic equilibrium at which the gradients of parameters are small and the velocities are low. Further development of the statistical derivation of the hydrodynamic equations was directed toward eliminating these restrictions. Richardson [14] and Piccirelli [15] developed a solution to the Liouville equation using a projection operator technique, and have obtained relationships between the dissipative fluxes and the hydrodynamic forces valid for arbitrary hydrodynamic gradients. The integral kernels in the non-local expressions for the dissipative fluxes are connected with the time correlation functions of the flux densities and formally contain an averaging with the local equilibrium distribution function. These kernels are a generalization of the Green–Kubo formulæ for essentially nonequilibrium states. Later [16], general transport relations were derived by the Mori method [17]. However, the calculation and approximation of the integral kernels including Green functions for the Liouville equation are as difficult as solution of the Liouville equation itself.

A rather fruitful method for analyzing problems of nonequilibrium statistical mechanics was developed by D.N. Zubarev [18,19]. This method is called the nonequilibrium statistical operator method. It is built on obtaining solutions to the Liouville equation that depend on canonical variables through a reduced set of parameters characterizing a problem.

The constitutive relationships between the dissipative fluxes J (viscous stress tensor, heat conduction vector) and conjugated forces X (velocity gradient tensor, temperature gradient) derived by statistical mechanics methods are non-linear, non-local, and retarding.

For structureless particles within a volume V, we shall proceed from the form

$$J(\mathbf{r}, t) = \int_{-\infty}^{t} \int_{V} \Re(\mathbf{r}, \mathbf{r}', t, t') X(\mathbf{r}', t') \, d\mathbf{r}' \, dt' \tag{6.1}$$

of the governing relationships. Here, $\Re$ represents the relaxation transport kernels extending the transport coefficients to essentially nonequilibrium conditions.

All cross effects and the initial effects in the relationships of Eq. 6.1 are neglected. Spatial non-locality is caused by non-local molecular interactions, statistical effects due to the interaction of small volume elements of the medium, and retention of the spatial correlations.

By neglecting the non-locality and memory effects in the relationships of Eq. 6.1 we are left with the linear and local relationships

$$J(\mathbf{r}, t) \approx \Re_0(\mathbf{r}, t) X(\mathbf{r}, t)$$

$$\Re_0(\mathbf{r}, t) \equiv \int_{-\infty}^{t} \int_{V} \Re(\mathbf{r}, \mathbf{r}', t, t') \, d\mathbf{r}' \, dt'. \tag{6.2}$$

In these equations, $\Re_0$ represents the usual transport coefficients.

Speaking more generally, if a process is described in terms of a chosen set of variables and there are some other processes that affect the described process but are not included in the description, then the chosen set of variables should necessarily be governed by the non-local equations both in space and time. In non-equilibrium statistical mechanics it has been proven that the governing relationships between the fluxes and gradients of the chosen variables should include averaging on the scale of the statistical correlation length. In equilibrium, where all particles at the microscopic scale level are moving chaotically, the averaging law is the normal distribution. In chaos, there is no correlation between motions of different particles, the typical correlation length is zero, and the normal distribution becomes the Dirac δ-type function. This means that a high-rate and high-gradient process referred to the nonequilibrium ones cannot be correctly described without taking into account the non-local effects.

In the derivation of the non-local relations it was shown that, in general, $\mathfrak{R}$ represents nonlinear functionals of the hydrodynamic gradients. Now we have already realized that any calculations of the relaxation transport kernels or the time correlation functions required to solve the generalized hydrodynamic set simultaneously with the particle dynamic equations is practically impossible for real systems. Therefore, the functional dependence of the transport kernels is usually specified in the form of additional assumptions or by modeling based on known or expected asymptotics for large or small scales. When the transport kernels or the correlation functions are considered as given functions of time and space, the particle dynamics are discarded, whole set separates, and the hydro-dynamic equations become self-contained. This approach can be applied to get different governing relationships for real systems using adequate approximations for the integral transport kernels. Taking different explicit expressions for the kernels, we can get governing relationships for media with different properties and construct convenient models for practical problems. However, attempts to introduce non-local hydrodynamics as a statistico-mchanical background for calculating hydrodynamics and applied mechanics have only recently become common because of the large gap in description levels and approaches.

It is very important to point out that the nonequilibrium statistical operator method and all other methods proceeding from the Liouville equation are valid only for isolated systems. However, it is nonequilibrium stationary states main-tained by the imposed fixed boundary conditions that are of prime practical interest. At present, analyzing the evolution of open systems actively interacting with their surroundings is not practical within the scope of nonequilibrium statistical mechanics.

6.3. Self-Consistent Non-Local Models for High-Rate Transport in Media

6.3.1. A New Hydrodynamic Approach to Modeling Relaxation Transport Kernels

This chapter is a continuation of series of works [8,21–23] in which a new theory of non-local hydrodynamics based on a construction of self-consistent models for the relaxation transport kernels is being developed. The theoretical background of the new approach involves non-local hydrodynamic equations obtained by the nonequilibrium statistical operator method as shown in Sec-tion 6.2.3. Herewith, it is supposed that the boundary effect connected with the interaction of an open system with its surroundings may become involved as an additional element of modeling for the relaxation transport kernels. Insofar as such modeling can be treated immediately at the level of hydrodynamic description, the proposed approach falls into a third group of approaches.

In high-rate flows and at large hydrodynamic gradients, a medium begins to exhibit effects of internal structures characterized by additional scales of length and time. Classical hydrodynamic (differential) equations are invalid under these nonequilibrium conditions. Balance equations for hydrodynamic densities that imply constitutive relationships between thermodynamic forces X and fluxes J of Eq. 6.1 are non-local in space and time. These equations are integro-differential and incorporate integrals over the whole volume of a medium and over the history of the medium deformation. In the general case, the relaxation kernels are unknown functionals of hydrodynamic densities and depend on space and time parameters connected with the internal structural length λ and time τ

If deviations from equilibrium are small, the scales λ and τ can be neglected compared with typical flow scales L and θ: $\lambda/L \to 0$ and $\tau/\theta \to 0$. The relaxation kernels reduce to transport coefficients and relationships between forces and fluxes become local and linear,

$$J(\mathbf{r}, t) = \Re_0(\mathbf{r}, t) X(\mathbf{r}, t) = J_0(\mathbf{r}, t), \qquad (6.3)$$

where the J_0 are fluxes in the linear and local approximation.

In the general case, bearing in mind that near equilibrium, i.e., as $\lambda/L \to 0$, $\tau/\theta \to 0$, the constitutive relationships become local, Eq. 6.3, and we can introduce parameters $\varepsilon_i = \lambda_i/L_i$, $i = 1, 2, 3$, and $\varepsilon_t = \tau/\theta$, which have meaning of non-locality parameters in space and time. Then the relationships 6.1 take the dimensionless form

$$J(\mathbf{r}, t) = \int_{-\infty}^{t} \int_{V} \overline{\Re}(\mathbf{r}, \mathbf{r}', t, t'; \varepsilon_i, \varepsilon_t) J_0(\mathbf{r}', t') d\mathbf{r}' dt', \qquad (6.4)$$

where the kernel $\overline{\Re}$ differs from $\Re$ by a factor $\Re_0$.

To obtain a closed set of hydrodynamic equations it is necessary to know the expressions for the relaxation transport kernels $\overline{\Re}$. In further considerations the overbar will be omitted. Because this does not lead to derivation of any nontrivial kinetics-based expressions for kernels applicable to practical problems, the only real possibility is to construct model expressions for the relaxation transport kernels. There exist three different ways to construct new models. The first way is to construct semi-kinetic models by using the model parameters to introduce some incomplete kinetic information about the interactions among internal structural elements. The second way is to construct new semiempirical models with parameters that can be found experimentally. It must be pointed out that, in each of these cases, the model parameters are determined from inverse problems using *a priori* incomplete kinetic information. This last procedure gives rise to new possibilities for processing experimental data. The third way consists in constructing space–time dependencies at the level of macroscopic description. According to the proposed approach, the model parameters are related to integral properties of the system through either integral relationships including such characteristics as a flow rate, total momentum, and total energy or by imposing additional boundary conditions. The last relationships with respect to the model parameters close a set of the non-local equations and make the formulation of a boundary-value problem self-consistent. The self-consistent property of the proposed approach is its special feature that leads to very important consequences.

6.3.2. Specification of Boundary-value Problems within the Scope of the New Self-consistent Approach

The other special feature of this approach is preservation in the generalized hydrodynamic equations of integral information about a system in the description of its local properties. This circumstance leads to a radical change in the concept of boundary-value problems in a non-local theory.

Consider two different materials separated by a boundary Γ. Each of the materials, with its internal structures in an essentially nonequilibrium state, disturbs the other to a finite depth. This gives rise to near-boundary regions D_{Γ_1} and D_{Γ_2} in materials 1 and 2, respectively. The material in each of these regions interacts with that in the other region and with undisturbed internal regions D_{0_1} and D_{0_2}. If all parts of each material influence the other material, there are no undisturbed regions D_{0_1} and D_{0_2}. In order to specify a boundary-value problem, we must solve the problem jointly at the macroscopic description level for both interacting regions D_{Γ_1} and D_{Γ_2}. In the general case, this is practically impossible. Then we have two alternative ways to specify boundary problems. In the case in which one of the near-boundary regions, for instance D_{Γ_1}, is thin, we can find boundary conditions for the undisturbed region D_{0_1} by analyzing a microscopic problem for the region. Neglecting the small depth of the boundary layer, we can get effective boundary conditions for the whole region D_1. Such a procedure makes sense only for thin boundary layers and for slight deviations from local equilibrium. For highly nonequilibrium states the near-boundary regions D_{Γ_1} and D_{Γ_2} arise, replacing the discrete boundary Γ. In this case, the concept of boundary conditions requires reconsideration. The proposed approach is to incorporate the boundary effects immediately into the macroscopic model as an additional model element. We shall follow this procedure in constructing self-consistent models including the parameters connected with some additional boundary conditions. Unlike the classical continuum models and those of the gradient type, the self-consistent non-local models are uniformly valid up to the boundaries. Thus, solutions provided by these equations can satisfy real boundary conditions considered to be the continuity conditions for hydrodynamic fields. This means that, on solid boundaries, we can use non-slip conditions even for highly nonequilibrium flows when the classical continuum models lead to discontinuities on or near boundaries.

6.3.3. Construction of Space-Dependent Relaxation Kernels

Non-local models must correspond to general invariance and asymptotic principles and depend on a minimal number of parameters. The model kernels meet the following general requirements:

1. Invariance with regard to the expansion transformations.

2. Invariance with regard to the displacement and rotation transformations only far from boundaries; δ-type kernels correspond to the condition

$$\Re(|\mathbf{r}\text{-}\mathbf{r}'|,\varepsilon) \to \delta(|\mathbf{r}\text{-}\mathbf{r}'|), \ \varepsilon \to 0. \tag{6.5}$$

   Near solid boundaries in the presence of large gradients in the regions D_Γ, polarizing effects arise that make the kernels asymmetrical because of correlations. In this case, they are not invariant with regard to displacement and rotation.

3. Uniformity of the limiting transition to classical hydrodynamics,

$$J(\mathbf{r}, t; \varepsilon) \to J_0(\mathbf{r}, t), \text{ as } \varepsilon \to 0, \tag{6.6}$$

   determines the uniform validity of the non-local description throughout the flow region.

These principles define a Dirac δ-type class of model relaxation-transport kernels depending on parameters. The physical meaning of the parameters is examined by means of a test problem for which it is possible to compare results with those obtained in the kinetic theory and with experimental data [8]. The internal structural scales determine the model parameters connected with the influence of an internal structure on hydrodynamics as a whole. To account for the spatial non-locality along the x direction, a simple model of the δ-type kernel valid for boundary problems is proposed:

$$\Re(x, x'; \varepsilon) = \frac{1+\alpha(x,\varepsilon)}{\varepsilon} \omega\left\{ \frac{|x'-x-\sigma(x,\varepsilon)|}{\varepsilon} \right\}. \tag{6.7}$$

The model of Eq. 6.7 corresponds to the condition 6.6 if the order of the limiting transition is fixed: $\varepsilon \to 0$, $\sigma \to 0$. The expression 6.7 is easily generalized to the three-dimensional case taking into account the fact that the non-locality scales along different directions can also be different.

The model of Eq. 6.7 includes the following parameters:

i) ε is a non-locality parameter or a relative spatial correlation scale for hydrodynamic densities (for gases $\varepsilon \sim Kn$);

ii) $\omega(|\xi|;\varepsilon) \geq 0$ is a δ-type function of $|\xi|$ depending on ε as a parameter and defining a rate of spatial relaxation of hydrodynamic correlations;

iii) $\sigma(x, \varepsilon)$ is a shift parameter connected with polarization effects excited by large hydrodynamic gradients and near-boundary interaction; owing to the boundary relaxation parameter, $\sigma(x, \varepsilon) \to 0$ decays in a distance from boundaries;

iv) $\alpha(x, \varepsilon)$ is a parameter characterizing the average structural effects such as a relative effective transport coefficient, $\alpha \to 1$, when $\varepsilon \to 0$.

As the form of ω-parameters we can use, for instance, the Gaussian form for the model kernels employing the shifted argument. It is worth remarking that, according to the chosen controlled characteristics, the self-consistent model can be adequate even when the real form of the transport kernel is far from the Gaussian form.

According to the self-consistent approach to the construction of the model relaxation transport kernels, the model parameters are related to integral properties of a system either through integral relationships including such characteristics as a flow rate, total momentum, and total energy, or by imposing additional boundary conditions

$$\Phi_i(S, \sigma, \varepsilon) = 0, \quad i = 1,2,3. \tag{6.8}$$

These additional relationships for the model parameters complete a set of the non-local equations and make the formulation of the boundary problem self-consistent. The self-consistency of the proposed approach is its special feature and results in very important consequences.

The essential property of the proposed approach is preservation, in the generalized hydrodynamic equations, of the integral information about a system in the description of the local hydrodynamic fields. This circumstance changes the concept of a boundary-value problem in the non-local theory in a radical way. Unlike the classical continuum models, the self-consistent non-local model is uniformly valid up to boundaries. Thus, solutions to these equations can satisfy real boundary conditions considered as continuity conditions for hydrodynamic fields. This means that, on solid boundaries, one can use the non-slip condition even for high-rate flows when the classical continuum model leads to discontinuities on or near boundaries.

In as much as the additional boundary conditions closing the self-consistent model can be imposed rather arbitrarily, this approach allows prediction of the conditions for formation of spatial structures with *a priori* predicted properties. This is a very important advantage of the self-consistent non-local theory that can be used in a wide range of technological applications.

6.3.4. Mathematical Basis of the Self-consistent Non-local Approach

During the past decade new results in the theory of nonlinear operator sets have been obtained and applied in the mechanics of resonance systems by Vavilov [24,25]. Recall that the case $\varepsilon \approx 1$ corresponds to the resonance situation where the external field scale and scale of the internal structure are the same. The self-consistent formulation for the non-local equations can be reduced by standard methods to a special type of nonlinear operator set,

$$u = F(u;\xi), \qquad \Phi_i(u;\xi) = 0, \; i = 1,\ldots,n \,, \tag{6.9}$$

with respect to an unknown element $u \in E$ in Banach space and n internal model parameters $\xi \in R^n$, F is a nonlinear operator: $E \times R^n \to E$, and Φ_i are nonlinear functionals.

In order to analyze the solvability problem for the operator set of the type of Eq. 6.9, two methods have been developed: a geometrical method based on a Galerkin approximation scheme and a comparison method based on a simple iteration process. Both methods establish universal mathematical grounds for examining a wide class of physical situations. By using the methods developed, the solvability conditions for the Eq. 6.9-type set have been formulated as a set of nonlinear inequalities with respect to the external loading parameters. The algorithms based on the methods for constructing approximate solutions to the Eq. 6.9-type problems are well established. If the solvability conditions are satisfied, the iteration procedures converge to a precise solution. These methods also allow successive analysis of the branching solutions to nonlinear problems, whereas complicated practical problems cannot always be examined by the classical methods.

6.4. Non-local Generalization of the Maxwell Model for a Medium with Internal Structure

6.4.1. Introduction

In series of experimental studies performed by Yu.I. Mescheryakov et al. [33–35] on the shock loading of solid materials, it has been found that the stress relaxation was followed by formation of a new internal structure. These structures are of the mesoscopic scale level, which lies between the atom/dislocation scale level and the macroscopic scale. The mesoscopic structural elements, which can be different from grains of the material, begin to play the role of new carriers of deformation. A new branch of mechanics describing the deformation and fracture properties from the point of view of the multiscale hierarchy of carriers of deformation is called *mesomechanics* [33]. Mesoscopic processes are responsible for the macroscopic behavior of solids. At present, it is clear that collective effects occurring during nonequilibrium processes that take place under high-rate loading conditions cause formation of mesoscopic structures. Unlike the case of quasistatic straining, the most important feature of high-strain-rate deformation of solids is the emergence of a space–time correlation among the elementary carriers of deformation. The collective effects lead to formation of new structural elements of larger scale than that of those initially present. However, the approach developed by Panin et al. [33] cannot form the basis of a satisfactory theory of mesoscopic effects because collective interactions must be included in the description. It is understood that relaxation in a shock-compressed solid cannot be described within the framework of the traditional elastic–plastic theory. On the other hand, the liquid-like behavior of the medium that is observed cannot be described using classical hydrodynamics of an ideal liquid or the Navier–Stokes equations for a viscous liquid. These equations are local, valid only for structureless media, and do not predict the changes of flow regime that are observed to occur over a wide range of imposed conditions. For this reason, it becomes necessary to extend classical hydrodynamics to high-rate processes in media with finite-sized structural elements. This presupposes that the needed theory, being hydrodynamic, would be capable of predicting both rearrangement of structural scales and change of kinematic mechanisms of flow during dynamic straining. For example, it must provide a transition from laminar to turbulent flow. Therefore, the extended hydrodynamic theory should be non-local to take into account collective effects and self-consistent to introduce a feedback mechanism into the system.

6.4.2. Conclusions from Maxwell Relaxation Theory

The Maxwell relaxation theory agreed upon unites elastic, plastic, and viscous medium properties in the simplest additive way. According to the theory, the viscous relaxation defined by the typical relaxation time begins when the shear stress reaches a threshold value. Hooke's law describes the elastic properties of the medium. The Newtonian viscous-fluid model represents its viscous behavior. Both models are valid only for rather small hydrodynamic gradients or for low strain rates. This means that the Maxwell model can't be used for high-rate dynamic processes.

In a one-dimensional case, the total stress component in the x direction, σ_1, consists of spherical, P, and deviator, S_1, parts:

$$\sigma_1 = P - S_1. \tag{6.10}$$

The well-known Maxwell model of a medium can determine the deviator stress S_1:

$$\frac{\partial S_1}{\partial t} = \frac{4G}{3}\frac{\partial u}{\partial x} - \frac{S_1 - S_1^*}{t_r(S_1)}, \tag{6.11}$$

where u is the mass velocity in the x direction, G denotes the elastic shear modulus, and $t_r\,(S_1)$ is the relaxation time. From the continuity equation we have

$$\frac{\partial u}{\partial x} = -\frac{1}{\rho}\frac{\partial \rho}{\partial t},$$

where ρ is the mass density. The medium is assumed to have a well-defined critical shear stress, S_1^*. Here, instead of the strain, the velocity u is chosen in our description as a variable that can be measured in experiments on shock compression of materials. Equation 6.11 as a linear first-order ordinary differential equation, when constrained by the initial condition $S_1(0) = 0$, integrates to the expression

$$S_1 = S_1^*\left\{1-\exp\left[-\int_0^t\frac{dt'}{t_r}\right]\right\}+\frac{4Gt_r}{3}\int_0^t\left\{\exp\left[-\int_{t'}^t\frac{dt''}{t_r}\right]\right\}\frac{1}{t_r}\frac{\partial u(x,t')}{\partial x}dt'. \quad (6.12)$$

Equation 6.12 shows that the elastic, plastic, and viscous medium properties can be specified only relative to the loading time or to the time typical for the process compared to the relaxation time. This is the very important consequence of the Maxwell theory.

For large times compared to the relaxation time $(t \gg t_r)$, in the limiting case of complete relaxation, Eq. 6.12 yields the viscous–plastic medium model with a Newtonian viscous term:

$$S_1 \to S_1^* + \frac{2}{3}\eta\frac{\partial u}{\partial x}; \quad \eta = 2Gt_r \quad t_r \to 0.$$

Here $\eta = 2Gt_r$ is an effective shear viscosity of the medium. In the absence of dissipation, the last term can be neglected and, over the threshold S_1^*, an ideal fluid flow takes place. This is the ideal plasticity model.

In the case of frozen relaxation for short time intervals $(t \ll t_r)$, the elastic solid model is obtained:

$$S_1 \to \frac{4}{3}Ge_1, \quad t_r \to \infty$$

$$e_1 = \int_0^t\frac{\partial u}{\partial x}dt, \qquad (6.13)$$

where e_1 is the total (elastic plus plastic) deviator strain component in the x direction.

For time intervals comparable to the relaxation time $(t \sim t_r)$, the first term in Eq. 6.12 is growing in time from zero and reaches the value S_1^* only for large time. This means that the agreed-upon classification for elastic, plastic, viscous or viscous–elastic, elastic–plastic media can be used only for rather narrow fixed loading conditions. If the loading conditions vary rapidly, this classification becomes invalid. Within the scope of the theory, even the main medium properties such as solidity or fluidity can be correctly determined only with respect to the character of the loading applied to the medium. Besides, the additively constructed model doesn't allow any interaction mechanisms except viscosity in the intermediate region between the solid and fluid states ($t \sim t_r$). At present, it has been found that the mesoscopic structures occur in this region.

In the case where the relaxation time t_r is a constant, Eq. 6.12 takes the form

$$S_1 = S_1^* \left\{ 1 - \exp\left[-\frac{t}{t_r} \right] \right\} + \frac{4Gt_r}{3} \int_0^t \left\{ \exp\left[-\frac{t-t'}{t_r} \right] \right\} \frac{1}{t_r} \frac{\partial u(x,t')}{\partial x} dt' . \tag{6.14}$$

For a perturbation propagating in the x direction at the constant velocity C, Eq. 6.14 can be rewritten as follows:

$$S_1 = S_1^* \left\{ 1 - \exp\left[-\frac{Ct}{Ct_r} \right] \right\} + \int_0^{Ct} \frac{4Gt_r}{3} \left\{ \exp\left[-\frac{Ct-Ct'}{C} \right] \right\} \frac{\partial u}{\partial x'} \bigg|_{x'=Ct'} \frac{dCt'}{Ct_r}$$

$$= S_1^* \left\{ 1 - \exp\left[-\frac{x}{\varepsilon} \right] \right\} + \int_0^\infty \left\{ -\exp\left[-\frac{x-x'}{\varepsilon} \right] \right\} \frac{2\eta}{3} \left(\frac{\partial u}{\partial x} \bigg|_{x=x'} \right) \frac{dx'}{\varepsilon} , \tag{6.15}$$

where $\varepsilon = Ct_r$ denotes a typical relaxation length which is assumed to be constant, as t_r is constant. The upper limit in the last integral can be taken infinite because, over the interval (Ct, ∞) on the x axis, the medium remains unperturbed.

So, it is possible to say that the Maxwell model in the integral form of Eq. 6.15 for the time compared to the relaxation time incorporates either memory or non-local effects. The integral relationship 6.15 corresponds to the general constitutive relationship 6.6 where the shear stress S_1 presents a component of the momentum flux tensor J and J_0 is equal to the viscous stress

$$J_0 = \frac{2}{3}\eta \frac{\partial u}{\partial x}.$$

The integral kernel in this relationship is a simple exponential with the relaxation time as a parameter. It becomes clear that all the restrictions concerning the validity region of the Maxwell theory are connected to the integral kernel form. The more the integral kernel differs from the simple exponential, the wider the validity range of the model and the more high-rate processes it can describe.

6.4.3. Non-local Generalized Model of a Viscous Elastic and Plastic Medium with Mesoscopic Structures

A new step resulting from the non-local hydrodynamic approach proposed by the author [26–29] consists in generalization of the relaxation model of Eq. 6.15 for nonlinear dynamical processes:

$$S_1 = S_1^* \int_0^\infty \frac{1}{\varepsilon} N(x, x'; \varepsilon)\, dx' + \frac{2}{3}\eta \int_0^\infty N(x, x'; \varepsilon)\frac{1}{\varepsilon}\frac{\partial u}{\partial x'}\, dx'. \qquad (6.16)$$

An integral relaxation kernel $N(x, x'; \varepsilon)$ that differs from the simple exponential, introduces the non-local properties (or collective effects) of the medium under conditions far from equilibrium. The general relaxation model of Eq. 6.16 describes the spatial relaxation due to irreversible formation of a new internal structure at the mesoscopic scale level that develops during high-rate straining of media. This model kernel should satisfy the same asymptotic conditions as that in Eq. 6.12. In the case where $\varepsilon \to \infty$ or $t_r \to \infty$, $N(x, x'; \varepsilon) \to 1$,

$$S_1 \to S_1^* \int_0^\infty \frac{dx'}{\varepsilon} + \frac{2\eta}{3}\int_0^\infty \frac{\partial u}{\partial x'}\frac{dx'}{\varepsilon} = \frac{4Gt_r}{3}\int_0^{Ct}\frac{1}{\varepsilon}\frac{\partial u}{\partial x'}dCt'$$

$$= \frac{4G}{3}\int_0^t \frac{\partial u}{\partial x}dt' = \frac{4G}{3}\frac{U_0 - u}{C} = \frac{4G}{3}\frac{\rho - \rho_0}{\rho} = \frac{4G}{3}e,$$

one obtains the elastic solid model. In the opposite case, where $(1/\varepsilon)N(x, x'; \varepsilon) \to \delta(x - x')$ as $\varepsilon \to 0$ or $t_r \to 0$ after the internal structural relaxation has already been completed, we obtain a viscous–plastic unstructured medium:

$$S_1 \to S_1^* + \frac{2}{3}\eta\frac{\partial u}{\partial x}.$$

In the intermediate case, the non-local model describes the relaxation in a medium with mesoscopic internal structures. As a matter of fact, S_1^* is a result of the internal structural relaxation already completed whereas, during the structural relaxation $(\varepsilon \sim 1)$, there is no threshold. Thus, we see that the new non-local model generalizes the Maxwell model of an elastic, plastic, and viscous medium for high-rate straining and, in addition, introduces a relaxation due to the formation structural elements at the mesoscopic scale level characterized by additional finite relaxation lengths. By neglecting the structural effects when the typical relaxation time is characterized only by the atom/dislocation scale and, for the simple exponential integral kernel, the model of Eq. 6.16 is reduced to the Maxwell model of Eq. 6.15 of an elastic, plastic, and viscous medium without internal structure.

In accordance with the proposed approach [26,27], the representation in the form 6.7

$$N(x, x'; \varepsilon) = (1+\alpha)\exp\left[-\frac{\pi}{\varepsilon^2}(x'-x-\sigma)^2\right]$$

for the spatial relaxation integral kernel can be used. Then, instead of Eq. 6.16, we have the constitutive equation

$$S_1 = S_1^*(1+\alpha)k_0(x, t; \varepsilon, \sigma)$$

$$+\frac{2}{3}\eta(1+\alpha)\int_0^\infty \left\{\exp\left[-\frac{\pi}{\varepsilon^2}(x'-x-\sigma)^2\right]\right\}\frac{1}{\varepsilon}\frac{du}{dx'}(x', t)dx'. \tag{6.17}$$

The zeroth-order coefficient of the integral kernel is

$$k_0(x, t; \varepsilon, \sigma) = \int_0^\infty \frac{1}{\varepsilon}\exp\left[-\frac{\pi}{\varepsilon^2}(x'-x-\sigma)^2\right]dx'$$

$$= \frac{1}{2}\left\{\mathrm{erf}\left[\frac{\sqrt{\pi}}{\varepsilon}(Ct-x-\sigma)\right]+\mathrm{erf}\left[\frac{\sqrt{\pi}}{\varepsilon}(x+\sigma)\right]\right\}\begin{vmatrix}\xrightarrow[\varepsilon\to 0]{}1\\[2mm]\xrightarrow[\varepsilon\to\infty]{}0,\end{vmatrix}$$

where

$$\mathrm{erf}(z) = \frac{2}{\sqrt{\pi}}\int_0^z \exp(-\zeta^2)d\zeta .$$

The non-local model of Eq. 6.17 includes three internal parameters α, σ, and ε having the following physical interpretations:

1. ε is a typical radius of non-local correlation or a typical relaxation length connected with a typical scale of the internal structural elements;

2. $(1+\alpha)$ is a relative effective viscosity correcting the value of viscosity η for the case of material with mesoscopic internal structure; $\alpha \to 0$ as $\varepsilon \to 0$;

3. σ is a polarization parameter along the gradient direction due to boundary conditions; $\sigma \to 0$ as $\varepsilon \to 0$; there is a close connection between σ and existence of rotations of internal structural elements because σ makes the shear stress tensor asymmetrical.

All the model parameters are time-dependent functions because the typical correlation length is expected to change with time.

Besides the internal model parameters connected with the characteristic structural relaxation time t_r, there is another external characteristic time,

$$t_R = \left| \frac{\partial u}{\partial x} \right|_{max}^{-1}$$

related to the typical strain-rate conditioned by the time of the external loading. For the case where $t_R \ll t_r$ the mass velocity is considered to be a step profile of height U_0, propagating at the velocity $V(t)$ along the x axis, and the velocity gradient has the Dirac δ-type form. Substitution of this gradient into the relaxation model 6.17 yields the result:

$$S_1 = S_1^*(1+\alpha)k_0(x, t; \varepsilon, \sigma)$$

$$+ \frac{2}{3}\eta(1+\alpha) \int_0^\infty \left\{ \exp\left[-\frac{\pi}{\varepsilon^2}(x'-x-\sigma)^2 \right] \right\} \frac{1}{\varepsilon} U_0 \delta(x' - V(t)t)\, dx'$$

$$= S_1^*(1+\alpha)k_0(x, t; \varepsilon, \sigma) + \frac{2}{3}\eta \frac{1+\alpha}{\varepsilon} U_0 \exp\left[\frac{\pi}{\varepsilon^2}(V(t)t - x - \sigma)^2 \right].$$

$$(6.18)$$

Equation 6.18 shows that, owing to the structural relaxation at the mesoscopic scale level, a step-wise velocity profile becomes smooth; its typical width grows with ε. The non-steady front velocity $V(t)$ differs from the smoothed front velocity by the value $\sigma(t)$ characterizing a retardation of the smoothed front maximum from the step-wise front due to the mesostructural relaxation: $V(t) - \sigma(t)/t = V^m(t) \leq V(t)$ as the parameter σ for the positive half-space has the definite sign $\sigma \geq 0$. Then it is possible to say that the relaxation at the mesoscopic scale level leads to a retardation of the velocity profile during high-rate loading and this retardation can be connected to the shift parameter σ.

In the opposite case where $t_R \gg t_r$, Eq. 6.17 yields the ideal plasticity model, $S_1 \approx S_1^*$ just as in the case of large characteristic time $t \gg t_r$.

So, it becomes possible to identify two ranges of conditions where one can use the agreed-upon classification:

- for viscous–plastic media with fluid-like behavior ($t_r \ll t \ll t_R$);
- for elastic media with solid-like behavior ($t \ll t_R \ll t_r$).

In all other cases, the classification can't be used. This concerns the new non-local model 6.17 that works only in the intermediate region ($t \sim t_R \sim t_r$), which is typical of high-rate dynamic processes.

6.4.4. Self-Consistent Determination of the Model Parameters

In the previous section, a new model describing the macroscopic behavior of a medium with mesoscopic internal structure has been proposed. However, during high-rate straining of the medium the mesoscopic structure can evolve rapidly. Therefore, it becomes necessary to include a feed-back reaction of the medium deformation upon the medium internal structure. Then, the model parameters connected to the mesoscopic structure evolution should be determined by the changing integral values.

According to the self-consistent approach [26–29], the internal model parameters are unknown functionals of macroscopic gradients that can be written in the form of Eq. 6.9. In order to determine the parameters, three functional relationships, derived from non-equilibrium boundary conditions should be used:

$$\Phi_i\left[u(x, t), \frac{\partial u(x, t)}{\partial x}; x_r; \varepsilon(t), \sigma(t), \alpha(t)\right], \quad i = 1, 2, 3. \tag{6.19}$$

These integral dispersion relationships together with the model constitutive equation 6.17 determine a spectrum of the internal time-dependent structural parameters $\varepsilon(t)$, $\sigma(t)$, $\alpha(t)$. Near the limiting cases where the elasticity theory $(t \ll t_R \ll t_r)$ or the viscous–plastic model $(t_r \ll t \ll t_R)$ prevail, the spectrum is continuous, the internal structure scales don't influence the macroscopic behavior of the medium, which becomes structureless. In the intermediate case $(t \sim t_R \sim t_r)$ the dispersion relationships due to their nonlinearity have non-unique solution; they determine a discrete spectrum of the parameters $\varepsilon_\kappa(t)$, $\sigma_k(t)$, $\alpha_\kappa(t)$ that characterize the scales and types (translation or rotation) of mesoscopic structures.

It must be emphasized that the relationships 6.19 can define the times when branching of their solutions occurs. Then it becomes possible to define when and under what external loading new internal structure scales can arise. The changing of the medium structure or the new structure formation at the instant when the branching occurs is considered a structural transition. This means that the proposed approach allows predicting failure of materials as a structural transition to large-scale structures.

In the limiting cases some model parameters tend to the constant values corresponding to the usual near-equilibrium material characteristics such as the shear modulus and viscosity:

$$\frac{\eta(1+\alpha)}{\varepsilon}(t) \xrightarrow[t_r \to \infty]{} \frac{2G}{C}$$

$$\eta(1+\alpha)(t) \xrightarrow[t_r \to 0]{} \eta.$$

So, according to the results above, the dynamical behavior of a medium is characterized not by constant values characterizing the medium properties, but by some functionals (integrals) of macroscopic fields depending on the history of the dynamic deformation in the entire material volume including boundary conditions. At the same time, being dynamical integral characteristics, the model parameters are closely connected with internal structural characteristics of the medium, such as the internal structural scales $\varepsilon_i(t)$. For steady-state straining, or for structureless media, the parameters become constant, being interpretable as the material modulus and transport coefficients.

This conclusion can be immediately confirmed by the experiments on the sound propagation in materials under laser high-rate loading in the range of pressures $5-150$ MPa and with a period 3×10^{-8} s performed by Yu. Sudienkov, et al. [30–32]. It had been found out that in a thin layer $\sim 10^{-3}$ m near the loading surface the sound velocity differs essentially from its equilibrium value. For a wide class of materials the relaxation is characterized by the typical relaxation time 0.12×10^{-6} s independent of the initial pressure. However, the character of the relaxation, being different for different types of materials, is determined by their internal structure.

6.5. Shock-Wave Propagation in a Relaxing Medium

6.5.1. Non-Local and Self-Consistent Formulation of the Plane Wave Propagation Problem

Plane shock wave propagation in a material half-space is considered. The mass and momentum balance equations are written in Lagrange's coordinates in the one-dimensional case where the x axis is the shock-wave propagation direction:

$$\frac{\rho_0}{\rho^2}\frac{\partial \rho}{\partial t} + \frac{\partial u}{\partial x} = 0 \tag{6.20}$$

$$\rho_0 \frac{\partial u}{\partial t} + \frac{\partial}{\partial x}(P - S) = 0 , \tag{6.21}$$

where the coordinate is

$$x = \frac{1}{\rho_0} \int_0^x \rho(x',t)dx' .$$

The deviator part, S, and spherical part, P, of the stress tensor should each be divided in turn into two parts: $S = S^e + S^m$, $P = P^e + P^m$. The first of these parts, S^e and P^e, correspond to the elastic shear and elastic compression, respectively, representing the elastic properties of the medium: $S^e = (4/3)G(\rho - \rho_0)$, $P^e = (\rho - \rho_0)C^2$. The second parts, S^m and P^m, are assumed to be connected with the mesoscopic effects. Herewith, S^m can be determined by the non-local relaxation model of Eq. 6.17 for $S - S^e$ but P^m remains to be determined.

With this, the mass and momentum balance equations 6.20 and 6.21 can be reduced to the wave-type equation

$$\frac{\partial^2 u}{\partial t^2} - a_l^2 \frac{\partial^2 u}{\partial x^2} = \frac{1}{\rho_0} \frac{\partial^2}{\partial t \partial x}(S^m - P^m). \tag{6.22}$$

In this equation, $a = C\rho/\rho_0$ is the Lagrangian wave propagation velocity and $a_l^2 = a^2 + [4\rho^2/(3\rho_0^3)G]$ is the squared longitudinal sound velocity in Lagrangian coordinates. Here the velocity a_l is taken to be constant. The source in the right-hand part of Eq. 6.22 connected with the mesoscopic relaxation effects leads to a more gently sloping front.

By using the known Green function for the wave operator in the case where the wave front propagates in the positive direction along the x axis, Eq. 6.22 can be reduced to the integral d'Alembert form

$$u(x, t) = u_0(x - a_l t)$$

$$+ \frac{1}{a_l \rho_0} \int_0^t \int_{x - a_l(t-\tau)}^x \left[\frac{\partial^2 (S^m - P^m)}{\partial \tau \partial \xi} \right] d\xi \, d\tau. \tag{6.23}$$

The initial and boundary conditions $u(x, 0) = \partial u(x, 0)/\partial t = 0$; $u(0, t) = U_0 \Omega(t)$ are taken. The zeroth-order solution has the form of a simple wave $u_0(x - a_l t) = U_0 \Omega(x - a_l t)$ propagating without changing its form and the second term defines the changing form of the wave front.

The second item in Eq. 6.23 can be calculated disregarding the change of the front propagation for the risetime t_R:

$$\frac{1}{a_l \rho_0} \int_0^t \int_{x - a_l(t-\tau)}^x \frac{\partial^2 (S^m - P^m)}{\partial \tau \partial \xi} d\xi \, d\tau = \frac{1}{a_l \rho_0} \Delta_x (S^m - P^m)(x - a_l t, t). \tag{6.24}$$

The right-hand part of Eq. 6.24 corresponds to the waveform evolution for the risetime due to the influence of the mesoscopic stress inside the front itself. It can be rewritten in a reference system connected to the wave front:

$$\Delta u(t) = -\frac{1}{\rho_0 a_l} \Delta_x (P^m - S^m)(t), \tag{6.25}$$

where the value Δu defines the form of the plastic wave front during shock loading. Equation 6.24 is considered adequate for small risetimes or far from the loading surface. However, supposing that the wave propagation velocity is related to the point inside the wave front where the velocity gradient is maximum, the deviation of the front propagation velocity $\Delta a(t)$ due to the mesoscopic effects can also be calculated using the rate of the evolution of the model parameter $\partial\sigma(t)/\partial t$ (see Eq. 6.18).

In the case of high strain rates ($t_R \to 0$) corresponding to the long characteristic time $t \gg t_R$, and when the mesoscopic relaxation has been completed ($\varepsilon \to 0$ or $t_r \to 0$), the source $S^m - P^m = 0$. However, Eq. 6.22 at the constant velocity a_l is not valid for nonlinear and non-steady strong shock wave propagation corresponding to the Hugoniot adiabat. It describes only shock-wave propagation of moderate intensity. In the case of low strain rates ($t_r \ll t \ll t_R$) Eq. 6.22 corresponds to acoustic wave propagation in liquids at the bulk sound velocity a:

$$\frac{\partial^2 u}{\partial t^2} - a^2 \frac{\partial^2 u}{\partial x^2} = 0 . \tag{6.26}$$

In the case of low strain-rates ($t_R \to \infty$) or small typical times $t \ll t_R$, when the mesoscopic processes have not yet been initiated, that is, for the frozen relaxation ($\varepsilon \to \infty$ or $t_r \to \infty$), the limiting case $t \ll t_R \ll t_r$, Eq. 6.22 governs acoustic wave propagation in solids at the longitudinal velocity a_l:

$$\frac{\partial^2 u}{\partial t^2} - a_l^2 \frac{\partial^2 u}{\partial x^2} = 0 . \tag{6.27}$$

So, in all limiting cases, the value Δu characterizing the waveform changing and the deviation of the wave propagation velocity Δa due to mesoscopic effects doesn't arise.

In the intermediate case for the characteristic time $t \ll t_R \ll t_r$, the values Δa, Δu that arise result non-stationary wave front propagation. It is possible to say that the value Δu describes plastic waveform evolution. It corresponds to the kinetic energy changing inside the wave front. If the value Δu at the pulse plateau is zero, the kinetic energy isn't changed for the rise-time. However, in the general case a wave amplitude loss occurs due to the irreversible character of the mesoscopic structure relaxation. This amplitude loss at the pulse plateau has been measured experimentally during the shock loading of metals before the dissipation into heat begins (Mescheryakov et al. [33–35]).

6.6. Nonequilibrium Thermodynamics of High-Rate Processes

6.6.1. Multiscale and Multi-Stage Energy Exchange During High-Strain-Rate Loading of Solids

Mescheryakov et al. [33–35] found experimentally that stress relaxation in the velocity range 500–1000 m/s was followed by formation of mesoscopic structures. Transfer of macroscopic kinetic energy to density and velocity fluctuations at the mesoscopic scale level is initiated within the wave front under high-rate loading conditions. The quantity characterizing the velocity fluctuations that can be experimentally measured and immediately included in the balance equations is the mass-velocity dispersion, $D^2(x, t) = \langle (v - u)^2 \rangle$, originating because of the macro–meso energy exchange. Then $D(x, t)$ has a meaning of the mean fluctuation velocity determining typical scales at the mesoscopic level. As experiments show, the velocity dispersion is generated at a constant temperature. It can be seen from a simple evaluation for the thermal conductivity, which is ~ 10^{-4} m^2/s for copper. For the typical risetime $t_R \sim 10^{-7}$ s the heat propagates for a distance ~ 10^{-6}–10^{-5} m. Compared to the typical front width ~ 10^{-4}–10^{-3} m this is too small. This means that the energy exchange between the macroscopic and mesoscopic scale levels during high-rate straining begins before the dissipation into heat. A part of the macroscopic energy goes to the mesoscopic scale level, but not immediately to the microscopic or atom/dislocation scale levels. First, the mass-velocity dispersion grows. This means that isotropic velocity fluctuations are generated. During high-rate straining, the fluctuations can change their scale. While they remain isotropic they are reversible. They are not heat fluctuations and therefore are not subject to the second law of thermodynamics. However, it is clear that velocity dispersion plays the role of a temperature for mesoscopic fluctuations and should be introduced into the equations of state for media deformed at a high rate.

When mesoscopic fluctuations become anisotropic, they contribute into the macroscopic mass velocity and deviator stress component that begins to relax either by means of new structure formation at the mesoscopic scale level or much later by viscous dissipation into a heat. The experiments show that the velocity dispersion can decrease, but doesn't always return to its initial value. (See Chapter 5 in this volume.) If the mesoscopic fluctuations related to the velocity dispersion have time to decay inside the wave front and there is no other structure formation, the macro–meso energy exchange is considered reversible. In this case there is no irreversible energy loss at the pulse plateau, the macroscopic velocity at the plateau has its initial value, and the part of macroscopic kinetic energy participating in the macro–meso energy exchange has come back to the macroscopic scale level but at a retardation conditioned by the risetime growth. This means that, inside the wave front, relaxation occurs but can be neglected for the entire wave propagation. In the other case, where the mesoscopic structure formation is followed by irreversible structural transitions, there is an irreversible loss of the macroscopic velocity at the pulse plateau. This value has been measured experimentally and, in just these cases, microstructure investigations show formation of new structures such as meso-rotations and shear bands at the mesoscopic scale level. It is very important that the velocity dispersion and the wave amplitude loss that characterize macro–meso energy exchange can be measured, and the phenomenon of mesostructure formation can be observed in experiments on the shock loading of materials.

Different behavior of the velocity dispersion experimentally measured by Yu. Mescheryakov and presented in Chapter 5 [35] can be seen in his Figs. 5.14, and 5.23. Later they, along with Figs. 6.1 and 6.2, will be predicted theoretically for different situations.

The thermodynamic properties of a system with excited mesoscopic fluctuations are introduced by equations of state. The spherical part of the stress tensor connected with the isotropic velocity fluctuations at the mesoscopic scale level, P^m, can be called the mesoscopic fluctuative pressure, and can be postulated to be

$$P^m = \rho_0 \, a \, D \,, \tag{6.28}$$

where a is the Lagrangian sound velocity in a given medium and D is the mean velocity fluctuation at the mesoscopic scale level. The stress deviator, $S^m = S_1 - S^e = \rho_0 R$, related to relaxation at the mesoscopic scale level, is determined by the non-local model of Eq. 6.17.

The velocity dispersion, D, should be determined from the internal energy balance. The general balance equation for the specific internal energy takes its usual form

$$\frac{\partial E}{\partial t} + (P - S)\frac{\partial u}{\partial x} = 0 \,. \tag{6.29}$$

As with both parts of the stress tensor, the internal energy density, E, should also include two components: the elastic energy, E^e, of elastic compression and shear, and the energy, E^m, of the mesoscopic degrees of freedom, respectively:

$$E = E^e + E^m \,. \tag{6.30}$$

With this, Eq. 6.29 can be separated into two parts related to the different scales and stages of the relaxation:

$$\frac{\partial E^e}{\partial t} + (P^e - S^e)\frac{\partial u}{\partial x} + \frac{\partial E^m}{\partial t} + (P^m - S^m)\frac{\partial u}{\partial x} = 0 \,. \tag{6.31}$$

The mesoscopic component is determined by the equation of state

$$E^m = c_m \rho_0 \, a \, D + E^{ms}, \quad \text{where} \quad c_m = \frac{\partial E^m}{\partial P^m} \,. \tag{6.32}$$

The coefficient c_m is the energy capacity of the mesoscopic fluctuations. The full energy at the mesoscopic scale level, E^m, consists of two parts: kinetic energy of the mesoscopic fluctuations, $c_m \rho_0 \, a \, D$, and potential energy, E^{ms}, accumulated in the mesoscopic structures due to irreversible formation of new structures. For example, the formation of new, rotational structures changes this energy.

At the initial stage of relaxation $t \ll t_r$ when the mesoscopic fluctuations have not yet become excited, $c_m \to \infty$, the mesoscopic scale level should be considered frozen. Equation 6.31 yields the result $D = D(t = 0)$. As the mesoscopic level has already relaxed $c_m \to 0$, velocity fluctuations become associated with heat and Eq. 6.31 describes the last stage of the relaxation for large typical times $t \gg t_r$. This is the irreversible dissipation stage of relaxation when the compression and high-rate deformation cause heating of the medium:

$$\frac{\partial E^T}{\partial t} + (P^T - S^T)\frac{\partial u}{\partial x} = 0 . \tag{6.33}$$

In this equation, the term $P^T - S^T$ refers to the bulk and shear viscous stress components near the state of complete structural relaxation $(t_r \to 0)$.

At the intermediate stage, when the material properties are characterized by finite values of c_m, the structural relaxation at the mesoscopic scale level plays the main role. This relaxation goes by means of the energy exchange between different degrees of freedom at the macro-, meso-, and microscale levels. The macro–meso energy exchange can be either reversible or irreversible depending on whether or not $\partial E^{ms}/\partial t = 0$. This means that the irreversible macro–meso energy exchange corresponds to a structural transition. The micro–meso energy exchange also can be reversible in isotropic mesofluctuations. The macro–micro energy exchange is always irreversible according to the thermodynamic laws. So, the mesoscopic scale level plays a role of energy buffer between macro- and mesoscale levels that energy passes during high-rate processes.

Substitution from Eqs. 6.32 and 6.28 into the balance Eq. 6.31 yields the result

$$\frac{\partial(c_m \rho_0 \, a \, D + E^{ms})}{\partial t} + (\rho_0 \, a \, D - S^m)\frac{\partial u}{\partial x} = -\frac{\partial E^e}{\partial t} - (P^e - S^e)\frac{\partial u}{\partial x} . \tag{6.34}$$

For the typical strain rate $\partial u / \partial x = t_R^{-1}$ the following integration with respect to inside-front time in a reference moving with the front results in the integral form of the energy balance (6.1.7):

$$c_m \rho_0 \, a \, D - \int_0^t \rho_0 \, a_l \Delta u \frac{1}{t_R} d\tau = -\Delta E^{ms} - \Delta E^e . \tag{6.35}$$

Here $\Delta(\rho_0 \, a \, D - S^m) = -\rho_0 \, a_l \Delta u$, according to Eq. 6.25, ΔE^e denotes a changing of the elastic energy inside the wave front due to the nonequilibrium misbalance at the microscopic level $-\partial E^e / \partial t - (P^e - S^e)(\partial u / \partial x) \neq 0$ and ΔE^{ms} is energy expenditure due to the structure transition at the mesoscopic scale level.

At the initial stage of macro–meso energy exchange the right-hand part of Eq. 6.35 is equal to zero. This is the case of reversible macro–meso energy exchange when mesostructures are not changing, i.e., when $\Delta E^{\mathrm{ms}} = 0$, the elastic energy balance is satisfied separately, $\Delta E^{\mathrm{e}} = 0$. In this case of dynamic equilibrium Eq. 6.34 describes the dynamics of mesofluctuations and determines the velocity dispersion $D(x, t)$. As the mean velocity fluctuation is always positive, the integral should be negative. This means that there exists an energy balance between the kinetic energy of mesofluctuations and the deviations of the macroscopic kinetic energy inside the wave front.

This stage, however, may be irreversible for materials with small energy capacity of mesofluctuations. Then the left-hand part of Eq. 6.35 can become negative and the equality is satisfied only on account of the elastic energy balance $\Delta E^{\mathrm{e}} > 0$. In this case energy can be transferred to the mesoscopic level from the potential energy of the crystal lattice as a result of resonance effects. The possibility of the lattice potential energy changing had been predicted theoretically [37]. This case may concern to the brittle materials failure.

At the intermediate stage, the difference on the left-hand side can be either negative or positive. The case where it is positive is very important for materials with large energy capacity of mesoscopic fluctuations. The energy difference $\Delta E^{\mathrm{ms}} < 0$ goes to new, more complicated, structure formation. This structural transition involves new degrees of freedom and new scales. In the case of the negative left-hand part of Eq. 6.35, the structural transition leads to an increase of the energy at the mesoscopic scale level, $\Delta E^{\mathrm{ms}} > 0$, and simpler mesoscopic structures. This is an irreversible stage of structural relaxation.

The last stage is an irreversible dissipative process that takes place after all previous ones, if the loading was very strong. Then mesoscopic structures have already relaxed and macroscopic energy loss goes to the viscous relaxation with the heating of the medium

$$\Delta E^{\mathrm{T}} = \int_0^t \frac{\rho_0 \, a_1 \Delta u}{t_{\mathrm{R}}} d\tau .\tag{6.36}$$

Here ΔE^{T} denotes the increase of the heat energy resulting from the complete relaxation.

6.6.2. Entropy Production During the Macro–Meso Energy Exchange

The entropy production inside the wave front related to the irreversible energy transfer at the mesoscopic scale level is given by the equation

$$\Pi^{\mathrm{m}} = J^{\mathrm{m}} X^{\mathrm{m}} = \frac{\partial E^{\mathrm{ms}}}{\partial t}(P^{\mathrm{m}} - S^{\mathrm{m}})\frac{\partial u}{\partial x}.\tag{6.37}$$

The high-rate form changing $(P^m - S^m)\partial u / \partial x = X^m$ $(P^m - S^m = -\rho_0 a \Delta u)$ is the thermodynamic force that induces irreversible energy flux $\partial E^{ms} / \partial t = J^m$ related to the internal structural transitions. Both the thermodynamic force and the irreversible energy flux can change their signs due to the macro–meso energy exchange inside the wave front. The sign of the entropy production depends on the relaxation stages.

On the initial reversible stage, the entropy production is zero: $\Pi^m = 0$. This stage corresponds to the dynamical equilibrium macro–meso energy exchange, where no structural transitions take place. During the intermediate stage, the sign of the entropy production depends on the character of the irreversible processes inside the wave front. If $\partial E^{ms} / \partial t < 0$ and $X^m > 0$, the potential energy of mesoscopic structures decreases because of a structural transition with the formation of new structures connected with the new degrees of freedom and new structural scales. This is self-organization in the medium under high-rate loading. In this case, the entropy production at the mesoscopic scale level is negative: $\Pi^m < 0$. If $\partial E^{ms} / \partial t > 0$, the potential energy of the mesostructure increases due to the structural transition that results in a simpler structure than that initially present. In this case, the entropy production is positive, $\Pi^m > 0$. This structural transition can lead to material failure.

The rate of the entropy production has a minimum in a state of dynamic equilibrium:

$$\frac{d\Pi^m}{dt} = \frac{\partial^2 E^{ms}}{\partial t^2}(P^m - S^m)\frac{\partial u}{\partial x} + \frac{\partial E^{ms}}{\partial t}\frac{\partial}{\partial t}\left[(P^m - S^m)\frac{\partial u}{\partial x}\right]$$

$$\frac{\partial}{\partial X^m}\left[\frac{d\Pi^m}{dt}\right] = \frac{\partial^2 E^{ms}}{\partial t^2} = 0.$$

(6.38)

In this state, the entropy production is zero, $\Pi^m = 0$, and has a point of inflection (see Fig. 6.1). It should be noted that, unlike the case of steady-state deformation, the rate of the entropy production, but not the entropy production itself, has a minimum in dynamic equilibrium. It must be emphasized that Prigogine's theorem on the minimal entropy production in nonequilibrium steady states is not valid for dynamic processes.

Mescheryakov [35] has found that the experimental curve of spall strength has the maximum at the same conditions at which dynamic equilibrium takes place and the rate of the entropy production has a minimum (see Figs. 5.20 and 5.21). This means that real mechanical properties of materials are entirely defined by different stages of the nonequilibrium macro–meso energy exchange.

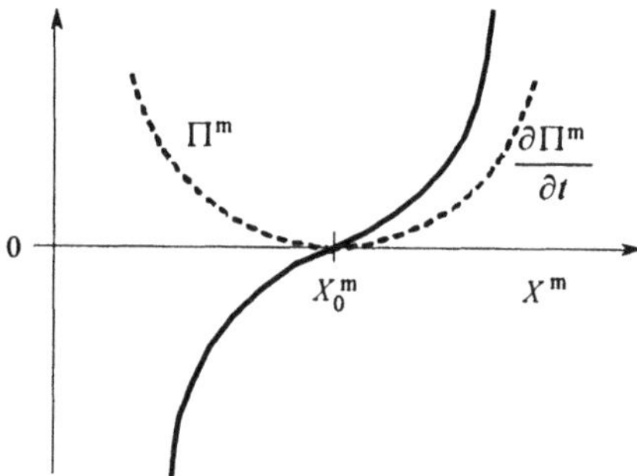

Figure 6.1 The entropy production Π^m and the entropy production rate $\partial\Pi^m/\partial t$ versus the thermodunamic force X^m.

6.6.3. Dynamical Equilibrium Macro–Meso Energy Exchange

In the dynamic equilibrium for the typical strain rate, $\partial u/\partial x = t_R^{-1}$ Eq. 6.35 yields

$$c_m \rho_0 a D - \int_0^t \frac{\rho_0 a_1 \Delta u}{t_R} d\tau = 0.\tag{6.39}$$

Equation 6.39 presents dependence between the waveform evolution and the velocity dispersion behavior inside the wave front. If the velocity dispersion, for example, is measured experimentally, the velocity profile in the dynamic equilibrium is determined.

At the pulse plateau when $t = t_R$, it is found that, at the zero mean velocity fluctuation $D(t = t_R) = 0$, the integral also becomes zero:

$$\int_0^{t_R} \Delta u \, dt = 0.\tag{6.40}$$

Then Δu is odd function of time inside the front and the mean velocity fluctuation, being always non-negative, is a symmetrical function having its maximum in the middle of the front. During the first half of the front $\Delta u < 0$ and the velocity dispersion is growing. During the second half, to the opposite, $\Delta u > 0$ while the dispersion decreasing. It is clear that a part of the kinetic energy of the front is going to the fluctuations at the mesoscopic scale level. This is the macro–meso energy exchange forming the waveform. Equation 6.40 means that dynamic equilibrium macro–meso energy exchange inside the wave front corresponds to symmetrical waveforms. During the first part of the front, the kinetic energy goes from the macroscopic scale level to the mesoscopic level and the entropy production is positive $\Pi^m > 0$. During the second half of the front, the energy from the mesolevel returns to the macroscopic kinetic energy at the negative entropy production $\Pi^m < 0$. The self-organization at the mesoscopic scale level can occur in a resonance way under high-rate external loading.

The full velocity deviation in the front can be divided into central symmetrical and antisymmetrical parts: $\Delta u = \Delta^s u + \Delta^{as} u$. Only the antisymmetrical part contributes into the integral over the risetime. All asymmetrical waveforms originate due to irreversible structure formation. The more steeply sloping the front becomes, the more intensive become the irreversible processes occurring inside the front:

$$\int_0^t \rho_0 a \, \Delta u \, \frac{dt}{t_R} = \Delta E^{ms} . \tag{6.41}$$

If there exists a point, t^*, inside the wave front where the velocity profile sharply changes its turning angle, it means that at this instant the irreversible relaxation stage begins. Such points had been observed experimentally (see Fig. 5.23). In this case the value ΔE^{ms} can be easily calculated from Eq. 6.41:

$$\int_{t^*}^{t_R} \rho_0 a_l \, \Delta^{as} u \, \frac{dt}{t_R} = \Delta E^{ms}(t_R) .$$

At the pulse plateau for the time $t = t_R$, Eq. 6.41 determines the irreversible loss of the wave amplitude due to the irreversible structure formation inside the wave front at the mesoscopic scale level:

$$\Delta U = \int_0^{t_R} \Delta u \, \frac{dt}{t_R} . \tag{6.42}$$

Then Eq. 6.41 is approximately replaced by the criterion

$$c_m D(t_R) - \Delta U = -\frac{\Delta E^{ms}(t_R)}{\rho_0 a_l} . \tag{6.43}$$

Here in the numerator on the right-hand side stands for the energy loss due to the structural transitions occurring during the risetime. As Eq. 6.42 shows, for materials with small c_m the structural transitions occur predominantly when $\Delta E^{ms} > 0$. On the other hand, for materials with large c_m when $\Delta E^{ms} < 0$, destructive processes would prevail.

If $D(t_R) = 0$ and $\Delta E^{ms} > 0$, the irreversible structure transition takes place at the condition $\Pi^m > 0$ when the destructive processes prevail. Then the value $\Delta U < 0$ results the irreversible wave amplitude loss that can be observed experimentally and is presented in Figs. 5.18–5.21. In the opposite case where $\Pi^m < 0$, the self-organization leads to the positive value $\Delta U > 0$. It is very interesting that such situations are also had been found out in experiments (See Fig. 5.24).

In the case of dynamic equilibrium where structural transitions are absent and $\Delta E^{ms} = 0$, the maximal mean velocity fluctuation is in the middle of the front. The maximal velocity deviations inside the front take place in the middle of each half. Then Eq. 6.42 yields the result

$$c_m D^{\max} \approx \frac{1}{2} \Delta u^{\max} . \tag{6.44}$$

This means that the level of mesoscopic fluctuation is compared to the level of the macroscopic velocity deviation. It is well known that, in equilibrium, the difference between the scale levels decreases and they can't be separated. In dynamic equilibrium, only the two scale levels participate and the condition 6.44 is necessary in order that the macro–meso energy exchange be reversible.

Mescheryakov [34] had experimentally obtained this condition as a criterion of the maximal spall strength (see Figs. 5.20 and 5.21). From the point of view of the present theory, this criterion corresponds to dynamic equilibrium macro–meso energy exchange followed by reversible fluctuative structure formation.

6.6.4. Velocity Dispersion at Dynamic Equilibrium

At dynamic equilibrium, macro–meso energy exchange Eq. 6.34 on the condition $\delta E^{ms} = 0$ defines only the velocity dispersion $D(x, t)$:

$$c_m \frac{\partial D}{\partial t} + \left(D - \frac{R}{a} \right) \frac{\partial u}{\partial x} = 0 . \tag{6.45}$$

Equation 6.45 with respect to the velocity fluctuation D integrates as a first-order ordinary differential equation in time to the expression

$$D(x, t) = D(x, 0) \exp\left[- \int_0^t \frac{1}{c_m} \frac{\partial u}{\partial x} dt' \right] + \int_0^t \exp\left[- \int_{t'}^t \frac{1}{c_m} \frac{\partial u}{\partial x} dt'' \right] \frac{R}{a} \frac{1}{c_m} \frac{\partial u}{\partial x} dt' . \tag{6.46}$$

It must be noted that, according to the homogeneous equation 6.45 where $R = 0$, the initial value of the mean velocity fluctuation $D(x, 0)$ (the same for the velocity dispersion) falls exponentially during the shock compression. Furthermore, the initial dispersion value is assumed to be zero: $D(x, 0) = 0$.

The rate of the dispersion generation depends strongly on the strain rate. In the case of dynamic equilibrium, the velocity dispersion is growing for the first half of the wave front, then the second term in Eq. 6.45 changes its sign and the velocity dispersion falls. If irreversible structural transitions occur inside the front, they can depress the velocity dispersion or, to the opposite, initiate its explosive increasing. The last situation is considered as a material failure.

By introducing a typical risetime

$$t_R = \left(\frac{\partial u}{\partial x}\right)^{-1}_{max},$$

and typical strain-rate t_R^{-1}, Eq. 6.46 inside the wave front can be rewritten approximately as follows:

$$D(x, t) \approx \int_0^t \exp\left[-\frac{t-t'}{c_m t_R}\right]\frac{R}{ac_m}\frac{dt'}{t_R}. \tag{6.47}$$

Equation 6.47 shows that the velocity fluctuations are generated by the shear stress connected to the changing of the waveform.

In the limiting case of large strain rates for a step-wise waveform with a risetime $t_R \to 0$, the integral kernel becomes of the Dirac type. In this case, the velocity dispersion is entirely determined by the structural relaxation R:

$$D(x, t)\xrightarrow[t_R \to 0]{} \int_0^t \delta(t-t')\frac{R}{a}dt' = \frac{R}{a}. \tag{6.48}$$

Herewith, in the case where the structural relaxation is already completed ($\varepsilon \to 0$ or $t_r \to 0$) the medium should be considered an ideal fluid with $R = 0$. The same occurs for the frozen relaxation ($\varepsilon \to \infty$ or $t_r \to \infty$). Then,

$$D(x, t)\xrightarrow[t_R \to 0,\, t_r \to 0]{} 0,$$
$$D(x, t)\xrightarrow[t_R \to 0,\, t_r \to \infty]{} 0. \tag{6.49}$$

These situations can occur for the large times $t \gg t_R \sim t_r$ and for times $t_R \ll t \ll t_r$, respectively.

In the opposite case of small strain rates, $t_R \to \infty$ and, in both the cases of completed structural relaxation ($\varepsilon \to 0$ or $t_r \to 0$) and frozen relaxation ($\varepsilon \to \infty$ or $t_r \to \infty$), no velocity dispersion is generated:

$$D(x, t) \to \int_0^t \frac{R}{ac_m}\frac{dt'}{t_R}\xrightarrow[t_R \to \infty]{} 0. \tag{6.50}$$

Such situations occur for small times $t \ll t_R \sim t_r$ and for times $t_R \gg t \gg t_r$, respectively.

Therefore, velocity dispersion originates only in the intermediate case for real waveforms characterized by finite risetimes $t_R \sim 1$ during structural relaxation at the mesoscopic scale level $(\varepsilon \sim 1)$ for the time $t \sim t_R \sim t_r$. Depending on the relation between the two typical times t and t_r, different situations can occur: the velocity dispersion for $t_R = t_r$ is presented in Fig. 6.2 and Fig. 5.14b, and for $t_R < t_r$ in Fig. 6.3 and Fig. 5.14a.

This means that the value D and the radius of non-local correlation ε are closely connected. The influence of the mesoscopic parameter c_m on the dispersion and velocity profiles becomes essential also in the intermediate case. For solids, the dispersion originates beyond the scope of the linear elasticity theory due to anharmonic vibrations of the atomic lattice. It means that solids under high-rate straining conserve their elastic properties at much larger stresses than during quasi-static loading. That is why the dynamic properties of materials can considerably exceed the quasi-static properties.

The high-rate form changing and the history of the relaxation (plastic effects) produce the velocity dispersion.

6.7. Non-Steady Waveforms in the Dynamical Equilibrium

6.7.1. Analysis of the Solution to the Problem

In dynamic equilibrium, the set of equations 6.22 and 6.34 describes plane shock wave propagation followed by energy exchange between macroscopic and mesoscopic scale levels and by reversible structure formation. The non-local formulation of the problem allows immediate connection between macroscopic field variables and characteristics of mesoscopic structure without introducing any kinetic interaction mechanisms at the mesoscopic scale level. Owing to the

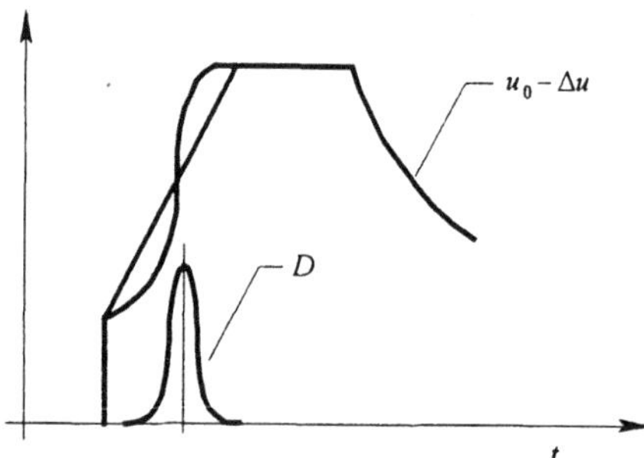

Figure 6.2. Shock wave velocity and mean velocity fluctuation profiles in the dynamic equilibrium case.

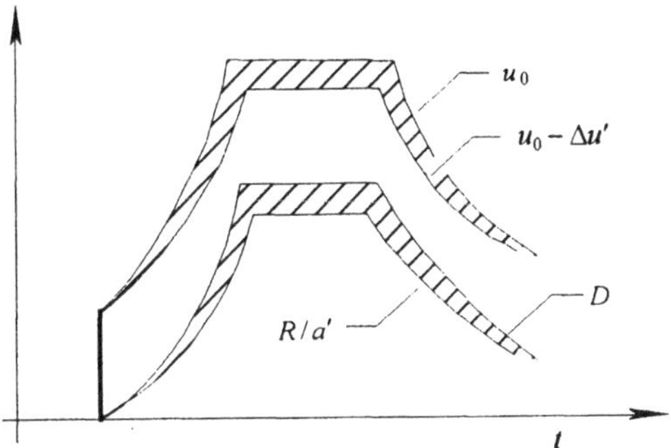

Figure 6.3. Shock wave velocity and mean velocity fluctuation profiles in the non-equilibrium case.

self-consistent non-local relaxation model, this formulation of the problem takes into account the reverse effect of mesostructural formation on the shock-wave characteristics.

In the high-strain-rate limit ($t_R \to 0$) expansion of the velocity dispersion in Eq. 6.47 yields

$$aD \to R + t_R \frac{\partial R}{\partial t}$$

$$R - aD \to t_R \frac{\partial R}{\partial t},$$

$$(6.51)$$

where the exponential terms are neglected.

In the case of the already relaxed mesoscopic structures at $t \gg t_R \sim t_r$, the viscous relaxation proportional to the strain rate is completely irreversible:

$$R \to v\frac{\partial u}{\partial x} = \frac{v}{t_R} > 0$$

$$R - aD \to t_R \frac{\partial}{\partial t}\left(\frac{v}{t_R}\right).$$

$$(6.52)$$

Here the wave amplitude loss at the pulse plateau, $\Delta U > 0$, $\Delta U \to 0$, becomes zero. The more general situation $t \sim t_R \sim t_r$ can also be referred to the dynamic equilibrium state if the relaxation has an opportunity to proceed to completion within the wave front. Otherwise a structural transition would be irreversible, and a state of dynamic equilibrium would not develop. The typical dynamic equilibrium situation is presented in Fig. 6.2.

6.7.2. Approximate Solution of the Problem of Non-Steady Wave Propagation with Non-Equilibrium Macro–Meso Energy Exchange

In the case of high strain rates for $t_R \ll t \ll t_r$, inside the wave front of amplitude U_0, the relaxation and velocity dispersion according to Eqs. 6.18–6.48 are Gaussian:

$$
\begin{aligned}
D &\approx \frac{R}{a} \approx \frac{2}{3}\frac{v(1+\alpha)}{a} \int_0^\infty \exp\left[-\frac{\pi}{\varepsilon^2}(x'-x-\sigma)^2\right] U_0\, \delta(x'-(a_l-\Delta a)t)\frac{dx'}{\varepsilon} \\
&= \frac{2v(1+\alpha)U_0}{3a\varepsilon}\exp\left[-\frac{\pi}{\varepsilon^2}((a_l-\Delta a)t - x - \sigma)^2\right].
\end{aligned}
\tag{6.53}
$$

In the reference frame connected to the front, the mean velocity fluctuation according to Eq. 6.53 depends on the parameter $\sigma(t)$ evolving in time:

$$
D = \frac{2v(1+\alpha)U_0}{3a\varepsilon}\exp\left[-\frac{\pi}{\varepsilon^2}\sigma^2\right].
\tag{6.54}
$$

In this case, an asymptotic expansion of the function D near the limit of Eq. 6.18 at $t_R \to 0$ results the difference $R/a - D$ being nonzero:

$$
R/a - D \to t_R \frac{\partial}{\partial t}\left\{\frac{2v(1+\alpha)U_0}{3a\varepsilon}\exp\left[\frac{\pi}{\varepsilon^2}\sigma^2\right]\right\}.
\tag{6.55}
$$

Unlike the limiting situation, the near-limiting relaxation is not completed within the risetime of the wave, and the macro–meso energy exchange decreases the wave amplitude by the amount $(R/a - D)(t_R)$. Just the same can be seen for the other near-limiting situation of low strain rates at $t_R \to \infty$.

Substitution $S^m - P^m$ from Eq. 6.55 into the solution 6.23–6.25 yields the front evolution in the near-limiting cases.

In the limiting case of complete structural relaxation, velocity dispersion appears only inside the Dirac-type disruption

$$
D = \frac{2v(1+\alpha)U_0}{3a\varepsilon}\exp\left[-\frac{\pi}{\varepsilon^2}\sigma^2\right] \xrightarrow[\varepsilon\to 0]{} \frac{2v(1+\alpha)U_0}{3a}\delta(\sigma).
\tag{6.56}
$$

Here one has a step-wise wave front moving at the relative velocity $\sigma(t)$. For the frozen relaxation, the mean velocity fluctuation presents an elastic wave of amplitude that can be considerably more than usual:

$$D \xrightarrow[\varepsilon \to \infty]{} \frac{4GU_0}{3a^2} \, .$$

(6.57)

In the case where $t \sim t_R \sim t_r$, the mean velocity fluctuation is generated over the whole time from the beginning. Its profile accumulates information concerning the entire history of all relaxation processes that have occurred inside the front during its propagation in the medium.

In general, non-equilibrium macro–meso energy exchange can result in a wide assemblage of different situations with respect to the behavior of velocity dispersion and wave amplitude loss. It has been calculated that the wave amplitude loss due only to energy transfer to the fluctuations for copper in the range of impact velocities 100–300 m/s is about 10% in the case where the velocity dispersion decreased to zero and about 40% in the case where it did not reach zero at the plateau of the compressive pulse. Therefore, the remaining part of the macroscopic energy loss is conditioned by the structural transitions. This case is presented on Fig. 6.3. The non-steady front evolution is given in Fig. 6.4.

6.8. General Scheme of the Problem

The whole problem consists of two parts: direct and inverse. The direct problem is calculating of the non-steady wave front evolution for the given dynamical structure characteristics. The inverse problem is determining the dynamical structure characteristics on the basis of incomplete data for the wave front evolution. First, the inverse problem should be solved. There are two integral relationships 6.25 6.46 connected with the experimentally measured values $\Delta u(t)$ and $D(t)$:

$$\Delta u(x, t) = a_1^{-1} \left[R\left(\frac{\partial u}{\partial x}, v, \varepsilon, \sigma \right) - aD\left(\frac{\partial u}{\partial x}, v, \varepsilon, \sigma, \delta E^{\mathrm{ms}}, c_{\mathrm{m}} \right) \right](x,t)$$

(6.58)

$$D(x, t) = \int_0^t \exp\left[-\int_{t'}^t \frac{1}{c_{\mathrm{m}}} \frac{\partial u}{\partial x} dt'' \right] \frac{R(v, \varepsilon, \sigma) - \delta E^{\mathrm{ms}}}{a} \frac{1}{c_{\mathrm{m}}} \frac{\partial u}{\partial x} dt' \, .$$

At a fixed point $x = l$ (l equals to the target's thickness), these relationships determine the unknown structural characteristics $v, \varepsilon, \gamma, \delta E^{\mathrm{ms}}, c_{\mathrm{m}}$ and the risetime t_R also can be included here if the strain rate, $\partial u / \partial x$, is given. Choosing the appropriate initial approximation for the velocity profile u_0, one can get an approximate value of the velocity gradient. But, for the determination of the structural characteristics, six relationships are needed. The deficit relationships can be obtained from three independent measurements for three different thicknesses or for three different impact velocities. Then the six relationships 6.58 at $x = l_i$ ($i=1,2,3$) determine the six spectra of the time-dependent functions $\{ v, \varepsilon, \sigma, \delta E^{\mathrm{ms}}, c_{\mathrm{m}}, t_R \}$.

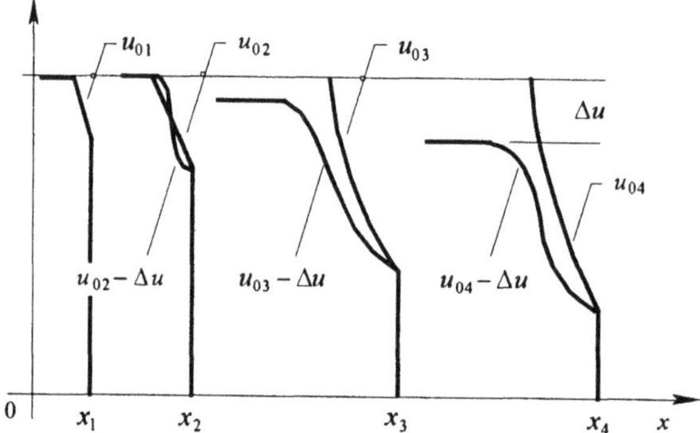

Figure 6.4. Nonstationary waveform evolution.

After that, the direct problem is solved for the given structural characteristics $v, \varepsilon, \gamma, \delta E^{\mathrm{ms}}, c_{\mathrm{m}}, t_{\mathrm{R}}$, and the waveform evolution during the whole wave propagation can be calculated:

$$\Delta u(x, t) = a_1^{-1}\left[R\left(\frac{\partial u}{\partial x}, v, \varepsilon, \sigma\right) - aD \right](x,t). \tag{6.59}$$

Equation 6.59 defines the velocity profile evolution for the known structural evolution. Both problems are referred to the special operator formulation of the type of Eq. 6.9 that presents a mathematical basis for solving such problems. An iteration procedure can be constructed on the basis of Eqs. 6.58 and 6.59. Under conditions when the procedure converges, the solution can be obtained with any needed precision.

6.9. Conclusions

1. The present theory shows that problems of high-rate loading of materials and non-steady shock wave propagation cannot be formulated correctly without introducing the dynamical formation of internal structural features at the mesoscopic scale level. Formation of mesoscopic structural elements is closely related to collective interaction effects and should induce non-local properties of the system as a whole. The mesoscopic effects following high-strain-rate loading of materials completely define the stress relaxation. The relaxation model of the medium must be valid through the range of conditions from those in which the material responds as an ideal elastic solid to those in which it responds as an ideal fluid. Any real medium should be considered as a medium with changing internal structure when subjected to high-rate loading. In addition, the model should incorporate a feedback mechanism. That is why the present theory is based on the non-local and self-consistent relaxation model of a medium, which, unlike the well-known Maxwell model, involves internal

structural parameters that are functionals of macroscopic fields. In limiting cases, some of these parameters tend toward constant values that are the usual elastic moduli or transport coefficients. Owing to its non-linear relaxation properties, the model determines a spectrum of spatial and temporal scales corresponding to a hierarchy of structural relaxations. The bifurcation points determine the structural transitions.

2. It has been determined experimentally that energy dissipation into heat begins only after macro–meso energy exchange has taken place. The experimentally measured characteristics of the macro–meso energy exchange involved in the theory are the velocity dispersion and the wave amplitude loss at the plateau of a compressive pulse. The high-rate form changing generates fluctuations at the mesoscopic scale level that induce increasing velocity dispersion. If the velocity dispersion returns to its initial value for the risetime, no structural transitions occur. In the case where the relaxation time exceeds the rise time of the wave, the velocity dispersion does not return to its initial value and a wave amplitude loss occurs at the plateau of the compressive pulse. A sharp change of the velocity front and dispersion profile is evidence of a possible structural transition.

3. Dynamic properties of materials are defined by the thermodynamics of macro–meso energy exchange. For materials with small energy capacity of mesoscopic fluctuations c_m, a part of the macroscopic kinetic energy of a shock wave goes into new degrees of freedom during structural transitions. Owing to these transitions, the entropy production decreases. For materials with large energy capacity, the energy deficit is compensated by the potential energy of the lattice. The large-scale structures that arise lead to material destruction, which induces growth of entropy production. If the macroscopic energy transfer to the mesoscopic scale level goes entirely into the mesoscopic fluctuations, dynamic equilibrium is reached. This state of macro–meso energy exchange is characterized by a minimum in the rate of the entropy production. The difference between the non-equilibrium steady state characterized by a minimum of the entropy production and the dynamic equilibrium with the minimal rate of the entropy production must be emphasized. It is very important that the experimental curve of spall strength has a maximum at this state of dynamic equilibrium. The criterion of the maximal spall strength of materials under high-rate loading obtained experimentally by Mescheryakov [38] is formulated as the scale equality 6.44.

4. An algorithm (split-type technique) for calculation of non-steady wave front evolution in dynamically deformed media is constructed with account being taken of the dynamics of structures at the mesoscopic scale level. The whole problem is split into two parts: the direct problem is a

calculation of the wave front evolution for the known mesoscopic structural dynamics and the inverse problem consists in the reconstruction of the mesoscopic structure dynamics via some independently measured velocity and dispersion profiles. The inverse problem should be solved first, after which the direct problem is solved. Approximate analytical solutions are obtained for one step of this procedure.

References

[1] S. De Groot and P. Mazur, *Nonequilibrium thermodynamics*, North-Holland, Amsterdam (1963).

[2] T.K. Shapovalov, "To the initial conditions for equations of rarefied gas hydrodynamics," *Aeromekhanika,* pp. 304–306, (1976) (in Russian).

[3] W. Garen, R. Synifzik, and G. Wertberg, "Experimental investigation of weak shock waves in noble gases," in: *Rarefied gas dynamics, 10th Int. Symp,* (ed. G. Potter), pp. 519–528 (1978).

[4] M.N. Kogan, *Rarefied gas dynamics.* Nauka, Moscow, (1967) (in Russian).

[5] L.C. Woods, "Transport processes in dilute gases over the whole range of Knudsen numbers. Part 1. General theory." *J. Fluid Mech.* **93**, pp. 585–607 (1979).

[6] C.R. Doering, M.A. Burshka, W. Horsthenike, "Fluctuations and correlations in a diffusion–reaction system: Exact hydrodynamics," *J. Stat. Phys.* **65**, pp. 953–970 (1991).

[7] D.N. Zubarev and S.V. Ticshenko, "Nonlocal hydrodynamics with memory," *Physica* **59**, pp. 285–304 (1972).

[8] B.V. Filippov and T.A. Khantuleva, *Boundary problems of nonlocal hydrodynamics*, Leningrad, Leningrad State Univ. (1984) (in Russian).

[9] M. Bixon, J.R. Dorfman, and K.C. Mot, "General hydrodynamic equations from the linear Boltzmann equation," *Phys. Fluids* **14**, pp. 1049–1057 (1971).

[10] N.N. Bogolyubov, *Problems of dynamic theory in statistical physics*, Gostekhizdat, Moscow (1946). (in Russian).

[11] L.P. Kadanoff and P.C. Martin, "Hydrodynamic equations and correlation functions," *Ann. Phys.* **24**, pp. 419–460 (1963).

[12] C.H. Chung and S. Yip, "Generalized hydrodynamics and time correlation functions," *Phys. Rev.* **182**, pp. 323–338 (1965).

[13] N. Ailavadi, A. Rahman, and R. Zwanzig, "Generalized hydrodynamics and analysis of current correlation functions," *Phys. Rev.* **4a**, pp. 1616–1625 (1971).

[14] J.M. Richardson, The hydrodynamical equations of a one-component system derived from nonequilibrium statistical mechanics. *J. Math. Anal. and Appl.* **1**, pp. 12–60 (1960).

[15] R. Piccirelli, "Theory of the dynamics of simple fluid for large spatial gradients and long memory," *Phys. Rev.* **175**, pp. 77–98 (1968).

[16] K. Kawasaki and J.D. Ganton, "Theory of nonlinear transport processes: nonlinear shear viscosity and normal stress effects" *Phys. Rev. A* **8**, pp. 2048–2064 (1973).

[17] H. Mori, "Transport, collective motion and brownian motion," *Progr. Theor. Phys.* **33**, pp. 423–454 (1965).

[18] D.N. Zubarev, "Statistical operator for nonequilibrium systems" *Doklady Akad. Nauk SSSR* **140**, pp. 92–95 (1961) (in Russian).

[19] D.N. Zubarev, "Modern methods of statistical theory of irreversible processes," *Itogi nauki i tekhniki. Ser. Sovremennye problemy matematiki* **15**, VINITI, Moscow, pp. 128–227 (1980) (in Russian).

[20] V.Ya. Rudyak, *Statistical theory of dissipative processes in gases and liquids*, Nauka, Novosibirsk (1987) (in Russian).

[21] T.A. Khantuleva, "Modern hydrodynamical problems on the basis of nonlocal hydrodynamical equations," *Modely mekhaniki sploshnoj sredy*, Vladivostok-Novosibirsk, pp.158–173 (1991) (in Russian).

[22] A.G. Vershinin and T.A. Khantuleva, "To a nonlocal description of flows with shock waves," *Mekhanika reagiruyuschikh sred i eye prilojeniya*, Nauka, Novosibirsk, pp. 89–96 (1989). (in Russian)

[23] A.G. Vershinin and T.A. Khantuleva, "Nonlocal hydrodynamical model of the shock wave front," *Fizicheskaya mekhanika* **6** , Leningrad State Univ., Leningrad, pp. 21–31 (1990) (in Russian).

[24] S.A. Vavilov, "Geometric methods of studying the solvability of a class of operator equations" *Russian Acad. Sci. Dokl. Math.* **45**, pp. 276–280 (1992).

[25] S.A. Vavilov, "A method of studying the existence of nontrivial solutions to some classes of operator equations with an application to resonance problems in mechanics," *Nonlinear Analysis* **24**, pp. 747–764 (1995).

[26] T.A. Khantuleva and Yu.I. Mescheryakov, "Nonlocal theory of the high-strain-rate processes in structured media," *Int. J. Solids and Structures* **36**, pp. 3105–3129 (1999).

[27] T.A. Khantuleva and Yu.I. Mescheryakov, "Kinetics and non-local hydrodynamics of mesostructure formation in dynamically deformed media," *Phys. Mesomechanics* **2**, pp 5–17 (1999).

[28] T.A. Khantuleva, "Non-local theory of high-rate processes in structured media," in: *Shock Compression in Condensed Matter—1999* (ed. M.P. Furnish, L.C. Chhabildas, and R.S. Hixson), American Institute of Physics, New York, pp. 371–174 (2000).

[29] T.A. Khantuleva, "Microstructure formation in the framework of the non-local theory of interfaces," *Mater. Phys. Mech.* **2**, pp 51–62 (2000).

[30] Yu. V. Sud'enkov, "Relaxation of the elastic constants in Al near the loading surface," *J. Tech. Phys. Letters.* 9, pp. 1418–1422 (1983) (in Russian).

[31] O.D Baizakov and Yu. V. Sud'enkov, "Relaxation phenomena in materials near the surface of the elastic submicrosecond loading," *J. Tech. Phys. Letters.* **11**, pp. 1433–1437 (1985) (in Russian).

[32] Yu. V. Sud'enkov, "Special features of the shoch wave propagation in solids near the surface of high-rate loading" In: *Problems of dynamical processes in heterogeneous media* ed. Kalinin Univ., pp. 120–126 (1987) (in Russian).

[33] Yu.I. Mescheryakov and A.K. Divakov, "Multi-scale kinetics of microstructure and strain-rate dependnence of materials," *DYMAT Journal* **1**, p. 271 (1994).

[34] Yu.I. Mescheryakov, A.K. Divakov, and N.I. Zhigacheva, "Role of mesostructure effects in dynamic plasticity and strength in ductile steels," *Mater. Phys. Mech.* **3**, pp. 63–100 (2001).

[35] Yu.I. Mescheryakov, "Mesoscopic effects and particle velocity distribution in shock compressed solids," in: *Shock Compression in Condensed Matter—1999* (ed. M.P. Furnish, L.C. Chhabildas, and R.S. Hixson), American Institute of Physics, New York, pp 1065–1070 (1999) .

[36] *Physical mesomechanics and computer construction of materials* (ed. V.A. Panin), Nauka, Novosibirsk (1995).

[37] A.I. Olimskoj and V.A. Petrunin, *Izvestiya Vuzov "Fizika"* **2**, pp. 82–117 (1987).

[38] Yu.I. Mescheryakov, *this volume,* Chapter 5.

CHAPTER 7

Non-Equilibrium Evolution of Collective Microdamage and Its Coupling with Mesoscopic Heterogeneities and Stress Fluctuations

Y.L. Bai, M.F. Xia, Y.J. Wei, and F.J. Ke

7.1. Introduction

In proposing a workshop to discuss "Shock Dynamics and Non-Equililbrium Mesoscopic Fluctuations in Solids", it was pointed out that: "The existence of mesoscale inhomogeneities and stress fluctuations has certainly been recognized by experimentalists and theoretical analysts. However, the issue of heterogeneous and non-equilibrium shock front dynamics on the mesoscale, has largely been ignored, in spite of the fact that these must strongly influence the phenomena such as fracture and phase transitions." The following specific questions were posed: (1) "What experimental data are available and what are their implications?" (2) "Are there new mesoscale theories for shock dynamics?" (3) "How do the theories affect the existing fracture and phase transition paradigms?" and (4) "What kinds of new computational and materials models are needed? " [1]

This statement reminds us of some earlier appeals. For instance, McDowell [2] wrote "rigorous treatment of non-uniformly distributed defects requires tools not yet fully developed in continuum damage mechanics. Weighing the influence of distributed damage at the microscale on the collective macroscale stiffness and evolution of damage is a challenge."

In engineering practice, distributed microdamage as a kind of mesoscopic inhomogeneity often plays a critical role in failure analysis. Time-dependent spallation [3–5], failure waves [6,7], and deflagration to detonation transition in propellants and explosives [8,9] are notable examples of nonequilibrium evolution of mesoscopic damage. More broadly, an emerging science of microstructural engineering is being developed to address the question of why, although a piece of blackboard chalk and a clam shell are chemically almost identical, the chalk will snap far more easily [10]. Based on the knowledge of microstructure and its evolution, this science is intended for use in design and in developing new advanced materials.

We would like to trace briefly the development of relevant theories and to evaluate the state of the art. Perhaps Budiansky's review papers [11] and Mura's book on micromechanics [12] are representative of early works dealing with microstructures. At that stage, the paradigm was to apply traditional continuum mechanics to the analysis of typical microstructural processes such as void collapse. Barenblatt [13] later argued whether this legitimizes micromechanics as a new branch of science. He proposed that, to determine the governing influence of the variations of the material microstructure on the macroscopic behavior of bodies, the macroscopic equations of mechanics and the kinetic equations of microstructural transformations form a unified set that should be solved simultaneously. Departing from common continuum mechanics, he suggested approaches like scaling methods (renormalization group theory) and damage accumulation. Additionally, he noted the significance of the Deborah number, the ratio of the characteristic time of a mesoscopic process to the imposed macroscopic time scale, in the new branch of mechanics. Another unified formulation, recently proposed by Panin and his co-workers, is physical mesomechanics. In their theory, there are two mesoscopic levels. On level I, vortex plastic flow is characterized by the scheme "shear + rotation"; whereas on level II new defects appear irrespective of crystallographic orientations [14]. Parallel to these, some encouraging approaches were formulated in China [15–17], in which either intrinsic mesoscopic rate process or statistical features of heterogeneous media, or both, were emphasized.

In summary, it seems that the tools of continuum mechanics are insufficient to deal with the evolution of mesoscopic inhomogeneities. This leaves the question: What is the exact difficulty in establishing new effective formulations? Principally, we are facing a problem across three levels: the microscopic, mesoscopic, and macroscopic levels. Moreover, we are usually concerned with trans-scaled phenomena.

Therefore, the challenge we meet is to look for a trans-scale formulation (taking rate effects into account) that reveals the effects of mesoscopic inhomogeneities on macroscopic behaviors of engineering significance. In the course of this work, we encounter two major obstacles. One is how to properly close the formulation, because this kind of trans-scale formulation usually leads to an endless hierarchy from macroscales to microscales, as with BBGKY [18]. The other obstacle is that the averaging we are used to in continuum mechanics is still a necessity but it usually obscures the effects of mesoscopic details that may be magnified and become critical on the macroscale. These, we suppose, are the real challenges we face.

We are greatly interested in discussion of this emerging and immature topic. To meet these challenges, this chapter reviews some of our thinking on the subject and approaches to the problems. In particular, we review the framework of statistical microdamage mechanics and stress redistribution (SRD) models with

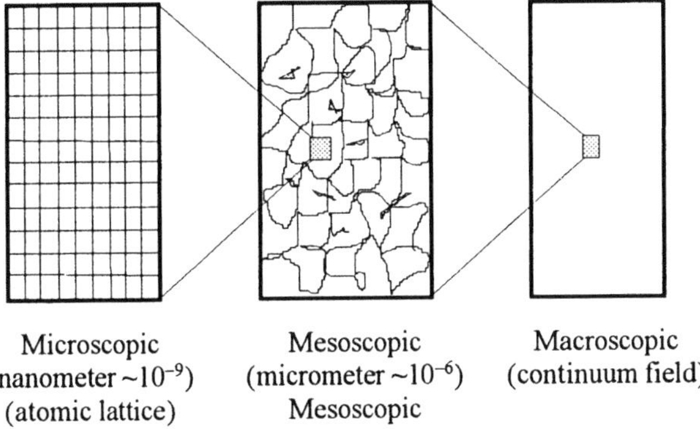

Figure 7.1. Schematic illustration of length scales [19].

stress fluctuations, as well as their application to wave-induced failure and some distinctive features of catastrophe transition.

7.2. Two Approaches to Non-equilibrium Evolution in Heterogeneous Media and Stress Redistribution Models

Real media are usually highly heterogeneous. However, in classical continuum mechanics, collective effects of heterogeneity are usually incorporated into constitutive relations as macroscopic averages. At most, some internal variables and their macroscopic empirical evolution laws are introduced as a complement to the constitutive relations. Obviously, this is insufficient to capture the complexity of behaviors of heterogeneous media. In these situations, a comprehensive theoretical framework is needed to describe the formation and evolution of microstructures at appropriate length and time scales and to establish a connection with the corresponding macroscopic properties (see Fig. 7.1) [19].

On the other hand, for dislocations on the atomistic scale, one could resort to quantum mechanics. However, for tangled dislocations and mesoscopic defects and damage, the descriptions in present-day theories are hardly complete. Large-scale computer modeling can provide insight into the behavior of solids as data are averaged at different scale levels; however, the conclusions are case-specific [20].

Above all, for the cases with mesoscale inhomogeneities and stress fluctuations, some new approaches are badly needed. These new approaches should be neither too delicate on the atomic level nor too coarse on the continuum level. In other words, the approaches must still rely on some averaging on mesoscale levels but they should be able to capture the significant effects of some sensitive details on lower levels and to reveal some mesoscopic links.

One additional important issue is that most behaviors of such media are governed by mesoscopically progressive processes evolving under external stresses.

Therefore, stress redistribution plays a key role in the non-equilibrium evolution. Updating the stress field according to the progression of distributed damage depending on inhomogeneities is the core of these approaches.

We proposed two stress redistribution (SRD) approaches to the problem with non-equilibrium fluctuations in mesoscopically heterogeneous media. One is statistical microdamage mechanics, the other involves SRD models with stress fluctuations.

The approach of statistical microdamage mechanics is based on a macroscopically local mean field approximation. Variables are all averages over a macroscopically small volume including a number of grains, microcracks, and microvoids (see Fig. 7.1). The key dependent variable in the approach is the number density of microdamage elements in phase space, rather than continuum variables. The theory is analogous to the Boltzmann integrodifferential equation of the non-equilibrium theory of dynamical systems, but the statistical representation is the number density of microdamage elements, rather than density of probability distribution of particles. The reason for this modification is that the total microdamage does not remain constant, as does the total number of particles in the case that Boltzmann discussed. This is because the aim we pursue is to determine damage evolution. Moreover, the number density of microdamage elements can provide much more information than continuum damage and provides the link between mesoscopic dynamics and continuum variables.

Because the foregoing approach still obscures effects of some sensitive inhomogeneities, SRD models with fluctuations were proposed. To some extent, statistical microdamage mechanics is still a macroscopic deterministic theory. In order to understand sample-specificity due to non-equilibrium mesoscopic fluctuations, one has to examine a number of macroscopically identical samples to obtain a statistical description and prediction of material response. To minimize the computational complexity due to the huge ensemble of samples with mesoscopic inhomogeneities, stress redistribution mapping must be limited to some simple rules instead of realistic constitutive relations and continuum or quantum equations. Nevertheless, these models still should extract the essence of phenomena and provide fundamental and universal features resulting from sensitive mesoscopic inhomogeneities. Here we considered local stress concentration as well as non-local coupling models. By means of these models, some significant features, like evolution induced catastrophe, sample-specificity, trans-scale sensitivity, and critical sensitivity are revealed.

For the future, we are looking toward combining the above two approaches, in order to predict the outcome of realistic problems in heterogeneous media without taking the risk of missing crucial effects of sensitive mesoscopic inhomogeneities.

7.3. Statistical Microdamage Mechanics

For heterogeneous materials, distributed microcracks or microvoids are produced under impact loading. Therefore, the evolution laws of damage and critical damage to failure should be based on the non-equilibrium essence of collective microdamage evolution. There are two major aspects worthy of special attention. The first is that there are at least two scales: macroscale and the mesoscale, and different mesoscale rate processes, like nucleation and growth, especially under strong dynamic loading. The second is the strategy for closing the trans-scale formulation used because of the coupling of mesoscopic and macroscopic processes contributing to the dynamics of damage.

Roughly speaking, microdamage is formed at mesoscale inhomogeneities. For example, particulates in alloys can become the sites of microdamage. The nucleation of microdamage usually has the size of particulates or grains, on the order of micrometers, and the total number of such microdamage elements on the surface of metals is in the range of $10^2 - 10^4 / \text{mm}^2$ [3–5]. The microdamage may grow and coalesce leading to eventual failure. Hence, the main issues in microdamage evolution are nucleation, growth, and coalescence of microdamage elements [3–5]. This means that a dual scale (from mesoscopic to macroscopic) understanding of damage evolution is badly needed.

In this section, we review the basic concepts and framework of statistical microdamage mechanics. As an example, its application to wave motion and spallation under one-dimensional impact loading is reviewed.

7.3.1. Basic Equations and the Dynamic Function of Damage on the Continuum Level

In the light of the Boltzmann equation of the non-equilibrium theory of dynamical systems, the fundamental equation is the evolution equation of microdamage number density, n, in phase space $\{p_i\}$ (see Fig. 7.2),

$$\frac{\partial n}{\partial t} + \sum_{i=1}^{I} \frac{\partial (n \cdot P_i)}{\partial p_i} = n_N \quad , \tag{7.1}$$

where t is generalized time [21]. The quantities $P_i = \dot{p}_i$ are the rates of the variables p_i, and n_N is the nucleation rate, respectively. For example, the phase space of microdamage can be $\{c, \mathbf{x}\}$, which consists of the current size c of a microdamage element and the macroscopic position, $\mathbf{x}$, of a representative volume where microdamage is present. For instance, $n(c, \mathbf{x}) = 10/(\mu\text{m} \cdot \text{mm}^3)$ means that a volume of $1\ \text{mm}^3$ at the macroscopic point $\mathbf{x}$ contains 10 microdamage elements with sizes ranging from c to $c + 1\ \mu\text{m}$.

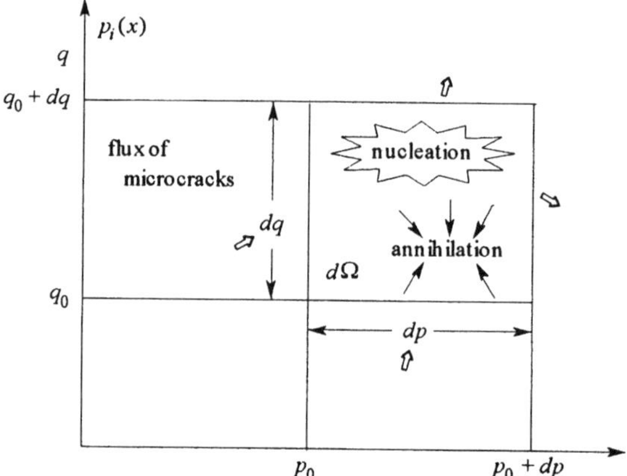

Figure 7.2. Schematic of microdamage in phase space, showing the derivation of the fundamental evolution equation of microdamage number density n.

According to the definition of microdamage number density, we can define continuum damage D,

$$D(t,\mathbf{x}) = \int n(t,\mathbf{x},c)\cdot\tau\cdot dc , \tag{7.2}$$

where τ is the failure volume of an individual microdamage element with size c. For example, $\tau \sim \pi c^3/6$ for a spherical microdamage element. For the number density of microdamage elements, $n = n(t,\mathbf{x},c)$, in the definition of Eq. 7.2, Eq. 7.1 takes the following simple form

$$\frac{\partial n}{\partial t} + \frac{\partial(n\cdot A)}{\partial c} + \frac{\partial(n\cdot\mathbf{v})}{\partial\mathbf{x}} = n_N , \tag{7.3}$$

where A is the growth rate of microdamage elements with current size c and $\mathbf{v}$ is the velocity of the macroscopic representative volume.

Multiplying Eq. 7.3 by the microdamage failure volume τ and integrating the resulting equation with respect to microdamage size, c, from zero to infinity, yields the continuum damage field equation [22]

$$\frac{\partial D}{\partial t} + \frac{\partial(D\cdot\mathbf{v})}{\partial\mathbf{x}} = f , \tag{7.4}$$

where

$$f = \int_0^\infty n_N(c;\sigma)\cdot\tau\cdot dc + \int_0^\infty n(t,\mathbf{x},c)\cdot A(c,\sigma)\cdot\tau'\cdot dc , \tag{7.5}$$

with $\tau' = d\tau/dc$. In the derivation of Eq. 7.5, it can be verified that other terms arising in the manipulation from Eq. 7.3 to Eq. 7.4 are negligibly small when $c \to \infty$, for the usual size distribution of microdamage nucleation, i.e., the size distribution density of microdamage nucleation tends to zero when $c \to \infty$. The function f is the dynamic function of damage (DFD), which is an agent bridging microdamage and continuum damage. Equation 7.4 can be combined with traditional continuum equations, such as the continuity, momentum, and energy equations, to establish a complete formulation. To illustrate the framework, we examine the one-dimensional strain state in Lagrangian coordinates (T, Y) [22]. By a one-dimensional strain state we mean one in which all displacement and velocity components and spatial derivatives are zero, except the displacement component u and the particle velocity $v = du/dt$ in the y direction, and the derivative operator $\partial()/\partial y$. Hence, the only nonzero strain component is $\varepsilon = du/dY$. Also, the transformation from Eulerian (t, y) to Lagrangian (T, Y) coordinates is $t = T$ and $y = Y + u$, or

$$\frac{\partial}{\partial t} + v \cdot \frac{\partial}{\partial y} = \frac{\partial}{\partial T} \quad \text{and} \quad \frac{\partial}{\partial y} = \frac{\rho}{\rho_0} \cdot \frac{\partial}{\partial Y},$$

where ρ and ρ_0 are current and initial densities, respectively, and

$$\frac{\rho}{\rho_0} = \frac{\dfrac{\partial u}{\partial y}}{\dfrac{\partial u}{\partial Y}} = \frac{1}{1+\varepsilon}.$$

For simplicity, the energy equation is ignored. In this case, the system of evolution equations for the deformation and damage fields is

$$\frac{\partial D}{\partial T} + D \frac{\rho}{\rho_0} \frac{\partial v}{\partial Y} = f \tag{7.6}$$

$$\frac{\partial \rho}{\partial T} + \frac{\rho^2}{\rho_0} \frac{\partial v}{\partial Y} = 0 \tag{7.7}$$

$$\frac{\partial v}{\partial T} - \frac{1}{\rho_0} \frac{\partial \sigma}{\partial Y} = 0. \tag{7.8}$$

Clearly, this is a completely combined system of continuum equations and mesoscopic kinetics of microdamage, conforming to Barenblatt's suggestion [13]. Principally, as soon as the DFD is known, the problem of damage evolution is solvable. For instance, the DFD can be empirically expressed as

$$f = f(D, \sigma) = G(\sigma) \cdot \gamma \cdot (1 + \beta \cdot D^\mu), \tag{7.9}$$

where G is a function of stress and γ, β, and μ are material parameters. Compared to the definition of the DFD as a sum of nucleation and growth rates as in

Eq. 7.5, γ should be a parameter denoting the damage nucleation rate. More important, the dimensionless number β, which is proportional to the ratio of two rates, the growth rate over the nucleation rate, is much more significant in the transition to failure. We shall see its significance later in the discussion of damage localization. It should be noted that Davison and Stevens have named the two terms in Eq. 7.9 as simple and compound damages corresponding to nucleation and growth (as well as coalescence) of microdamage elements, respectively, and pointed out the significance of compound damage accumulation [4].

7.3.2. A Closed Trans-Scale Formulation and Intrinsic Deborah Number

We mentioned before that the core of the problem is how to close the formulation. For the system of equations 7.6–7.8, it is how to close the DFD formulation of Eq. 7.5. In fact, the difficulty in DFD is that it is inversely dependent on an unknown mesoscopic variable: the number density n in Eq. 7.5. The foregoing empirical DFD, Eq. 7.9, is the simplest closure on the continuum level for engineering practice, but without a clear link to mesoscopic kinetics. Therefore, for physical understanding instead of the empirical closure, an alternative closed trans-scale approximation to Eq. 7.5 is proposed.

Provided nucleation and growth rates are both time-independent kinetics $n_N(c_0; \sigma)$ and $V(c, c_0; \sigma)$, where c_0 is nucleation size of microdamage, instead of number density n in phase space $\{c\}$, we should examine number density of microdamage elements n_0 in phase space $\{c, c_0\}$. Therefore, we should examine another version of Eq. 7.1:

$$\frac{\partial n_o}{\partial t} + \frac{\partial (n_0 V)}{\partial c} = n_N(c) \cdot \delta(c - c_0), \tag{7.10}$$

where $\delta(c - c_0)$ is the Dirac δ-function having the same dimension as the reciprocal of its argument, i.e., the reciprocal of microdamage element size c. Note that we have taken Eq. 7.10 to be independent of the macroscopic spatial coordinates, $\mathbf{x}$.

For the solution to Eq. 7.10 and the corresponding calculations to obtain the closed DFD, we skip tedious manipulations and just write down some key points as follows (the details are given in [23,24]).

Equation 7.10, a first-order quasi-linear partial differential equation, has the characteristic ordinary differential equations

$$\frac{d(n_0 V)}{dc} = n_N \delta(c - c_0) \tag{7.11}$$

$$\frac{dt}{dc} = \frac{1}{V}. \tag{7.12}$$

Then, regarding the stress σ as a prescribed parameter, the solution to Eq. 7.10 is

$$n_0(t, c, c_0; \sigma) = \begin{cases} \dfrac{n_N(c_0; \sigma)}{V(c, c_0; \sigma)}, & \text{when } c < c_0 < c_f \\ 0, & \text{otherwise}, \end{cases} \tag{7.13}$$

with a moving front c_f:

$$t = \int_{c_0}^{c_f} \frac{dc}{V(c, c_0; \sigma)}, \tag{7.14}$$

where c_f is the moving microdamage front. The implication of the front $c_f = c_f(t, c_0; \sigma)$ is that the microdamage nucleated at time $t = 0$ with the initial size c_0 will extend to the current size c_f at time t (see Fig. 7.3).

By definition, the relation between number densities of microdamage elements n and n_0 is the integral of n_0 with respect to initial size c_0:

$$n(t, c; \sigma) = \int_0^c n_0(t, c, c_0; \sigma)\, dc_0. \tag{7.15}$$

Substitution of the solution of Eq. 7.13 into Eq. 7.15 gives an expression for the number density of microdamage elements, n, in terms of the kinetics of nucleation and growth of microdamage, n_N and V. Similarly, the growth rate, A, of microdamage elements can be expressed as

$$A(t, c; \sigma) = \frac{1}{n} \int_0^c V\, n_0\, dc_0. \tag{7.16}$$

Substitution of Eq. 7.15 for the number density n and Eq. 7.16 for the growth rate A into Eq. 7.5 leads to the closed DFD. To facilitate integration, the region

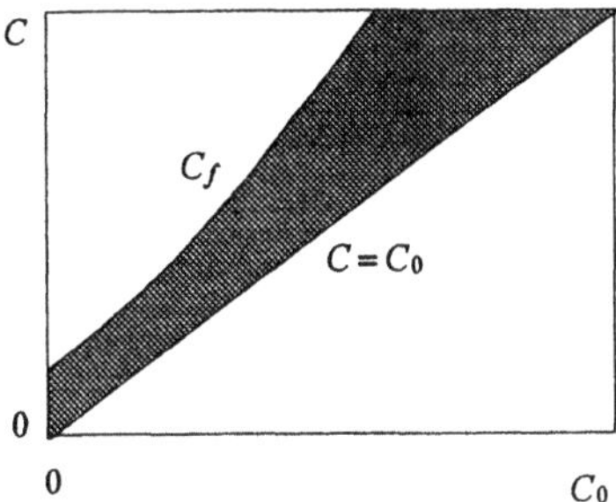

Figure 7.3. Schematic of the solution region in phase space $\{c, c_0\}$, where the non-zero solution of number density of microdamage n_0 exists (the shaded area). $c_f(t)$ is the time-dependent front of microdamage evolution moving upward.

of the resulting double integral with nonzero number density n_0 in Eqs. 7.13 and 7.14 is illustrated by the shaded area in Fig. 7.3. After exchange of the integrating order of the double integral, a closed DFD is expressed directly by two mesoscopic kinetic laws for the nucleation and growth rates of microdamage,

$$f = \left\{ 1 + \frac{\displaystyle\int_0^\infty \left[n_N(c_0) \int_{c_0}^{c_f} \tau'(c) \cdot dc \} \cdot dc_0 \right] dc_0}{\displaystyle\int_0^\infty n_N(c;\sigma) \cdot \tau(c) \cdot dc} \right\} \int_0^\infty n_N(c;\sigma) \cdot \tau(c) \cdot dc, \quad (7.17)$$

along with the microdamage front c_f, given by Eq. 7.14. In this way, we close the formulation of the DFD on the continuum level and directly combine it with the physical laws of microdamage on mesoscopical level by Eqs. 7.14 and 7.17. However, we have to confess that although continuum damage $D(t, \mathbf{x})$ is assumed to be a variable depending on macroscopic spatial coordinates $\mathbf{x}$ in Eqs. 7.6–7.8, the closed form of the DFD given by Eq. 7.17 is based on the approximate solution of number density of microdamage elements n, given by Eqs. 7.13 and 7.15, namely, independent of macroscopic spatial coordinates $\mathbf{x}$. This is in accord with the assumption of locality of constitutive theory. One more issue is that σ in Eqs. 7.6–7.8 is nominal stress, whereas the stress in Eqs. 7.13–7.14 and 7.17 was assumed to be the stress in the intact matrix. So, the closed approximation is valid for damage $D \ll 1$.

In spite of the complex integration in the DFD of Eq. 7.17, we can indicate that the second term in the braces should be characterized by a dimensionless number $V^*/[n_N^* \cdot (c^*)^5]$, where the superscript * denotes corresponding characteristic quantities. As a matter of fact, its inverse, $D^* = [n_N^* \cdot (c^*)^5]/V^*$, characterizes the damage rate ratio of two intrinsic processes: nucleation over growth. The effect of growth is stronger for lower values of this ratio. Although the significance of the Deborah number, a ratio of the intrinsic relaxation time over the imposed one, was proposed in micromechanics [13], notably, another intrinsic Deborah number D^* arises in analysis of damage evolution. The intrinsic Deborah number originates in multi-time-scale intrinsic processes, like nucleation and growth of microdamage in the present case. In this sense, the introduction of intrinsic Deborah numbers is very natural and universal in non-equilibrium processes in heterogeneous media. We will see later that this intrinsic number is truly the decisive factor in failure prediction. Incidentally, dimensionless numbers β in the empirical expression, Eq. 7.9, or D^* derived from mesoscopic kinetics of microdamage growth and nucleation given by Eq. 7.17 are both representations of the intrinsic Deborah number. Additionally, the second term of the trans-scale formulation of the DFD of Eq. 7.17 could be approximately fitted to a power law of damage D in some cases. This may greatly simplify the calculation of damage evolution.

7.3.3. Damage Localization

"Now everybody loves a localization problem!" [26]. "Localization of plastic deformation into shear bands in ductile metals is, indeed, a fascinating problem," Barenblatt said at the 18th International Conference on Theoretical and Applied Mechanics. But, he continued, "in principle, this problem is close to the problem of damage accumulation" [13]. In fact, failure of materials has been well known as a highly nonlinear process of dimension reduction, as when a roughly two-dimensional fracture surface forms in a three-dimensional body with accumulated damage. In the light of the damage field concept, the criterion

$$\frac{\partial\left(\dfrac{\partial D}{\partial Y}\right)}{\partial T} \Bigg/ \left(\frac{\partial D}{\partial Y}\right) \geq \frac{\left(\dfrac{\partial D}{\partial Y}\right)}{D} \tag{7.18}$$

predicts damage localization. This means the relative rate of the damage gradient becomes greater than the relative rate of the damage itself, as shown in Fig. 7.4. In terms of the damage field equations 7.6–7.8 and the approximate form of the DFD, $f = f(D, \sigma)$, an approximate criterion for damage localization,

$$f_D \geq \frac{f}{D} , \tag{7.19}$$

can be derived under the quasi-static assumption [22]. This is no longer a geometric description of damage localization as is Eq. 7.18, but is a critical dynamic condition, governed by the intrinsic nature of materials with damage accumulation, i.e., the DFD, f. To our surprise, the criterion of Eq. 7.19 looks like the well-known Chapman–Jouguet condition for a detonation front [22,27]. The condition is clearly interpreted in Fig. 7.5.

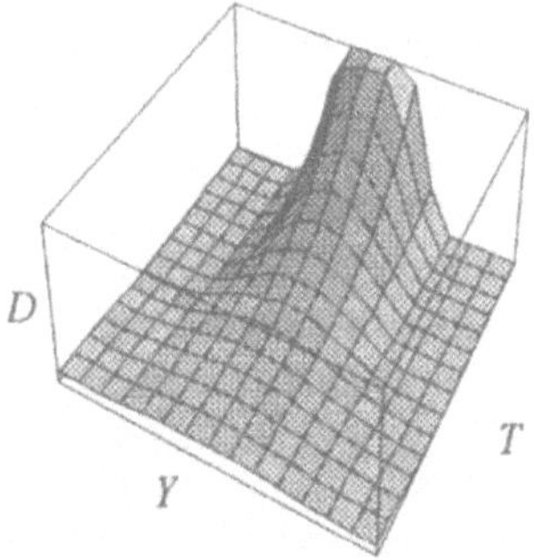

Figure 7.4. Pattern evolution of damage localization.

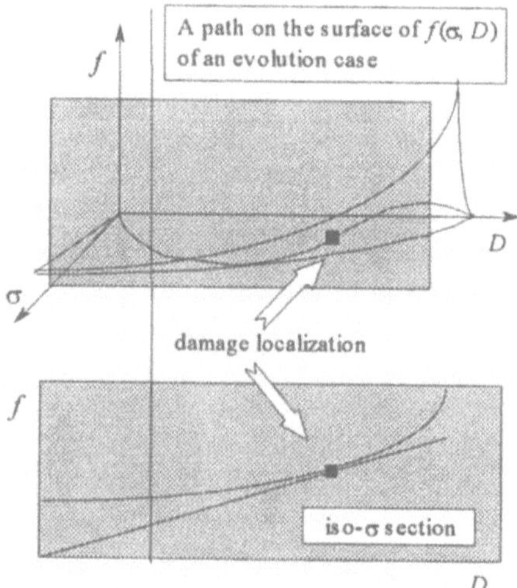

Figure 7.5. Interpretation of the criterion for damage localization $f_D > f/D$.

When the slope, f_D, of the tangent of the DFD on an iso-stress section in co-ordinates (f, D, σ) becomes greater than that of its secant, f/D , damage localization will occur. Taking the simplest DFD of Eq. 7.9 as an example, the critical damage for damage localization becomes

$$D_c = [(\mu - 1) \cdot \beta]^{-1/\mu} . \tag{7.20}$$

If the second term of the trans-scale formulation of the DFD of Eq. 7.17 can be fit by a power law of damage D, the important relation

$$D_c \sim D^* \tag{7.21}$$

arises. Thus, the dimensionless numbers D^* or β play a significant role in the prediction of damage localization, and therefore failure, in heterogeneous media. As a matter of fact, the dimensionless numbers are both intrinsic Deborah numbers, indicating the relation between nucleation and growth rates. From data fitting, it is known that μ is on the order of 10^0, like 2–4, but β may change from 10^1 in creep to 10^4–10^6 in impact. From mesoscopic measurement, D^* ranges from tenths to hundredths, or even thousandths. It might be interesting to compare these values to the transition thresholds of several tenths given by percolation theory (random nucleation only). The difference of the critical values is the reflection of the radical difference between the non-equilibrium evolution of microdamage to failure and the equilibrium phase transition (percolation). Furthermore, numerical simulations do show that the damage localization criterion can properly predict the collapse of media under loading (see Fig. 7.6).

In one word, the intrinsic Deborah number β or D^* dominates the critical transition from distributed microdamage to formation of a macroscopically localized failure surface.

7.3.4. Wave Dynamics and Spallation

Owing to the rate-dependent nature of spallation of materials subjected to impact loading, there have been various efforts to formulate this process, such as the integral criterion, microstatistical fracture mechanics [3], continuum measures of spallation [4], etc. Meyers [28] stressed that we still need quantitative/predictive models based on a continuum measure of spallation and nucleation and growth of microcracks. As mentioned above, the statistical microdamage mechanics and damage localization criterion (7.19), may be the proper framework in which to formulate the problem.

Equations 7.6–7.8 and dynamic function of damage (DFD), either the empirical macroscopic one of Eq. 7.9 or the trans-scale one of Eq. 7.17, together with proper boundary and initial conditions, form the complete formulation of the wave problem. Here, we ignore all computational details and just give an example of computed wave profiles with damage evolution in the tension phase to show the ability of the framework (see Fig. 7.7). This demonstrates that as long as the mesoscopic dynamics of damage, such as nucleation and growth rates, are known, it is possible to calculate the wave problem with continuous evolution of damage. Additionally, there are three time scales: the imposed one,

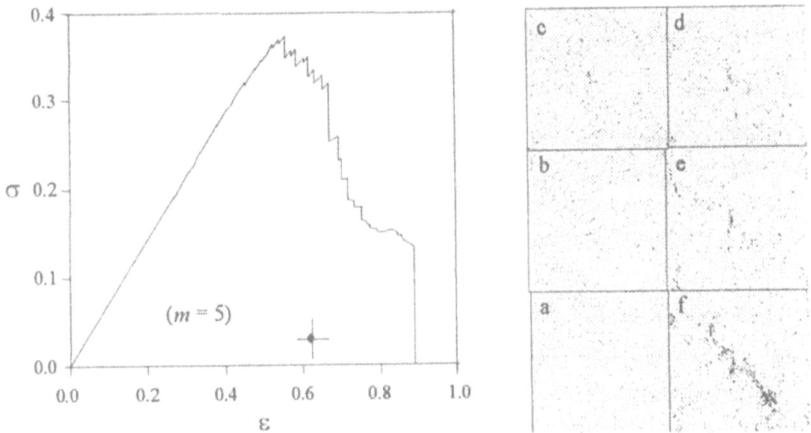

Figure 7.6. Simulated stress–strain relation of a heterogeneous model possessing Weibull distribution function with $m = 5$ and corresponding damage patterns. The cross (+) indicates the damage localization condition.

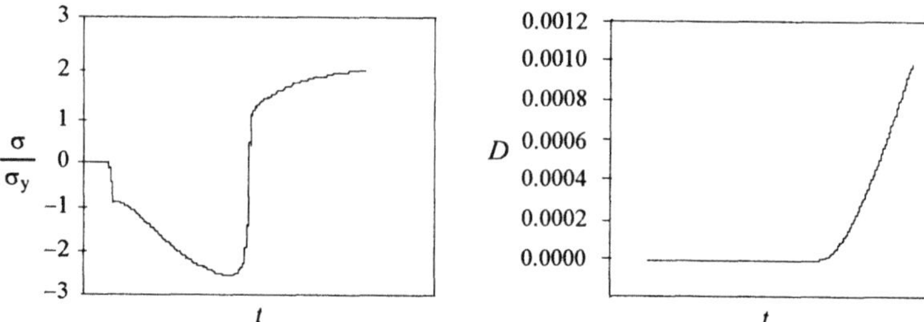

Figure 7.7. Simulation of stress and damage evolution for a sample subjected to boundary impact load. The quantity σ_y is the yield strength of the material.

L/a, where L and a are thickness and elastic wave speed, and the growth and nucleation time scales, c^*/V^* and $[n_N^*(c^*)^4]^{-1}$, respectively. Hence, two dimensionless numbers (Deborah numbers), i.e. the imposed number $De = a/[Ln_N^*(c^*)^2]$ and the intrinsic number $D^* = n_N^*(c^*)^5/V^*$ are involved in the governing equations of the wave propagation problem. We shall see that D^* indicates the characteristic critical damage, whereas De governs whether or not the sample can attain the critical threshold within the imposed time.

Now let us have a quick look at spallation. Mesoscopic experimental measurements of an aluminum alloy subjected to a loading of 1.4 GPa give $D^* \sim (10^{-3} - 10^{-2})$, see [5,17]. Data fitting on the continuum level gives $\beta \sim (10^5 - 10^6)$. If the damage localization criterion, Eq. 7.20, still applies, the condition leads to the critical damage to localization, $D_c \sim 4 \times 10^{-3}$, and a critical time of 0.7 µs. Compared to the loading duration of 0.68–0.85 µs for the appearance of incipient macro-cracking in tests, the agreement is very encouraging.

7.4. Catastrophe Transition in Stress Redistribution Models with Stress Fluctuations

Mechanical failure in heterogeneous media is a problem of scientific and technological importance and has been intensively studied for years. However, a number of fundamental questions are still open owing to its complexity. As stated at the very beginning of this chapter, meso-scale inhomogeneities and stress fluctuations can strongly influence phenomena such as fracture and phase transitions. Although the above approach of statistical microdamage mechanics can help us deal with evolution problems, it does not provide insight into catastrophe transition and adjoint sample specificity (SS) observed in the failure of samples that are mesoscopically inhomogeneous but macroscopically appear to be identical (the same distribution function), owing to its nature as a local mean field approximation. This demonstrates an urgent need of other new conceptual and computational models. We suppose that these new models should be stress redistribution models with mesoscopic stress fluctuations.

Actually, the catastrophe transition results from cascade of damage from small to large scales far from equilibrium. Clearly, this behavior is sensitively dependent on mesoscopic inhomogeneities and dynamical evolution of damage plays an important role in the catastrophe transition. In this process, mesoscopic fluctuations may become very strong near the catastrophe transition. This leads to another important feature: sample-specificity (SS). It has been found that samples with mesoscopic inhomogeneities having the same distribution function, i.e., macroscopically identical samples, exhibit universal scaling (as the mean field approximation describes), when damage remains small and independent. But with increasing damage, the samples show a transition from universal scaling to sample-specific behavior, namely significant differences from sample to sample, even though they are initially identical macroscopically. Meanwhile, the derivation of the controlling field from the scaling based on the mean field approximation is also enhanced.

In order to clarify the idea, let us look at a simple example, the deformation and failure of plane samples (the plane size is $\Omega \sim L^2$, where L is the linear size of the sample) with the same distribution function of mesoscopic unit strength $h(\sigma_c)$ and the identical unit elastic modulus k [17]. Mean-field approximation implies uniform strain, $\varepsilon = u/L$, and uniform stress, $\sigma = k\varepsilon$, in unbroken units, where u is the boundary displacement. Therefore, the unbroken units on a section of unit thickness is

$$L \int_{u/L}^{\infty} h(k\varepsilon_c)d\varepsilon_c$$

and the corresponding driving force

$$F = L \int_{u/L}^{\infty} h(k\varepsilon_c)d\varepsilon_c \cdot \sigma = k\,u \int_{u/L}^{\infty} h(k\varepsilon_c)d\varepsilon_c \,. \tag{7.22}$$

Clearly, Eq. 7.22 can be rewritten as the scaling law

$$F = k L \frac{u}{L} \int_{u/L}^{\infty} h(k\varepsilon_c)d\varepsilon_c = L\,\psi\left(\frac{u}{L}\right). \tag{7.23}$$

Equivalently, irrelevant to details of either sample size or inhomogeneity, there is a universal relation of macroscopic stress and strain of the samples

$$\sigma = \frac{F}{L} = k\varepsilon \int_{\varepsilon}^{\infty} h(k\varepsilon_c)d\varepsilon_c = \psi(\varepsilon). \tag{7.24}$$

Numerical simulations show (see Fig. 7.8) that the $\sigma-\varepsilon$ data do collapse onto the same curve in the initial regime. But, with increasing deformation, the deviations from the scaling become stronger and stronger and differences in failure of samples become distinct as well.

The transition of macroscopic behavior from scaling to sample specificity is a common feature in heterogeneous media. Obviously, the dynamic evolution leads to the strong amplification of initial mesoscopic differences of the samples as they approach eventual failure.

The difficulty involved in the establishment of new models to cope with the catastrophe transition is twofold [17]. First, there are three coupled patterns: inhomogeneity, stress, and damage. For the above one-dimensional elastic–brittle example, the three coupled patterns are the mesoscopic strength pattern $\Sigma_c = \{\sigma_{ci}\}$, the stress pattern $\Sigma = \{\sigma_i\}$, and the damage pattern $X = \{x_i, x = 1 \text{ or } 0\}$, where $x = 0$ or 1 denote intact or broken units respectively. The three coupled patterns rather than a unique governing field, like stress, greatly increase the complexity of the problem. More significantly, when approaching the catastrophe transition, the stress may switch quickly from one pattern to another in a way that is sensitively dependent on the damage pattern. Second is the computational complexity due to inhomogeneity. Taking a simple example $\Sigma_c = \{\sigma_{ci}, \sigma_c = 0 \text{ or } 1\}$, the total number of states for one dimensional chain is 2^N, where N is the total number of mesoscopic units. Clearly, this is a problem of great computational complexity because of the great number of samples, which increases exponentially with increasing N. Therefore, when we intend to find out the sensitive patterns of inhomogeneities in a huge ensemble and to obtain statistical description, we have to resort to simple stress redistribution rules, rather than realistic constitutive relations, because of its incredible computing consumption.

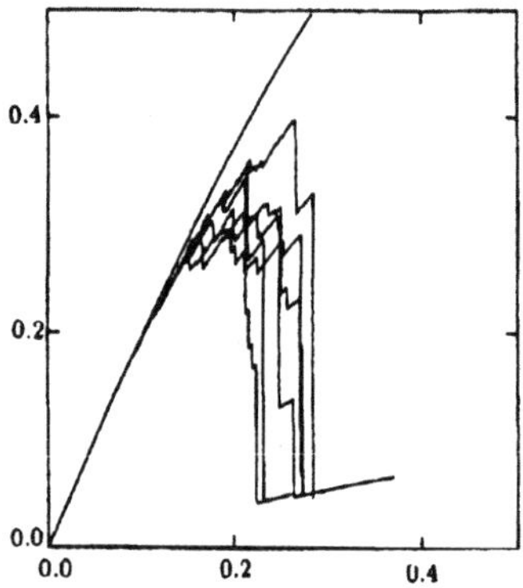

Figure 7.8. Stress and strain relation in a network model. The scaling law works in the weakly damaged regime for different samples. However, as damage proceeds, sample specificity appears for samples with the same macroscopic parameter.

In order to reveal the common features of the catastrophe transition, heterogeneous brittle media suffering from damage evolution to rupture are assumed to be a kind of heterogeneous and nonlinear threshold system.

We consider a system comprised of a great number of interacting and nonlinear mesoscopic units. The interactions between the mesoscopic units are assumed to be described by one of the following stress redistribution rules [29]:

1. Global mean field rule (GMF), in which external driving force is supported uniformly by all intact units.

2. Local mean stress concentration (LMSC) rule, in which the nominal driving force of a broken cluster is supported by the neighboring influence regions adjacent to the broken cluster. This rule takes stress concentration effect in the vicinity of micro-damage into account.

3. Cluster mean field (CMF) model, in which the nominal driving force of a broken cluster is uniformly shared by its two neighboring intact clusters. This rule takes the effects of stress concentration and nonlocal coupling between adjacent microdamages into account.

GMF is a rule without stress fluctuations but both LMSC and CMF take stress fluctuations into account.

The evolution dynamics or the algorithm is the same for all stress redistribution rules. For the j-th stress pattern Σ_j, the comparison between the two patterns of Σ_j and Σ_c gives a new damage pattern X_j. Then calculate a new stress pattern Σ'_j from damage pattern X_j according to the corresponding stress redistribution rules and compare the two patterns Σ'_j and Σ_c again until no more breaking occurs. Now, a new equilibrium state is attained. Then, add an increment to the nominal stress and repeat the above loop.

We do not want go into the details of calculations. What we would like to do now is to emphasize some distinctive aspects of the catastrophe transition in the stress redistribution models (SRD) with stress fluctuations.

7.4.1. Evolution Induced Catastrophe (EIC)

An example calculation using the CMF rule is shown in Fig. 7.9. Before the critical transition, nucleation, growth, and minor coalescence of microdamage happen here and there but the sample remains globally stable. In particular, there is no hint of a catastrophe transition in the damage pattern. However, for a certain pattern, a newly nucleated microcrack can initiate successive coalescence and more and more existing microcracks become involved in the coalescence and this leads to eventual fracture. This is why we call it evolution induced catastrophe (EIC) [30]. This is a non-equilibrium transition that is extremely sensitive to mesoscopic inhomogeneities and is very different from an equilibrium phase transition like percolation. To illustrate the idea of EIC clearly, Fig. 7.10 shows a proverb: the last straw breaks the camel's back, in Chinese it is said to break an ox's back. No matter camel or ox, the last slight straw plays a critical role at the catastrophe transition point. The challenging problem in EIC is predicting which straw will be the last one. Moreover, the last straw varies from one camel to another, though the camels look the same.

7.4.2. Sample Specificity (SS)

Sample specificity has the following implications [29,31]:

1. As damage proceeds from the initial regime to the catastrophic failure regime, there is a transition from universal scaling to diversity in failure of heterogeneous media. The transition is attributed to the serious deviation of the controlling field from the mean field.

2. Macroscopic failure may be significantly different from sample to sample, so a statistical rather than a deterministic description of failure should be introduced.

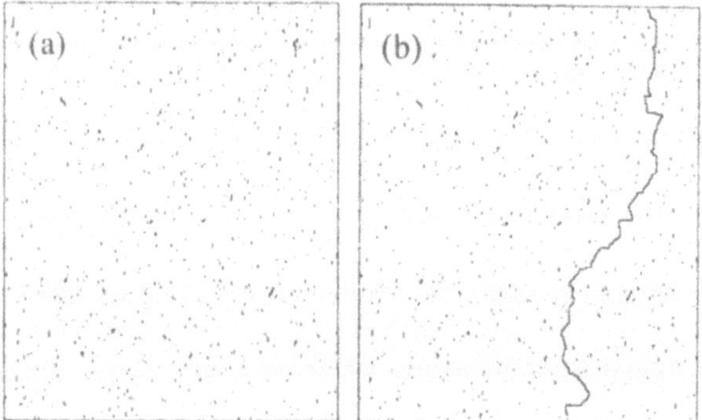

Figure 7.9. Simulation of microcrack patterns. (a) Pattern of microcracks just before EIC, (b) Pattern of EIC triggered by a newly nucleated crack in (a).

Figure 7.10. A cartoon for Evolution Induced Catastrophe: The last straw breaks the camel's back.

Figure 7.11 shows a statistical description: failure probability dependent on nominal stress and initial damage fraction. This figure clearly indicates that deterministic prediction of failure may not be made for moderate stress and damage. The statement on sample specificity (SS) may be sketched by the cartoon of Fig. 7.12. The young boys are sextuplets so it is hard to tell them apart although minor differences do exist. When they grow up (a nonlinear dynamic process), only one becomes a doctor. When we trace back to look for their early differences, we see that the future doctor had one less hair than the others!

7.4.3. Trans-Scale Sensitivity

As we have seen, failure may be very sensitive to mesoscopic details of the samples. As a nonlinear dynamical system, the final state is uniquely determined by the initial state, but the final differences between the samples can be strongly

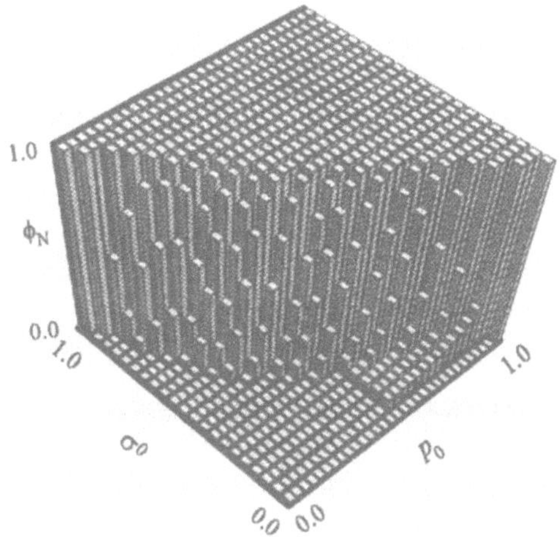

Figure 7.11. Failure probability $\Phi = \Phi(\sigma_0, p_0)$ for CMF and $N = 100$.

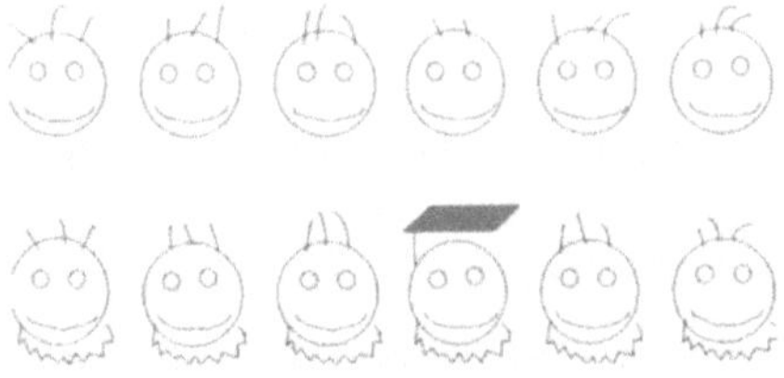

Figure 7.12. Sample specificity. Evolutions of samples with mesoscopic differences may present sample specificity in macroscale although they have the same macroscopic parameters and are in the same environment.

enhanced by the dynamical evolution. So, different failure behaviors of samples could not be simply attributed to initial mesoscopic differences between the samples and it is hard to predict EIC behaviors solely by recognizing initial mesoscopic inhomogeneities. Both initial differences and nonlinear dynamical evolution are inherent in the sample specificity of failure. This is what we call trans-scale sensitivity. Therefore, trans-scale sensitivity should be based on nonlinear dynamics [29,31]. As shown in Fig. 7.13, looking for and listing all ant holes in a long dike must be extremely tedious and is very far from the identification of the specific ant hole causing the dike collapse. The ideas of sample-specificity and trans-scale sensitivity can be illustrated by simple evolution of two chains with the same initial damage fraction but a minor difference in initial damage pattern (see Fig. 7.14).

Figure 7.13. Trans-scale sensitivity. One ant hole may cause the collapse of a thousand-mile dike.

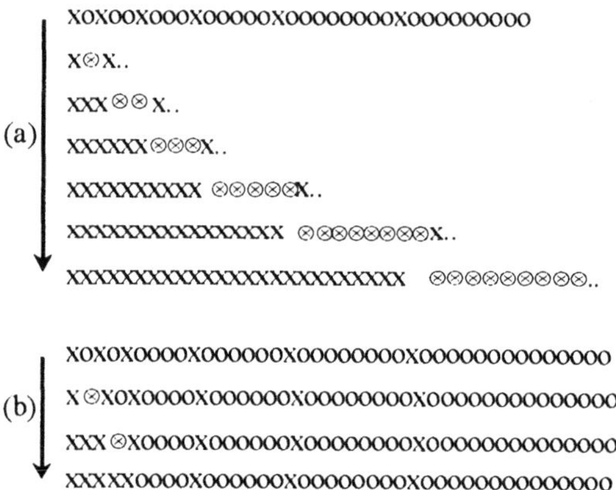

Figure 7.14. (a), Quasi-Fibonacci series (1,2,3,5,8,13,20, ...) of breaking is a sensitive microstructure for cascading, In (b), the cascading will stop at a stable status.

7.4.4. Critical Sensitivity (CRS)

Owing to sample specificity and trans-scale sensitivity, we can hardly obtain deterministic predictions of failure of inhomogeneous media, by either macroscopic averaging parameters or thorough examination of mesoscopic inhomogeneities. Then what can we do to predict failure with mesoscopic inhomogenieties? Perhaps, the concept of critical sensitivity can provide some help in coping with the problem.

Critical sensitivity S is defined as the sensitivity of energy release rate to an increment of a governing external variable [32,33],

$$S = \frac{\Delta E' / \Delta \sigma'}{\Delta E / \Delta \sigma},\tag{7.25}$$

where ΔE and $\Delta \sigma$ are the increments of energy release due to damage and governing stress, respectively, and $\Delta \sigma'$ is slightly greater than $\Delta \sigma$. It has been found that there is a significant increase in S when a sample is approaching its catastrophe transition (see Fig. 7.15). Hence, if the increments of energy release and a governing variable are both measurable, critical sensitivity may provide clues for catastrophe prediction. This concept has been applied to earthquake forecast and looks promising [32,33].

We do not yet know how to deal with the sample specificity in failure associated to shock dynamics in heterogeneous media, for instance in the case of the deflagration-to-detonation transition in powdered explosives subject to impact loading. For safety studies, such low probability events are very sample specific and trans-scale sensitivity plays a significant role.

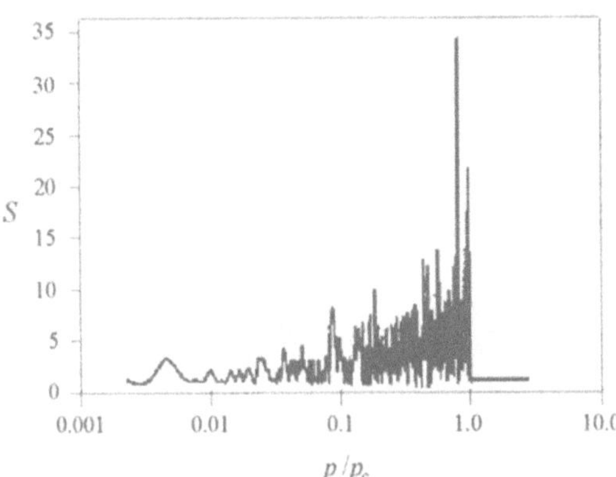

Figure 7.15. Critical sensitivity for CMF model, $N = 10000$, $m_c = 2$ and $\alpha = 0.001$.

7.5. Concluding Remarks

The challenge we face is to reach deep physical understanding of the non-equilibrium evolution of collective microdamage coupled with mesoscopic inhomogeneities and stress fluctuations and to develop manageable formulations that can properly extract the essence of the processes. This appeals strongly to new irreversible statistical mechanics and new tools to treat non-uniformly distributed microdamage.

So far what we know is:

1. Damage evolution and rupture of heterogeneous media can be described in terms of an associated system of continuum and damage field equations. The dynamic function of damage links mesoscopic kinetics of microdamage and continuum damage. Damage localization provides a proper threshold of failure forecast.

2. The intrinsic Deborah number is a significant representation of the effect of mesoscopic kinetics on macroscopic behavior of heterogeneous media. It is closely related to continuum damage localization.

3. Sample specificity (SS) is observed in evolution induced catastrophe in various stress redistribution (SRD) models with stress fluctuations, such as stress concentration and cluster mean field models.

4. Cascade of damage evolution magnifies the effects of initial inhomongeneities and induces trans-scale sensitivity. This is the essence of sample-specificity. For the sake of specific predictions for heterogeneous media, the concept of critical sensitivity seems to be promising.

What to do in the future? We suppose the questions cited at the very beginning of the paper are very much to the point, athough parts of these questions have been touched upon and discussed.

1. What experimental data are available and what are their implications?

2. Are there new meso-scale theories for shock dynamics?

3. How do the theories affect the existing fracture and phase transition paradigms?

4. What kinds of new computational and material response models are needed?

Acknowledgment

This work is granted by the National Natural Science Foundation of China (NSFC 19891180, 19972004, 19732060), Major State Research Project G200007735.

References

[1]   Y. Horie, private communication.

[2]   D.L. McDowell, *Applications of Continuum Damage Mechanics to Fatigue and Fracture*, ASTM STP 1315. pp. 1–3 (1997).

[3]   D.R. Curran, L. Seaman, and D.A. Shockey, "Dynamic failure of solids," *Physics Reports* **147**, pp. 253–388 (1987).

[4]   L. Davison and A.L. Stevens, "Continuum measures of spall damage," *J. Appl. Phys.* **43**, pp. 988–994 (1972).

[5]   Y.L. Bai, J. Bai, H.L. Li, F.J. Ke, and M.F. Xia, "Damage evolution, localization and failure of solids subjected to Impact Loading," *Int. J. Impact Engng.* **24**, pp. 685–701 (2000).

[6]   G.I. Kanel, S.V. Rasorenov, and V.E. Fortov, "The failure waves and spallation in homogeneous brittle materials," in *Shock Compression of Condensed Matter—1991* (eds. S.C. Schmidt, R.D. Dick, J.W. Forbes and D.G. Tasker), Elsevier Science Publishers B.V., Amsterdam, pp. 451–454 (1992).

[7]   R.J. Clifton, "Analysis of failure waves in glasses," *Appl. Mech. Rev.* **46**, pp. 540–546 (1993).

[8]   J.E. Field, G.M. Swallowe, and S.N. Heavens, "Ignition mechanisms of explosives during mechanical deformation," *Proc. Roy. Soc. London*, **A382**, pp. 231–244 (1982).

[9]   T.H. Zhang, *Damage of a Propellant and Its Stability of Combustion*, PhD Thesis, Chinese Academy of Sciences, Beijing (1999).

[10] P. Szuromi, "Microstructural Engineering of Materials," *Science* **277**, p. 1183 (1997).

[11] B. Budiansky, "Micromechanics," in *Advances and Trends in Structural and Solid Mechanics* (eds. A.K. Noor and J.M. Housner), Pergamon Press, Oxford, pp. 3–12 (1983).

[12] T. Mura, *Micromechanics of Defects in Solids*, Martinus Nijhoff Publishers, Hague (1982).

[13] G.I. Barenblatt, "Micromechanics of fracture," in *Theoretical and Applied Mechanics* (eds. S.R. Bodner, J. Singer, A. Solan and Z. Hashin), Elsevier Science Publishers B.V., Amsterdam (1992)

[14] V.E. Panin, "Overview on mesomechanics of plastic deformation and fracture of solids," *Theo. Appl. Frac. Mech.* **30**, pp. 1–11 (1998).

[15] L.W. Yuan, "The rupture theory of rheological materials with defects," in *Rheology of Bodies with Defects* (ed. Ren Wang), Kluwer Academic Publishers, Dordrecht, pp. 1–20 (1997)

[16] X.S. Xing, "The foundation of nonequilibrium statistical fracture mechanics," *Advances in Mechanics* **21** (in Chinese), pp. 15–168 (1991).

[17] M.F. Xia, W.S. Han, F.J. Ke, and Y.L Bai, "Statistical meso-scopic damage mechanics and damage evolution induced catastrophe," *Advances in Mechanics* **25** (in Chinese), pp. 1–40, pp. 145–173 (1995).

[18] L.E. Reichl, *A Modern Course in Statistical Physics*, University of Texas Press. Austin (1980).

[19] S.T. Pantelides, "What is materials physics, anyway?," *Physics Today*, Sept., pp. 67–69 (1992).

[20] G.C. Sih, "Micromechanics associated with thermal/mechanical interaction for polycrustals," in *Mesomechanics 2000* (ed. G.C. Sih), Tsinghua University Press, Beijing, pp. 1–18 (2000).

[21] Y.L. Bai, F.J. Ke, and M.F. Xia, "Formulation of statistical evolution of microcracks in solids," *Acta Mechanica Sinica* 7, pp. 59–66 (1991).

[22] Y.L. Bai, M.F. Xia, F.J. Ke, and H. L. Li, "Damage field equation and criterion for damage localization," in *Rheology of Bodies with Defects* 25, (ed. R. Wang), Kluwer Academic Publishers, Dordrecht, pp. 55–66 pp. 1–40 (1998).

[23] F.J. Ke, Y.L.Bai, and M.F. Xia, "Evolution of ideal micro-crack system," *Science in China, Ser. A* 33, pp. 1447–1459 (1990).

[24] Y.L.Bai, W.S. Han, and J. Bai, "A statistical evolution equation of microdamage and its application," ASTM STP **1315** pp. 150–162 (1997).

[26] A. Needleman and V. Tvergaard, "Analysis of plastic flow localization in metals," *Appl. Mech. Rev.*, **45**, pp. 243–260 (1992).

[27] G.I. Taylor and R.S. Tankin, "Gas dynamics of detonation" in: *Fundamentals of Gas Dynamics*, (ed. H.W. Emmons), Princeton Univ. Press, Princeton, NJ, Section G (1958).

[28] M.A. Meyers, *Dynamic Behavior of Materials*, Wiley, N.Y., (1994)

[29] M.F. Xia, Y.J. Wei, J. Bai, F.J. Ke, and Y.L. Bai, "Evolution induced catastrophe in a non-linear dynamic model of material failure," *J. Non-Linear Dynamics* **22**, pp. 205–224 (2000).

[30] Y.L. Bai, C.S. Lu, F.J. Ke, and M.F. Xia, "Evolution induced catastrophe," *Physics Letter A* **185**, pp. 196–200 (1994).

[31] M.F. Xia, F.J. Ke, J. Bai, and Y.L. Bai, "Threshold diversity and trans-scales sensitivity in a finite nonlinear evolution model of materials failure," *Physics Letters A* **236**, pp. 60–64 (1997).

[32] Y.J. Wei, F.J. Ke, M.F. Xia, and Y.L. Bai, "Evolution induced catastrophe of material failure," *Pure Appl. Geophy.* **157**, pp. 1945–1957 (2000).

[33] M.F. Xia, Y.J. Wei, F.J. Ke, and Y.L. Bai, "Critical sensitivity and trans-scale fluctuations in catastrophic rupture," to appear in *Pure Appl. Geophy.* (2001).

CHAPTER 8

Responses of Condensed Matter to Impact

John J. Gilman

The responses of liquids and solids to applied forces depend on time through the existence of viscosity. At sufficiently short times, liquids behave elastically, having insufficient time to flow. That is, they behave as if they were solid. Conversely, solids behave elastically at short times, but they flow at sufficiently long times, depending on how much force is applied to them. That is, they behave as if they were liquid. Between the two extremes lies plastic matter. Inside a plastic solid are small tubes (cores of dislocation lines) within which sliding can occur. This sliding is resisted by liquid-like viscosity and by fluctuating internal forces which cause energy dissipation.

There are two fundamental sources of response during impact: acoustic (mechanical) and electronic (chemical). These interact to create a large array of responses. Some responses have been misinterpreted. For example, descriptions of mechanical states, "elastic strains" and "plastic deformations", are physically distinct entities, yet both are commonly called "strains". Also, both are sometimes added together inappropriately. This has led to the false concept of a "plastic modulus"; to the assumption that plastic deformation is a continuous process; and other errors.

Phenomena at shock fronts are usually interpreted in terms of pressure changes. However, it is pointed out that because of the requirement for continuity, shock fronts in solids require the existence of shear strains in order to go from one pressure level to another. It is these shear strains that induce phase changes and chemical reactions at shock fronts.

8.1. The Impact Process

The face-on collision of two identical flat plates, as illustrated in Fig. 8.1, constitutes an idealized collision [1]. The plates move toward each other with equal but oppositely directed velocities $\pm v_0/2$ so that the relative velocity difference between the two plates is v_0. When the plates collide, their impacting surfaces are brought to rest. The effect of the collision is communicated to the interior of each of the plates by a wave that propagates away from the interface.

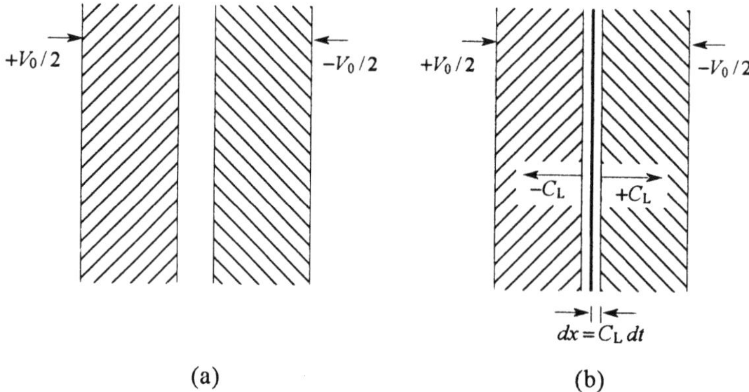

Figure 8.1. Impact of two plates: initial velocities: $(+v_0/2)$ and $(-v_0/2)$. (a) Just prior to impact. (b) Short time, dt after impact. Back surfaces of plates still have initial velocities. Front surfaces have $v = 0$. Longitudinal waves with velocities: $(+C_L)$ and $(-C_L)$ move into the plates the distances, Δx.

Let us focus our attention on the right-hand plate. Ahead of the wave in this plate, the material continues to move to the left with its initial velocity $-v_0/2$, so the change in particle velocity that results from the impact is $v_p = v_0/2$. The velocity difference between the stagnant material at the interface and the moving material ahead of the wave is accommodated by uniaxial compression of a layer of material between the wave and the interface. The compression wave moves into the uncompressed material at a velocity v_1 relative to the material particles ahead of the wave, or at a velocity C_L in the fixed laboratory coordinates. These velocities stand in the relation $v_1 = C_L + (v_0/2)$. In the time interval Δt following impact, the compression wave has engulfed a layer of material at the impact surface that had thickness $\Delta x = v_1 \Delta t$ in the uncompressed state. The kinetic energy of the moving plate is converted into potential, or strain, energy when the material stagnates at the impact interface.

In the time interval Δt after impact, the particles of the surface are displaced a distance $\Delta u = \varepsilon_{xx} \Delta x = \varepsilon_{xx} v_1 \Delta t$ toward the uncompressed particles ahead of the wave. The displacement can be expressed in the form $\Delta u = v_p \Delta t$. Thus, the strain in the layer is $\varepsilon_{xx} = \Delta u/\Delta x = (\Delta u/\Delta t)/(\Delta x/\Delta t) = v_p/v_1$. The transverse strains are constrained to vanish: $\varepsilon_{yy} = \varepsilon_{zz} = 0$. Applying the foregoing equations for v_p and v_1 yields the equation $\varepsilon_{xx} = v_0/2v_1$ for the strain in the compressed material.

The longitudinal speed of sound (isotropic medium) is given by

$$v_1 = \left(\frac{B + (4G/3)}{\rho} \right)^{1/2} , \tag{8.1}$$

where ρ = mass density, B = bulk modulus, and G = shear modulus. Thus, the disturbance consists of two parts: a pressure, and a shear. The pressure, P, is

$$P = B\frac{v_0}{2v_1} \tag{8.2}$$

and the maximum shear stress is

$$\tau_{max} = G\frac{v_0}{2v_1}. \tag{8.3}$$

These relationships hold only if $\varepsilon_{xx} < 1/10$. That is, if

$$v_0 < \frac{1}{5}v_1. \tag{8.4}$$

A typical value of v_1 is 6 km/s, so v_0 should be less than about one km/s for the relations of linear elasticity to hold.

The modulus, $B + (4G/3)$, in Eq. 8.1, multiplied by the square of the strain, yields the strain energy density. The modulus B measures the dilatational strain energy, while G measures the shear strain energy. Initially, during impact, these energies are roughly equal since $G \approx 3B/5$. However, as the longitudinal wavefront propagates, the shear stress is relieved. At first, the absorption occurs through plastic glide and twinning, and later through the fluctuation-dissipation mechanism. When the shear strain energy has been absorbed, only a small part of the longitudinal wave may be left. This is often called an elastic "precursor", but this is a misnomer [2]. It is not a precursor; it is the leading edge of the longitudinal wave.

Examples of experimental data that are consistent with the behavior outlined in Fig. 8.2 are presented in Fig. 8.3.

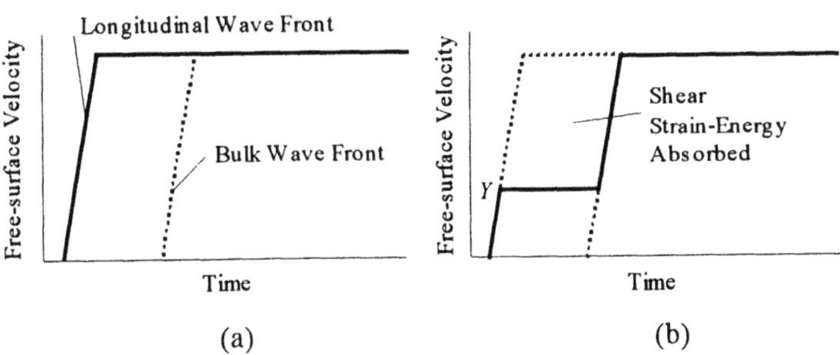

Figure 8.2. Particle velocities vs. time at back surfaces of impacted plates. (a) Schematic longitudinal elastic wave. (b) Longitudinal wave with part of the shear strain absorbed by plastic deformation so that the bulk part becomes delineated.

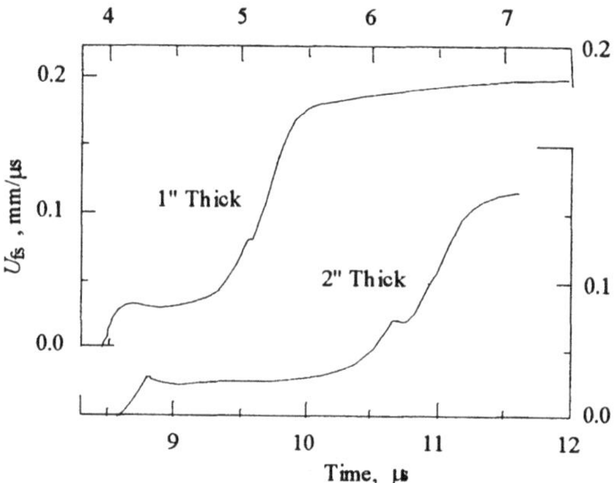

Figure 8.3. Examples from experiments of Taylor and Rice [1]. Back surface velocities of iron plates showing initial yield stress, followed by dissipative deformation at nearly constant stress, and the rise to the maximum hydrostatic stress where it is constant until the rarefaction wave arrives. Note that the longitudinal wave has a higher velocity than the bulk wave so the two fronts are separated more for the thicker plate.

8.2. Weak to Mild Impact

Figure 8.4 illustrates the general stages of deformation in an impacted plate [2]. First, uniaxial compression results in a combination of hydrostatic compression and shear deformation (Fig. 8.4a). Then the shear strains relax (Fig. 8.4b). Finally, a state of pure hydrostatic compression is achieved (Fig. 8.4c).

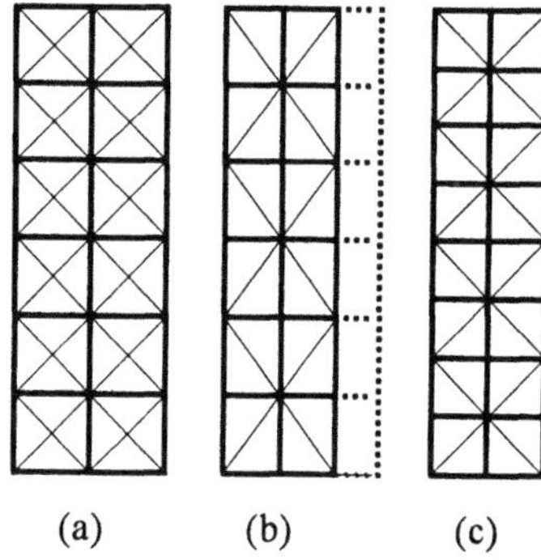

Figure 8.4. Schematic stages of the impact process: (a) Configuration prior to impact. (b) Configuration compressed in longitudinal wave immediately after impact. Note the combination of shear and volume change. (c) Shear relaxes with time, leaving only volume change. Wave becomes bulk type.

For relatively weak impacts, the shear strains may be relaxed by plastic deformation (glide or twinning). The rate of this deformation, $d\gamma/dt$ for a single glide system, is given by

$$\frac{d\gamma}{dt} = gb \int (\mathbf{v} \cdot \mathbf{n}) dl \ , \tag{8.5}$$

where g is a geometric factor of order unity, b is the Burgers displacement of the dislocation line, $\mathbf{v}$ is the average velocity vector along the line, and $\mathbf{n}$ is the outward normal to the line, lying in the glide plane. The line integration is over all of the mobile dislocation line length in the specimen. This is adapted from Orowan's equation [3]. For a low density of dislocations, it is assumed that $|\mathbf{v}|$, which is a function of the local shear stress, is constant. However, the line length continually increases. As the density increases, the average $|\mathbf{v}|$ decreases due to interactions, unless more stress is applied.

Conversely, plastic instability may occur at constant stress as the length l increases, so, with constant $\mathbf{v}$, according to Eq. 8.5, $d\gamma/dt$ increases, thereby allowing the local elastic stress to decrease. This leads to the drop in particle velocity that is often observed behind the elastic limit of the longitudinal wave front.

For loading in tension, the peak stress determines the fracture behavior (the peak being determined by the balance between the elastic rate of loading and the plastic deformation rate of Eq. 8.5). If the peak stress exceeds the local critical stress, fracture ensues [4]. If it does not, plastic deformation continues.

When the elastic strain exceeds about 10%, its effect on the electronic structure of the material becomes manifest, and crystallographic changes become likely. At constant structure, an insulating material may become metallic [5]. Or, because shear strain tends to close the gap in the bonding energy spectrum, a chemical decomposition/substitution reaction may occur [6]. Also, additional dislocation loops, or twins, may homogeneously nucleate.

8.3. Elastic Strain and Plastic Deformation

Both elastic and plastic deformations produce displacements within solid matter; i.e., changes of shape. However, they are distinctly different physical phenomena. Figure 8.5 is a reminder of the differences. Elastic deformation is usually temporary, well-ordered, and mostly affine. Plastic deformation is usually permanent, chaotic, and anything but affine.

Because the work needed to cause elastic deformation is stored in the deformed material in the form of changes of electronic structure (strain energy), it can be quickly recovered almost completely. Also, elastic deformation can

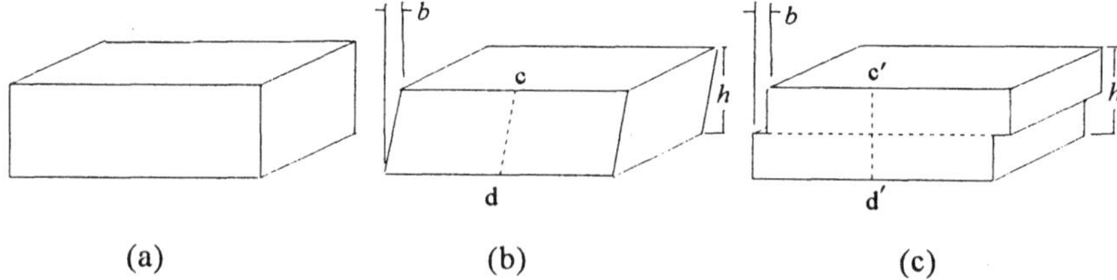

Figure 8.5. Difference between elastic strain, γ_e, and plastic deformation, γ_p. Both describe changes of shape, and $\gamma_e = \gamma_p = b/h$, but these quantities are distinct physically. A schematic undeformed specimen is shown at (a). After an elastic strain, γ_e, the specimen changes to (b). Internal lines like cd become strained, and elastic energy is stored in the chemical bonds along them. After a plastic deformation, γ_p, the specimen changes to (c). Internal lines like c'd' are undeformed so no elastic energy is stored. The work of deformation has been converted into heat (and/or internal defects).

propagate rapidly from one position to another. On the other hand, the work that causes plastic deformation is dissipated as heat, as well as configurational entropy, so almost none of it can be recovered, and the deformation can propagate only slowly, being limited by the viscosity at the cores of dislocations. An immediate conclusion is that elastic strains are state variables. They depend only on the values of other state variables (+/– pressure, shear stresses, temperature, electromagnetic fields, gravitational fields, etc.). Whenever these are given a particular set of values, the elastic strains (relative to a standard state) acquire corresponding values. Thus, the state variables are connected in a definite way by an equation of state and each specific state has a corresponding set of state variables. In passing from one state to another an elastic material may lose, or gain, small amounts of energy due to anelastic effects that convert small amounts of elastic energy into entropy, and thus are not recoverable. These losses may be associated with specific mechanisms, and their amounts can be calculated quantitatively (anelasticity theory) [7]. In contrast to the well-behaved case of elastic deformation, there is little about plastic deformation that is not chaotic, starting with the distribution of displacements that is associated with it.

For centuries, it was thought that, during plastic deformation, the distribution of plastic displacements (the plastic displacement field) was microscopically uniform [8]. Until the acceptance of the atomic theory of matter at the end of the 19th century, there was no reason to think otherwise. Localization of the deformation into shear bands was well known, but it was not realized that this fragmentation continued beyond microscopic dimensions for three to four orders of magnitude, down to atomic dimensions. Furthermore, the instrumentation that recorded "stress–strain curves" drew predominantly smooth lines with little change in passing from the elastic to the plastic regime (sometimes serrated curves were observed, but they were not the norm). It is now known that plastic deformation is anything but uniform. It is heterogeneous all the way down to the atomic level and somewhat beyond. The somewhat beyond refers to the fact that the quantum-mechanical amplitude functions are disturbed in a non-uniform way at the kinks on dislocation lines. The contrast with the homogeneous nature of elastic deformation is striking. The latter is affine right down to atomic dimensions.

To illustrate the discontinuity that occurs at the "yield point" of an elastic–plastic material, consider cyclic loading and the corresponding "Q", or cyclic quality of the deformation. This is defined as the ratio of the total elastic work done on the material per cycle divided by the amount of energy lost per cycle. Qs up to about 10^7 are observed for elastic deformations up to the yield point. When yielding occurs, this drops immediately to as little as 10. Thus, a quite discontinuous change occurs. Ideally, the discontinuity can be even larger. If a dislocation line is present, a plastic deformation of the order of unity can occur in a volume of order b^3, which is to be averaged over the specimen (say 1 cm^3). Here b is an atomic dimension, or about 2×10^{-8} cm, and b^3 is about 10^{-23} cm^3, so the change in Q could be as much as 10^{-23}. In practice, this is much too large to be observed, but it makes the point that, in principle, plastic yielding is an exceedingly discontinuous process.

Small changes of the shapes of solid bodies can be conveniently described in terms of displacement gradients. That is, in terms of a field of displacements normalized by dividing them by a corresponding field of local gauges, yielding a field of displacement gradients. To provide a useful description these fields must be continuous and compatible. Furthermore, in the elastic case, they must be differentiable with respect to both time and space. But, plastic displacement gradients do not meet this last criterion because they are irreversible, and they are microscopically discontinuous (Fig. 8.6).

Unfortunately, for the historical reasons outlined above, the word "strain" has been used to describe both elastic and plastic deformations; sometimes differentiated by subscripts. But the physical bases of elastic and plastic deforma-

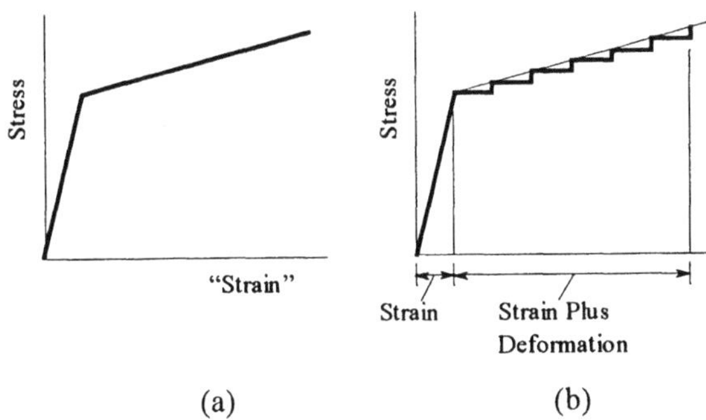

Figure 8.6. Comparison of elastic and plastic stress–deformation curves. (a) The elastic curve is continuous. (b) The plastic one is discontinuous on a fine scale.

tions are fundamentally different. Therefore, their names should be distinct. Also, it is improper to add, subtract, multiply, or divide them. In this paper, the word "strain", and its corresponding symbol, ε, are restricted to the elastic case.For the plastic case, the word "deformation", and its corresponding symbol, δ, is used.

It is especially important to recognize that quantities derived from ε and δ are also physically distinct. In particular, $d\sigma/d\varepsilon$, an elastic modulus, is physically very different from $d\sigma/d\delta$, which is not a modulus. The latter relates to the dissipation of stress, but not to its propagation. The slope $d\sigma/d\delta$ may legitimately be taken to be a strain-hardening coefficient, however. A formal distinction is that $d\sigma/d\delta$ may be either positive or negative, whereas $d\sigma/d\varepsilon$ must always be positive. Negative values of $d\sigma/d\delta$ are important because they produce a mode of plastic instability that results in localized plastic shear bands. The negative values are created whenever the rate of multiplication of dislocation lines increases the deformation rate more rapidly than is necessary to keep up with the deformation rate that the external system is trying to impose. The external system may be the loading machine, the strained material around the tip of a crack, etc. Physically, a displacement is the same geometric entity whether it results from an elastic strain or a plastic deformation. Therefore, it should be the variable of choice in dealing with elastoplastic problems. Since there are two kinds of displacement, although they are physically identical entities, they can be distinguished by means of subscripts, and they can appear together in equations. However, their space and time derivatives have different physical meanings, so the different types of derivatives cannot appear together in the same equation. Thus, most elastoplastic problems should be described by a set of at least two equations, each of which contains only one kind of derivative. Then the solution of the problem requires the simultaneous solution of the set (the elastic and the plastic differential equations). An additional reason why the elastic and plastic derivatives cannot be mixed is that the elastic displacements are continuous, whereas the microscopic plastic displacements are quantized in units of the Burgers displacement, b. To some extent these quantized displacements can be smoothed out in macroscopic problems, but this does not eliminate the other differences between the elastic and plastic displacements, i.e., the fact that one is conservative, whereas the other is nonconservative.

The second derivative of an elastic displacement with respect to position is an elastic modulus (either bulk or shear). But, this is not the case for the second derivative of the plastic displacement because the latter is not conserved. Thus, there is no "plastic modulus". This is important in interpreting elastic–plastic impacts, plastic zones at crack tips, local yielding, upper yield points, etc.

8.4. Mechanical Equation of State

If a general equation of state existed, it would be possible to approach any given point in a stress, σ, deformation, δ, temperature, T, space along an arbitrary path, and the state of the material at the point would always be the same. In other words, the material would obey an equation analogous to the state equation for gases (A = constant):

$$pV = AT . \tag{8.6}$$

This has been extensively tested by comparing its partial derivatives for various states (P, V, T points). It is obeyed as long as a material is elastic, but as soon as the yield point is reached, it fails to give consistent values. Equations of this type were extensively tested in the 1930–1950 time period. Some of the results are discussed by Tietz and Dorn [9]. Another type of state equation is one in which the rate of a reaction depends only on time and temperature if the stress, or the deformation, is held constant. This is true for simple chemical reactions in dilute systems. If it is also true for mechanical systems, the plastic behavior should depend only on the combination of time and temperature known as the Zener–Hollomon parameter [10]:

$$\text{rate} \sim t\left[\exp\left(-\frac{Q}{kT}\right)\right]. \tag{8.7}$$

Accordingly, deformation at a high rate and a low temperature should be the same as deformation at a low rate and a high temperature. To a considerable extent, this is true, but the substitution of time for temperature is not quantitative. Therefore, this parameter is useful for making interpolations, but it is not reliable for extrapolations.

8.5. Debris Production

The motion of dislocations through a solid (crystal, glass, or composite) causes plastic deformation. This was quantitatively verified in the 1950s by Gilman and Johnston [11]. One of the surprises of their direct observations was that there is a stochastic component of dislocation motion associated with the screw type. Screw dislocations are cylindrically symmetric and therefore not restricted to motion on one glide plane. As a result, they "double-cross-glide" in an apparently random fashion, perhaps in response to vibrational fluctuations [12]. In addition to enhanced dislocation multiplication, this creates "debris" in the wakes of moving screw dislocations. The debris consists of innumerable edge dislocation dipoles having a variety of dipole moments (distances between the individual dislocations). It is this debris that causes deformation hardening, and leads to fatigue cracks and other forms of structural degradation.

Dipole debris production occurs at a higher rate when the applied stress, and the deformation rate, is high, or the temperature is low. There is a long standing myth that most (90–95%) of the work that is done to produce plastic deformation is converted into heat during the deformation. This is entirely at odds with experimental measurements reviewed by Bever, Holt, and Titchener [13]. The partitioning of the work into heat vs. debris depends on the conditions of deformation. That is, on such factors as temperature, rate, prior deformation, initial perfection of the specimen, and so on. It can vary from a heat/defect ratio of less than 1/2 to about 9/10; and in the latter case most of the defects anneal out during the deformation. Therefore, definitive data are not readily available.

8.6. Heat Production

A dislocation moving through a nearly perfect pure metal produces very little heat. This is clear from the small amount of viscous damping that it experiences as indicated by internal friction measurements, and by direct velocity measurements. The measured viscosity values are of the order of 10^{-3} Poise, and less. Because the dislocation motion is of the "stick-slip" type, the maximum viscous losses occur during the fast slip events when the dislocation velocities reach about $v_{max} = v_s / \pi$, where v_s is the shear wave velocity $\approx 3.2 \times 10^5$ cm/s. Then the maximum deformation rate at the core of a dislocation is about $(d\delta/dt)_{max} = v_{max}/b \approx 3.2 \times 10^5 / 2.5 \times 10^{-8} = 1.3 \times 10^{13}$ /s; and the maximum drag stress is 1.3×10^{10} d/cm^2. From this, $U =$ the strain energy per atom $= (\tau^2/2G)b^3 \approx 3.2 \times 10^{-15}$ erg/atom. The maximum temperature rise, T, can be found by equating this to the thermal energy per atom $= kT$, where $k = 1.38 \times 10^{-16}$ erg/deg is Boltzmann's constant. The result is $T \approx 12$ K/atom distance moved. Even when the viscosity is much larger, the local temperature rise is small. Because heat production by individual dislocations is small, and the total heat production is large (for the parameters given above, the plastic work per atom is about $\tau b^3 \approx 2 \times 10^{-13}$ erg). It is obvious that this part of the plastic deformation process is poorly understood for the case of pure metals. For alloys containing hard particles, the heat production may well be located within the hard particles where there are covalent bonds to be broken irreversibly. In general, the Fluctuation–Dissipation Theorem of thermodynamics can be invoked [14], but this says nothing about the source of the fluctuations, so it is not very satisfying. The apparent drag caused by fluctuating internal stresses has been shown to be substantial by Chen, Gilman, and Head [15], but again this is rather formal theory.

8.7. Nonpropagation of Plastic Deformation

The wave equation for elastic shear strain (one dimension) is:

$$\rho \frac{\partial^2 u}{\partial t^2} = G \frac{\partial^2 u}{\partial x^2}, \tag{8.8}$$

where ρ = density, u = displacement, and G = shear modulus. Partly by analogy, Taylor [16], Rakhmatulin [17], as well as von Karman and Duwez [18] developed a similar equation for "plastic waves" with G replaced by a "plastic modulus". However, since plastic deformation is dissipative, such waves cannot exist, as pointed out by Gilman [19]. Taylor's analysis came first; the others perpetuated his mistake. He assumed that the right-hand side of Eq. 8.5 is differentiable, but it is not since the displacement due to dislocations is quantized. He also assumed that plastic stress–deformation curves are continuous and therefore differentiable. In reality, however, they consist of a series of steps because the deformation is dissipative (Fig. 8.6). That is, when the yield stress is reached, a small quantity of deformation occurs, absorbing some elastic strain energy, and thereby reducing the local stress slightly. The deformation then stops, waiting for the stress to rise back to the yield stress. This repeats if there is no deformation hardening, and the material is said to be elastic-perfectly-plastic. The deformation is discontinuous in time. Obviously, there is no propagation of the deformation, and the "plastic modulus" is zero (the horizontal lines in Fig. 8.6). If there is deformation hardening, the pattern is similar, except that the stress–deformation curve takes the shape of the rising staircase of Fig. 8.6. There is still no propagation of the deformation (Lüders band propagation is another matter).

8.8. Fluctuation–Dissipation Theorem

A fundamental theorem of statistical thermodynamics is the Fluctuation–Dissipation Theorem [14]. An important manifestation of it is the Einstein–Stokes Equation relating diffusivity and viscosity. This theorem is not intuitively obvious because it is a subtle consequence of the asymmetry of time under the assumption that the conservation laws for momentum and energy hold [20]. Consider a large particle, of mass M, moving through a sea of small particles, of mass m (in one dimension for simplicity). The large particle will experience both "front-end" and "back-end" collisions with the small particles. Front-end collisions will be more frequent than back-end ones when the large particle moves at constant velocity, v. Let η be the friction, or viscosity coefficient, and φ be a small random force, so the equation of motion (with both steady and random forces) is

$$F = M \frac{dv}{dt} = -\eta v + \varphi . \tag{8.9}$$

Denoting τ_c as the very short dwell-time of each collision, a standard thermodynamic deduction yields a connection between the average thermal energy $kT/2$, and the mean square fluctuating energy, $4\eta\tau_c <F^2>$, relating the fluctuating force and the viscosity:

$$\eta = \frac{\tau_c}{2kT} \left\langle F^2 \right\rangle . \tag{8.10}$$

Thus, fluctuating forces yield an apparent drag on the motion of the larger particles as a result of the asymmetry of the collisions between the small randomly moving particles and the large steadily moving ones. Rearranging Eq. 8.10:

$$T == \frac{\tau_c}{2k\eta} \left\langle F^2 \right\rangle . \tag{8.11}$$

This makes it apparent that, if the large particle is driven through the viscous sea of randomly moving small particles, the temperature will rise. The non-intuitive feature of this is the inverse dependence on the viscosity. The strong dependence of the temperature on the magnitude of the fluctuating forces might be expected. It may be helpful to note that this is related to the ancient rowboat puzzle in which there are two boats with equally strong rowers. They start at the same place on the bank of a river flowing with a velocity, v. One goes across the river whose width is, w, and comes back to the starting point. The other goes downstream a distance, w, and then comes back. Which one gets back first? One of the rowers represents motion through a constant background, and the other through a fluctuating background. In the case of dislocations, the most important fluctuations are those of shear strains, $\Delta\gamma$. Zero-point vibrations of the shear type are always present and play an important role in determining the specific heat. The average square fluctuation is given by [21]

$$\left\langle (\Delta\gamma)^2 \right\rangle \geq \frac{kT}{GV_0} , \tag{8.12}$$

where G = shear modulus, and V_0 = initial volume. A dislocation being driven through these fluctuations by an applied stress causes the temperature to rise.

8.9. Shear-Induced Instabilities

When solids are subjected to uniaxial compression (combined shear and negative dilatation), at critical values of the compression they may plastically deform, or they may undergo a phase transformation. According to conventional wisdom, the behavior is driven by pressure. In fact, shear strains are at least as important. For chemical reactions, this was discovered toward the end of the

19th century by Carey Lea [22]. More recently, it was deduced from the nature of phase changes [23] where it is bond bending, rather than bond compression, that leads to the phase transitions in semiconductors. Twinning is the most simple of all chemical changes, and it is obviously driven by shear strains. Next most simple are allotropic phase changes, because only the bond angles need to change. Isomerizations are the next in complexity, then decompositions, and so on. The evidence is that all of these are facilitated by shear strains as might be expected because they all involve changes of shape, whereas changes of size are a secondary factor. The conventional idea that pressure induces chemical reactions, disagrees with both the macro- and micro-observations. The macro observations of Lea [22], Bridgman [24], and many others, indicate that shear deformation is most important. Micro-observations show that bond angles change while bond lengths don't [23]. The reason for the large effect of shear on reactions is that shear profoundly affects the electronic structures of solids and molecules [25]. The effects are much larger than the effects of hydrostatic pressure, and are often of the opposite sign. In molecules, shear (bond bending) may be considered to be an inverse Jahn–Teller effect (closing the LUMO–HOMO gap). This way of looking at the effect emphasizes the role of symmetry. Unlike hydrostatic strain, shear breaks symmetry. For example, uniaxial compression converts cubic symmetry to tetragonal symmetry, or to rhombohedral symmetry, depending on the compression direction relative to the crystallographic axes. In solids, shear extends one direction while compressing a perpendicular direction, thereby reducing the minimum band gap [26]. Since the band gap, and the LUMO–HOMO gap, are measures of stability, reducing them leads to structural changes, decompositions, and metallization (conversion of insulators to metals). The shear strains near crystal defects cause local changes of electronic structure leading to chemical decomposition [27].

8.10. Shear at Strong Shock Fronts

There is a sharp change in the state of uniaxial strain at a shock front (Fig. 8.7). As mentioned above, uniaxial strain contains both shear and hydrostatic components. In pure metals, the shear strains are quickly dissipated by plastic deformation, but they *always* accompany the jump in deformation.

Electronic changes are very much faster than acoustic ones, so they *always* result from the shear strains and are present within a shock front. They vary in magnitude, of course, with the size of the jump in the uniaxial strain, and the front's thickness. In an exothermic substance, the change in electronic structure induced by a large shear strain initiates the chemical reaction; and the released chemical energy perpetuates the reaction, and thereby increases its velocity. As the magnitude of uniaxial compression increases, the ratio of the magnitude of the shear strain to the compressive strain increases without limit (Fig. 8.8). Therefore, at constant risetime, the shear strain rate also increases without limit. This is opposed, of course, by viscous drag, and eventually by compatibility con-

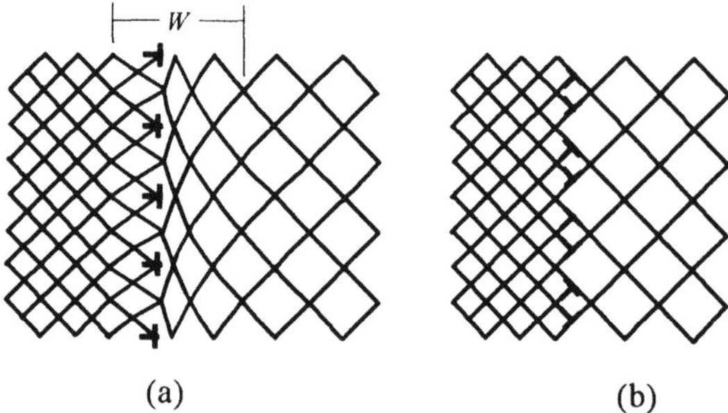

Figure 8.7. Schematic shears at strong shock fronts. (a) Relatively narrow shock front. "*W*" indicates region between arrival of longitudinal wave and arrival of bulk wave. The longitudinal wave arrives on the right. Shear strain increases rapidly and volume decreases simultaneously. Then, at the "dislocations" the shear strain becomes plastic deformation (heat and defects) leaving only volume change at the front of the bulk wave. (b) Very narrow shock front of approximately atomic thickness. This front travels at the bulk wave speed. The shear relaxation zone is atomically thin, consisting of dislocation cores only.

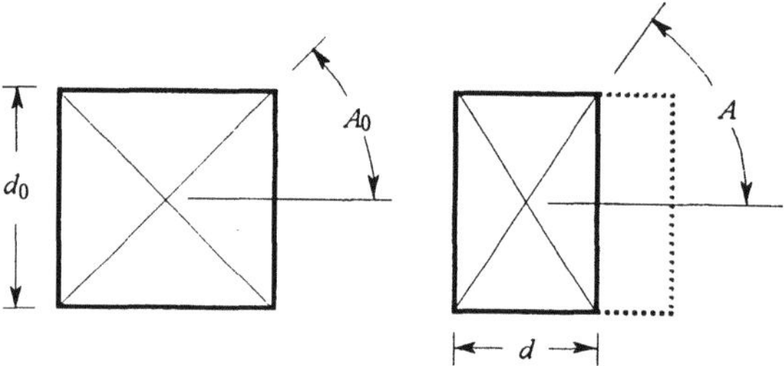

Figure 8.8. Ratio of shear deformation (shear angle) to compression deformation as the compression increases. The shear strain is: $\tan A - \tan A_0 = (d/d_0) - 1$; while the compressive strain, $\varepsilon = 1 - (d/d_0)$. Therefore, the shear strain $\gamma(\varepsilon) = \tan A - 1 = [1/(1-\varepsilon)] - 1$. Thus, $\gamma(\varepsilon)$ diverges as $\varepsilon \to 1$.

ditions that limit the overlapping of atoms. In the meantime, the shear forces can become very large as the product of the velocity gradient and the dissipation (viscosity) coefficient.

In the limit of the previous paragraph, energetic solids can decompose without thermal activation because the gap in the bonding energy spectrum (activation energy) is reduced to zero by the symmetry reduction. Thus, instead of heat causing the fast chemical decomposition of a deflagration, the shear-induced change in the electronic structure causes decomposition, which generates considerable heat. In other words, the chicken and the egg become interchanged. The process is now observed as a detonation, and a fast reaction becomes superfast.

8.11. Diagnosis of Shock Fronts

The experimental information that is used to diagnose the nature of shock fronts is very limited. There is no direct way to make observations because of the short time scale and because of the small sizes of the regions of interest. The last major advance in diagnostic methods was the invention of the velocity interferometer in about 1965 by Barker and Hollenbach [28]. This greatly improved the time resolution and accuracy of particle velocity measurements. However, these are external parameters that give no direct information regarding the internal state of a shocked material. Also, the improved time resolution is still far too long to observe important process rates. The best time resolution is about one nanosecond, but there are important events taking about 10 fs. Thus, there is a shortfall of about six orders of magnitude. Short laser pulses have begun to bridge the gap, but they are not quite short enough to do so completely. Perhaps plasmon spectroscopy can be developed. Because it probes electronic structure rather than atomic structure, its potential time scale is about 100 times shorter than current laser spectroscopy. Its time resolution would be in the 10 as range. It has been shown elsewhere [29] that numerous physical properties of solids are determined by the concentration of valence electrons, as is the plasmon frequency, ω_p. A local measurement of ω_p reveals the local valence electron density and therefore the local value of several properties of interest. For example, Fig. 8.9 shows that if the local value of the plasma frequency is known, then so is the bulk modulus; and a good estimate of the local pressure can be obtained from this, together with *dB/dP*.

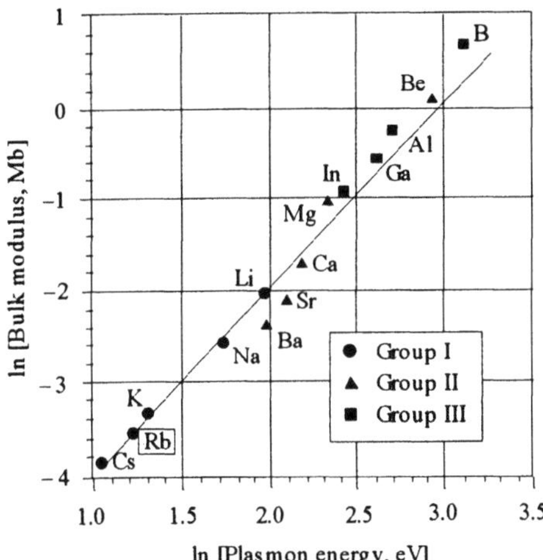

Figure 8.9. A ln–ln plot of bulk moduli vs. plasmon energies. For the simple metals from the first three Groups of the Periodic Table, the correlation line has a slope of two as expected from the theory of plasmons.

There are at least two ways of measuring plasmon frequencies. One is ultra-violet spectroscopy to determine the transparency edge (the original method of Wood); the other is EELS (electron energy-loss spectroscopy). In the context of shock fronts, a major problem is to find fast enough detectors.

8.12. Relaxation Times (Liquds, Solids, and Plastic Materials)

For all materials, their bulk moduli are nearly independent of time and have the same values for liquid, solid, or plastic specimens of the same composition. However, their shear moduli may differ considerably under standard conditions. As the strain rate, $d\varepsilon/dt$, is increased, and viscous processes create drag stresses, $D = \eta(d\varepsilon/dt)$, where η is the viscosity coefficient and $t =$ time. Under standard conditions, D is small for a liquid, but a liquid will support as much shear stress as a solid at high strain rates. That is, the maximum shear stress that a solid will support is $D = G/4\pi$, where $G =$ shear modulus, and

$$\frac{d\varepsilon}{dt} = \frac{G}{4\pi\eta}\,. \tag{8.13}$$

Then, for a liquid to support $G = 1\,\mathrm{Mbar} = 10^{12}\,\mathrm{d/cm^2}$, and $\eta = 1$ Poise, $d\varepsilon/dt \approx 10^{11}/\mathrm{s}$.

At room temperature, the viscosity of glycerin is about 1000 P. Thus, it will support a large drag stress, say 10 kbar, if the strain rate is $10^7/\mathrm{s}$ or greater. A 1-mm-diameter laser pulse lasting 1 ns can easily produce this strain rate.

The strain rate in a shock front of thickness 5Å, and velocity 10^6 cm/s is $2 \times 10^{13}/\mathrm{s}$. Therefore, even a small amount of viscosity, such as 5×10^{-4} P leads to drag stresses of the order of GPa.

8.13. Limiting Speeds of Dislocations

In a dislocation core, the local strain rate is approximately $= v/b$, where v is the dislocation velocity. For a metal with $b = 2.5$ Å, and

$$v \approx \frac{v_s}{2\pi} = \frac{1}{2\pi}\left(\frac{G}{\rho}\right)^{1/2}, \tag{8.14}$$

where $v_s = 3 \times 10^5$ cm/s , so the local strain-rate is $\approx 10^{13}/\mathrm{s}$.

In general, viscous drag limits the speeds of dislocations to values less than the speed of shear waves, v_s. With the mass density $= \rho = 5\,\mathrm{g/cm^3}$, and $G = 50$ GPa (copper), $v_s = 3 \times 10^5$ cm/s . The applied shear stress cannot exceed $\approx G/4\pi$. Setting this equal to the drag stress, $\eta v_s/b$, the required viscosity is about 3.3×10^{-3} P . This is comparable with the values directly measured for dislocations in metals.

For many years, it was asserted in the literature that inertia limits the speeds of dislocations to less than the speed of sound. This idea is incorrect for various reasons [30]. Fundamental among these are the facts:

a. The rest masses of dislocations are zero.

b. The standard inertial theory does not conserve angular momentum.

The idea that viscosity limits dislocation speeds is more consistent with experimental observations.

Another assertion in the literature is that dislocations can move at supersonic speeds. There is even a report that such motion has been observed in a computational simulation [31]. However, the behavior of the simulation was misinterpreted. Supersonic motion is not possible because the maximum shear stress that can be applied to a solid is about $G/4\pi$, and this is too small to accelerate the local atomic masses up to supersonic velocities in the time available when a dislocation is moving supersonically.

References

[1]   J.W. Taylor and M.H. Rice, "Elastic-Plastic Properties of Iron," *J. Appl. Phys.* **34**, p. 364 (1963).

[2]   J.W. Taylor, "Dislocation Dynamics and Dynamic Yielding," *J. Appl. Phys.* **36**, p. 3146 (1965).

[3]   P.P. Gillis and J.J. Gilman, "Dynamical Dislocation Theory of Crystal Plasticity. I. The Yield Stress," *J. Appl. Phys.* **36**, p. 3370 (1965).

[4]   J.J. Gilman, "Dynamic Criteria for Crack Nucleation and Growth," in *Proc. First Internat. Conf. Frac.*, Vol. 2 (ed.T Yokobori) Tohoku University, Sendai Japan. p. 733 (1966).

[5]   P.P. Edwards, T.V. Ramakrishnan, and C.N.R. Rao, "Metal–Insulator Transitions: A Perspective," in *Metal–Insulator Transitions Revisited* (eds. P.P. Edwards and C.N.R. Rao), Taylor & Francis, London, p. xv (1995).

[6]   J.J. Gilman, "Chemical Reactions at Detonation Fronts in Solids," *Phil. Mag. B* **71**, p. 1057 (1995).

[7]   C. Zener, *Elasticity and Anelasticity in Metals*, Univ. Chicago Press, Chicago, (1948).

[8]   S.P. Timoshenko, *History of the Strength of Materials*, Dover Publications, New York, (1983).

[9]   T.E. Tietz and J.E. Dorn, "The Effect of Strain Histories on the Work Hardening of Metals," in *Cold Working of Metals*, American Society for Metals, Cleveland, Ohio, p. 163, (1949).

[10] J.H. Hollomon and L.D. Jaffe, *Ferrous Metallurgical Design*, J. Wiley & Sons, New York, (1947).

[11] J.J. Gilman and W.G. Johnston, "Dislocations in Lithium Fluoride Crystals," in *Solid State Physics - Vol. 13* (ed. F. Seitz and W. Turnbull), Academic Press, New York, p. 147 (1962).

[12] J.J. Gilman, "Mechanism of the Koehler Dislocation Multiplication Mechanism," *Phil. Mag. A* **76**, p, 329, (1997).

[13] M.B. Bever, D.L. Holt, and A.L. Titchener, *Progr. Mat. Sci.* **17**, p. 1, (1973).

[14] G.H. Wannier, *Statistical Physics*, J. Wiley & Sons, New York, Chap. 22, (1966).

[15] H.S. Chen, J.J. Gilman, and A.H. Head, "Dislocation Multipoles and Their Role in Strain-Hardening," *J. Appl. Phys.* **35**, p. 2502, (1964).

[16] G.I. Taylor, "The Testing of Materials at High Rates of Loading"—The James Forest Lecture, *Jour. Institution of Civil Eng.*, #8, October 1945-46, p. 486.16, (1946).

[17] K.A. Rakhmatulin, "Propagation of a Wave of Unloading," *Appl. Math. & Mech.* **9**, p. 91 (1945).

[18] T. von Karman and P. Duwez, "Propagation of Plastic Deformation in Solids," *J. Appl. Phys.* **21**, p. 987 (1950).

[19] J.J. Gilman, "The Plastic Wave Myth," in *Shock Compression of Condensed Matter—1991* (ed.,S.C. Schmidt, J.J. Dick, J.W. Forbes, and D.G. Tasker), Elsevier Science Publishers B.V., New York, p. 387 (1992).

[20] P. Grassia, "Dissipation, Fluctuations, and Conservation Laws," *Amer. J. Phys.* **69**, p. 113 (2001).

[21] M. Parrinello and A. Rahman, "Strain Fluctuations and Elastic Constants, *J. Chem. Phys.* **76**, p. 2662 (1982).

[22] M.C. Lea, "Disruption of the Silver Halide Molecule by Mechanical Force," *Phil. Mag.* **34** (5th Series), p. 46 (1892).

[23] J.J. Gilman, "Shear-induced Metallization," *Phil. Mag. B* **67**, p. 207, (1993).

[24] P.W. Bridgman, "Effects of High Shearing Stress Combined with High Hydrostatic Pressure," *Phys. Rev.* **48**, P. 825 (1935).

[25] J.J. Gilman, "Shear-induced Chemical Reactivity," in *Metal-insulator Transition Revisited*, (ed. P.P. Edwards and C.N.R. Rao), Taylor & Francis, London, p.269 (1995).

[26] J.J. Gilman, "Mechanism of Shear-induced Metallization," *Czech. J. Phys.* **45**, p. 913 (1995).

[27] M.M. Kuklija and A.B. Kunz, "Electronic Structure of Molecular Crystals Containing Edge Dislocations," *J. Appl. Phys.* **89**, p. 4962 (2001).

[28] L.M. Barker and R.E. Hollenbach, *Rev. Sci. Instr.* **36**, p. 1617 (1965).

[29] J.J. Gilman, "Plasmons at Shock Fronts," *Phil. Mag. B* **79**, p. 643 (1999).

[30] J.J. Gilman, "The Limiting Speeds of Dislocations," *Met. & Mat. Trans. A*, **31A**, p. 811 (2000).

[31] P. Gumbsch and H. Gao, *Science* **283**, p. 965 (1999).

CHAPTER 9

The Discontinuous Shock—Fact or Fancy?

Ronald L. Rabie

9.1. Introduction

There is scientific interest in the use of shock waves to generate material conditions that are extreme states of matter. In the strongest shock waves commonly generated in the laboratory, pressures of hundreds of gigapascals and corresponding temperatures of an electron volt or two may be reached. In porous materials pressures are usually lower but the temperatures can be significantly higher. In some cases it has been argued, on the basis of empirical evidence and induction, that certain processes measured in shocked systems would be most easily explained if the shocks were essentially discontinuous changes in the state of the material, i.e., mathematical and physical discontinuities. Later in this section a practical definition of a "physical" discontinuity is provided. Clearly, in a material made up of atoms, one must pick a scale that is satisfactory to the notion of "physically" discontinuous for the problem at hand. It is easy to see that there is a huge difference in striking a diatomic molecule impulsively on one atom in a direction along the axis connecting the atoms and in pushing on the same atom in the same direction gently over a longer period of time to get the molecule to the same total center of mass energy. This conceptual difference lies at the heart of the interest in the structure of shock waves. Are shocks catastrophic (impulsive on the scale of atoms or molecules) or not? If shocks can be catastrophic, how does it happen, how is the structure maintained, and what is the dissipative mechanism if there is one? Finally, is the state at the end of the shock process actually an equilibrium state or does one simply hope that it is?

In this chapter, the meaning of shock-wave structures is examined and a recent experimental project aimed at gaining the basic understanding of shock processes at the molecular level is described. Before proceeding, some definitions are offered. The knowledgeable individual may skip this section. By an elastic material, one envisions a system of beads connected to one another by linear springs. This may be a one-, two-, or three-dimensional system. The force rule is that the force on any bead is linear in the relative displacement of the bead from the neighboring beads to which it is connected. When one refers to elastic behavior, one is referring to the behavior of such an elastic material as it undergoes deformation. An elastic material will return to its original shape exactly if deformed and released. The terms plastic and plasticity are also used in the following. The simple experience one has of plastic behavior in metals is when a bit of metal is permanently deformed by bending. Such deformations are called plastic deformations. The structure of shock waves is discussed extensively in the following. Some notion of the nature of such waves will be helpful in understanding the remainder of this chapter. If one were to sit on a shock wave and view the process that occurs as material flows through the shock, one would see the following. On one side of the shock, the view would be a stream of material arriving at the shock with a velocity in excess of the sound speed in the incoming material. At the shock, the density, pressure, temperature, internal energy, and entropy of the incoming material would be rapidly changed. If the shock is compressive, all of these variables would suffer rapid increases. On the other side of the shock, one would see this altered material flowing out at a speed less than the incoming velocity and less than the speed of sound in the compressed material. Shock waves are supersonic, faster than sound signals, in the material ahead of the shock and subsonic, slower than sound signals, in the material behind the shock. The statements made above may be quantified by use of the physical concepts of conservation of mass, momentum, and energy. When these are considered for the case of a continuum, the resulting expressions are known as Euler's Equations. When dissipative terms are added to Euler's momentum and energy equations, they are often referred to as the Navier–Stokes Equations.

9.2. Beginnings

A particular solution of the inviscid Euler Equations that describe the motion of a continuum is a propagating discontinuity in all thermodynamic state variables describing the material of interest. The Euler Equations are

$$\frac{\partial \rho}{\partial t} + \frac{\partial (\rho u)}{\partial x} = 0 , \tag{9.1}$$

$$\rho \left[\frac{\partial u}{\partial t} + u \frac{\partial u}{\partial x} \right] = -\frac{\partial p}{\partial x} , \tag{9.2}$$

and

$$\frac{dE}{dt} + P \frac{dV}{dt} = 0 . \tag{9.3}$$

The derivation of the particular discontinuous solution of interest follows. Equation 9.1 represents the principle of conservation of mass, Eq. 9.2 the principle of conservation of momentum, and Eq. 9.3 the principle of conservation of energy. The independent variables t and x are the time and Eulerian (Laboratory) position, respectively, and ρ, u, P, V and E are the mass density, particle velocity, pressure, specific volume (reciprocal of the mass density), and specific internal energy, respectively. The derivatives in Eq. 9.3 are material derivatives defined by the operator equation

$$\frac{d}{dt} = \frac{\partial}{\partial t} + u\frac{\partial}{\partial x}. \tag{9.4}$$

The transformation to a steady-wave frame having a velocity U is made by replacing the particle velocity in the above equations with

$$w = u - U$$

and setting the partial time derivatives to zero, which results in the set of ordinary differential equations

$$\frac{\partial(\rho w)}{\partial x} = 0, \tag{9.5}$$

$$\rho w\frac{\partial w}{\partial x} = -\frac{\partial P}{\partial x}, \tag{9.6}$$

and

$$\frac{\partial E}{\partial x} = -P\frac{\partial V}{\partial x}, \tag{9.7}$$

which can be integrated directly. The result of the integration is

$$\rho w = \rho_0 w_0, \tag{9.8}$$

$$\rho_0 w_0 (w - w_0) = -(P - P_0), \tag{9.9}$$

and

$$E - E_0 = \frac{1}{2}(P - P_0)(V_0 - V), \tag{9.10}$$

where, in writing Eq. 9.10 for the energy, we have used the first two conditions to get $P(V)$ prior to the integration. Nothing about the solution obtained thus far says that Eqs. 9.8–9.10 describe a discontinuity. The assumption we have made is that we want the conditions for a steady flow. Equations 9.8–9.10 are those conditions. How do we get to a discontinuous shock from here? The most direct route is to note that in the treatment immediately following this where viscosity and heat conduction are considered, it will be shown that the wave thickness approaches zero as the viscosity approaches zero, and the resulting equations are identical to those just given. It is important to note that both steady waves of arbitrary thickness and discontinuous shocks satisfy exactly the same set of equations.

Generally, we want values in the laboratory frame rather than in the shock frame so we substitute for w and w_0 the values $w = u - U$ and $w_0 = u_0 - U$. The subscript "0" refers to the state ahead of the shock while the unsubscripted variables refer to the shocked state. The equations in the form above are referred to as the Rankine–Hugoniot Jump Conditions or simply the Jump Conditions when applied to shocks. The Jump Conditions are three equations in five unknowns and require an Equation of State and a boundary condition for complete specification. It is immediately apparent that an Equation of State in the form $E(P, V)$ would be most useful in the solution of shock-wave problems in that such an EOS removes the difficulty of dealing with the introduction of additional thermodynamic variables such as the entropy, for instance, in a complete EOS of the form $E(S, V)$. Variables in the Jump Conditions are only defined in equilibrium states on the EOS surface for discontinuous shocks, thus the notion that one may eliminate particle velocity from the momentum jump condition in favor of specific volume and have a continuous $P(V)$ locus from the initial to the final state is in error. In fact, carrying out the suggestion above results in the expression

$$P_\mathrm{S} - P_0 = (\rho_0 U)^2 (V_0 - V_\mathrm{S}), \tag{9.11}$$

known as the equation of the Rayleigh Line. It is somewhat unfortunate that this term has been used in the case of discontinuous shocks, in that it encourages one to think of a continuum of states along this line with well defined specific volumes and corresponding pressures. It is, in fact, just an expression connecting the initial state in a discontinuous shock process with the final state and nothing more.

If the Euler Equations are modified to include viscous stresses or pressures and thermal conduction, the discontinuous nature of the change in state variables is removed and the wave across which the changes occur becomes spread in space and time. One may still look for steady-wave solutions of finite thickness, however. The changes through the wave constitute an equilibrium thermodynamic component with non-equilibrium dissipative term(s) added. The derivation from the Euler Equations with dissipation follows. Once again, in the frame in which the shock is stationary we have

$$\frac{d(\rho w)}{dx} = 0, \tag{9.12}$$

$$w_0 \frac{dw}{dx} + \frac{dP}{dx} - \frac{d}{dx}\left[\frac{4\mu}{3}\frac{dw}{dx}\right] = 0, \tag{9.13}$$

and

$$w_0\left[\frac{dE}{dx} + P\frac{dV}{dx}\right] = \frac{4\mu}{3}\left(\frac{dw}{dx}\right)^2 - \frac{d}{dx}\left[\kappa\frac{dT}{dx}\right]. \tag{9.14}$$

Note that when the viscosity coefficient, μ, and the thermal conductivity coefficient, κ, are set equal to zero, these equations reduce to the set of equations integrated above. The addition of a heat conduction term in Eq. 9.14 is interesting in its own right. By introducing such a term, one makes the tacit assumption that for the process one is about to describe, temperature will be defined. That is, one will have either thermodynamic equilibrium, even in the presence of large spatial gradients in the flow, or some suitable definition of a temperature in terms of the other flow variables. We may acquire first integrals of the above expressions as

$$\rho w = \rho_0 w_0 ,$$ (9.15)

$$P - \frac{4\mu}{3}\frac{dw}{dx} - P_0 = \rho_0 w_0 (w - w_0) ,$$ (9.16)

and

$$E + PV + \frac{1}{2}w^2 + \frac{1}{\rho_0 w_0}\left[-\kappa\frac{dT}{dx} - \frac{4}{3}\mu w\frac{dw}{dx} \right] = E_0 + P_0 V_0 + \frac{1}{2}w_0^2 .$$ (9.17)

We now specialize to a monatomic ideal gas. Note that the conservation of momentum, Eq. 9.16, is just the equation of the Rayleigh Line from the inviscid development above if one replaces P with $\sigma = P - (4/3)\mu(dw/dx)$. Furthermore, unlike the case of the discontinuous shock, the thermodynamic states on this line actually occur within the shock. The magnitude of the offset of the equilibrium point from the Rayleigh Line is just the viscous stress. Following Becker [1], and taking the Prandtl number, i.e., $\mu C_P / \kappa = 3/4$, allows one to convert the derivatives of temperature and velocity into a total derivative (recall we are discussing a monatomic ideal gas). That is,

$$-\kappa\frac{dT}{dx} - \frac{4\mu}{3}w\frac{dw}{dx} = -\frac{4\mu}{3}\frac{d}{dx}\left[E + PV + \frac{1}{2}w^2 \right].$$ (9.18)

It is clear from this expression that if, as x approaches positive infinity, the quantity in the derivative above is to remain finite, it must be independent of x. Thus,

$$E + PV + \frac{1}{2}w^2 = E_0 + P_0 V_0 + \frac{1}{2}w_0^2 .$$ (9.19)

Noting that for the ideal gas under consideration, $E = (3/2)PV$ and setting $\eta = \rho_0 /\rho$ one may write the expression above giving P as a function of V within the shock. The result [2, p. 472] is

$$P = P_0 \frac{4\eta_1 - \eta^2}{(4\eta_1 - 1)\eta} .$$ (9.20)

One notes immediately that this is not the equation of the Rayleigh Line, nor is it the Hugoniot. The expression above gives the equilibrium states (integral curve) the gas is assumed to go through in the shock process. It is important to recognize that for these states to be conventionally defined, the onset of equilibrium must occur on a time scale short relative to the shock-wave risetime or one must have non-equilibrium definitions of such variables as temperature and entropy. In the pressure volume plane this curve lies above the Hugoniot but below the Rayleigh Line and goes through both the initial and final states. At any specific volume between these values, the total stress acting is the sum of the pressure from the above equation and the viscous stress, given by $-(4/3)\mu(dw/dx)$. It is clear that the viscous stress is not the offset between the Hugoniot and the Rayleigh Line as is often depicted, but rather it is the offset between the integral curve and the Rayleigh Line. Replacing P with P_1 and η with η_1 (the subscript "1" denoting the final shock state) in Eq. 9.20 results in the equation of the Hugoniot—the curve of single shock end points in the EOS surface. Carrying this out, one finds

$$P_1 = P_0 \left(\frac{4 - \eta_1}{4\eta_1 - 1} \right),$$

where we note that the pressure on the Hugoniot of an ideal monatomic gas approaches infinity when the gas is compressed by a factor of 4.

Using the conservation of mass and momentum for the viscous flow given above, one may write an expression for the continuous change in velocity through the wave, noting that $\eta = w/w_0$. The result is

$$\frac{5}{3}\frac{\mu}{\rho_0 w_0}\eta\frac{d\eta}{dx} = (\eta - 1)(\eta - \eta_1). \tag{9.21}$$

The solution of this equation is an exponentially modified polynomial in η of order η_1 [2, p. 473]. The exponential term is the only one of interest to us in the discussion of shock thickness. It is approximately

$$\exp\left(\frac{(U/C_0)^2 - 1}{U/C_0} \right)\frac{x}{l_0}, \tag{9.22}$$

where C_0 is the initial state sound speed given by $[(5/3)P_0V_0]^{1/2}$ and

$$l_0 \sim \frac{\mu}{\rho_0}v_{ave}$$

is the mean free path of an atom of the gas and v_{ave} is the average thermal velocity of the atoms at energy E. It is clear from the argument of the exponential that the thickness of the shock is the order of

$$\delta = \frac{l_0 M}{M^2 - 1}, \tag{9.23}$$

where

$$M = \frac{U}{C_0} \tag{9.24}$$

is the Mach number. It is analyses such as this that lead to the observation that shocks in monatomic ideal gases are the order of a few mean free paths in thickness. For $M = 2$, we get that $\delta = 2l_0/3$. As M becomes larger, this particular development predicts that δ should approach zero. Here is a place where we come to the conclusion that *neither the Navier–Stokes nor the Euler Equations apply*. These equations are predicated on the assumptions that 1) the material is a continuum, i.e., has no finite scale such as a mean free path, and 2) when such a scale is present, we agree to ignore results when gradients in the flow occur on such scales. We have neither condition. It is a remarkable stroke of good fortune that the Navier–Stokes equations work as well as they do in predicting shock wave thickness under such abuse.

Other approaches to determination of shock structure in gases have been carried out. Notable among them are the attempts from the point of view of acquiring solutions to the Boltzmann equation for specific assumptions regarding the initial and final probability distributions. Generally, these distributions are weighted by a number fraction of atoms in one or the other of the distributions that are functions of the spatial coordinate in the direction of shock propagation. The problem, then, is to find these number fractions from which the entire flow may be determined. The first such calculation was by Mott-Smith [3] in which various moments of the velocity were calculated, namely, w^2 and w^3. This was followed by the work of Sakurai [4], Ziering and Ek [5], and Glansdorf [6]. In the end, the shock thickness was found to depend on the Mach number, Eq. 9.24, and transport coefficients and to be of the order of a few mean free paths (Glansdorf's results give a thickness of 11.4 mean free paths for strong shocks).

It is always possible to solve the Navier–Stokes Equations numerically for arbitrary forms of dissipation and arbitrary equations of state to determine steady-shock-wave structure in one dimension. When one resorts to this approach, there is no reason to avoid complex formulations of heat conduction, reactivity, or other dissipative mechanisms. In such solutions, it is commonly found that the approach to states in which the gradients in these dissipative mechanisms formally go to zero, i.e. thermal equilibrium prevails, occurs only as distance from the "shock" approaches infinity. It is likely that this is the case in virtually all shock-wave experiments on solid materials in which the dissipative mechanism is dominated by permanent deformation. Rearrangements or continued motion of shear bands, dislocations, etc., after the passage of the shock must surely continue as well as the transfer of energy by heat from locally hot regions of high dissipation to cooler regions where little dissipation occurred. If such is the case, one is forced to conclude that the shocked state in solids is not an exact state of thermodynamic equilibrium and that, in a mathematical sense at least, no such shocks are steady in finite time.

Almost limitless variety is possible by extending Euler's original framework. One may formulate the Euler Equations for multicomponent systems by adding mass source and sink terms to the species continuity equations, momentum source and sink terms to the species momentum conservation equations, and energy source and sink terms to the species energy conservation equations. One generally defines these added terms in such a way that, when the equations are summed over all constituents present (when one averages away the multispecies details), the sums over the interaction terms vanish so that one recovers the single species Euler Equations with only effective dissipative terms remaining. In the multispecies formulation one may make a wide range of assumptions regarding material properties and mixture rules appropriate to the system under study. For instance, one may imagine a nonlinear elastic porous material with interconnected pores filled with a gas or liquid that experiences no interaction, such as drag, with the elastic porous matrix. If both materials are inviscid and noninteracting, shocks will propagate individually in each component as though the other were not present. In the strong coupling limit, where the drag between the elastic porous matrix and the gas or liquid is infinite, they will share a common particle velocity and a simple mixture rule Equation of State (EOS) will involve the mass fractions of each constituent present. This EOS along with the jump conditions provide the changes in pressure, internal energy, specific volume, etc., across the single discontinuous wave. When viscosity, heat flow, pore collapse, or any other dissipative mechanisms are included in the description of the elastic porous medium and the gas or liquid interstitial phase, and mass, momentum, and energy exchange are allowed between phases, the dynamic situation within the shock may become quite complex (see, for example, [9] and [10]).

In the case of multicomponent systems, one has introduced into the continuum theory the possibility of thermodynamic nonequilibrium among the phases present. A particular flow history may result in a single phase being heated while others remain cold, for example. The interaction terms among the phases will establish equilibrium on a time scale governed by the rate constants (coupling constants) for the processes. This view will be expanded in later discussion of the catastrophic shock.

A great deal of shock wave work has focused on metals. We discuss this topic mainly because some of the best work in shock-wave structure has come from studying metals at relatively low strain rates. One of the principal results is that metals show the phenomenon of elastic–plastic behavior for loading stresses nominally less than the order of 1/3 of the material shear modulus. For most metals, this stress is at least 15–20 GPa. Thus, waves with final stress less than this value show both an elastic wave and a following slower plastic wave. Plastic waves and the stress states composing them are limited by a yield condition involving limits on the value of a function of the principal stresses [11]. Generally, one is considering isotropic materials and one-dimensional strain so that, with a plastic flow rule, enough information is available from experiment to determine unambiguously the stress state throughout the plastic wave. Swegle and Grady [12] proposed a simple model for the risetime of the plastic wave in metals. Simply put, the risetime goes as the inverse fourth power of the stress. In the case of a nearly overdriven plastic wave (when the plastic wave speed equals the elastic precursor speed), minimum measured risetimes of a nanosecond or so are observed. This is at stresses of 10–20 GPa. Extrapolating this to 50–60 GPa, one finds that rise times of a few picoseconds are predicted. Wallace [13] notes in his work that the risetime of strong shocks in metals should be in the range of a picosecond. For experiments at these time and space scales, the conventional shock and particle speeds often expressed in millimeters/microsecond can be more conveniently expressed as nanometers/picosecond. A typical lattice spacing in metals is the order of 1/2 nm, so that shock velocities of the order of 7–8 nm/ps suggest the wave thickness is only 15–20 atoms for strong shocks. The total strain through the wave might be 10–30% so that the strain rate would be the order of 10^{11} s^{-1}. The accompanying stress gradient would be the order of 1–2 GPa per lattice spacing.

Wallace's [13] argument for the risetime invokes the notion that heat conduction may be the operative dissipative mechanism in determining the steady-state shock thickness. In metals, heat conduction by free electrons supplements the lattice heat conduction of phonons in the presence of a thermal gradient and involves the electron thermal conductivity, $\mathbf{K}$, which is given generally as

$$\mathbf{K} = \frac{\pi^2}{3}\left(\frac{k_{\mathrm{B}}}{e}\right)^2 T\,\boldsymbol{\sigma},$$

where k_{B} is Boltzmann's constant, e is the electronic charge, T is the absolute temperature, and σ is the stress. This well-known result states that the electron thermal conductivity depends on the temperature and the electrical conductivity, that is, the heat transported by the free electrons in the metal through the shock thickness results from the presence of a thermal gradient and a flow of charge (electrons). An electric field opposing this flow of electrons will arise that will, in concert with the thermal gradient, generate a steady-state charge distribution through the shock thickness. This displacement of electrons in the direction of the thermal gradient must occur and should be measurable on a sufficiently short time scale with a technique sensitive to electron density or perhaps, at a free surface, electron work function. Such a measurement would provide some experimental evidence about the nature of the dissipation in strong shocks in metals.

Solids undergoing shock compression are fundamentally different than either gases or liquids. The analog of the gas-phase mean free path might be taken to be the average atom spacing (lattice dimension) in the solid. Solids are generally not homogeneous and isotropic, having orientational defects (grains) as well as point, line, and surface defects in abundance. When shocks are introduced into such structures, the nature of the shock becomes problematic. One begins to have visions of shocks within grains, shocks being diffracted, refracted, and reflected by defects, etc. Hayes and Grady [7], following on the work of Swegle and Grady [12], were able to show, with some rather simple assumptions about heating in shear bands, that shocks in shear banding materials ought to be the order of a few shear band spacings in width. Their argument is essentially this: if shear banding and the associated dissipation *is* the dissipation required for a given shock structure, then the shock must be steep enough (have high enough strain rate) to produce the shear bands but cannot be thinner than a few of them to get, from the resulting shock structure, the required dissipation. This is an important point. If one postulates a particular set of mechanisms that are intended to account for the dissipation required for a shock, then the space scale over which such dissipation is effective determines, to first order, the thickness of the shock. In the case of the monatomic ideal gas considered above in the Navier–Stokes approximation, the viscosity, directly proportional to the mean free path, is both the dissipative mechanism and establishes the scale of shock thickness. It is not the case that every recognizable internal structure will, in some way, contribute to the shock thickness. Distortion of a sharp shock by a macro- or mesoscale internal feature in a material, such as an object of hundreds of mean free paths extent suspended in a gas shock tube, is a different problem than the determination of the fundamental unperturbed steady-state shock structure. The averaging of shock structures by distortion at the macro- or mesoscales, if the intrinsic shock thickness is smaller than either, may be viewed as a statistical effect directly related to the defect structure.

Finally, sophisticated molecular-dynamics calculations potentially provide a more physically realistic alternative to continuum modeling in the sense that atomic scale structures are at the heart of the formulation from the start. Although the internal structure of atoms is not considered at present, some internal structure of at least simple molecules is. The trend in such work is to use the ever increasing computational capacity that is available to allow calculation of systems containing a few million atoms. Such systems have dimensions of a few hundred atoms on a side at most and are physically about 0.5 μm on a cube edge. Shocks in these systems are very thin—the order of tens of atoms at moderate pressures and thinner as pressures increase.

9.3. The Nature of the Shock

We have distinguished between the nature of steady waves and discontinuous waves. A truly discontinuous shock satisfying the Jump Conditions may change, in amplitude for instance, constantly in time, perhaps from running into a nonuniform initial state, while continuing to satisfy the Jump Conditions at each instant of time. A shock wave of finite thickness will also satisfy the Jump Conditions even though the change in state is not discontinuous, provided the wave is steady, i.e., not changing shape with time as has been demonstrated for the viscous heat conducting gas above. For the discussion that follows, we take discontinuous to mean a shock so thin as to be the order of a few molecules or lattice spacings in thickness. This is sufficient to the purpose at hand, namely, to distinguish between what have been termed benign and catastrophic shocks [14]. In the case of the benign shock, one envisions a shock as a gradual alteration of state from an equilibrium thermodynamic state initially, through a dissipative compression in which the departure from equilibrium is so slight (a single thermodynamic temperature is always defined) that one may still calculate the sequence of equilibrium states associated with each point in the wave, i.e., the integral curve. When a dissipative function is assumed that gives the stress offset

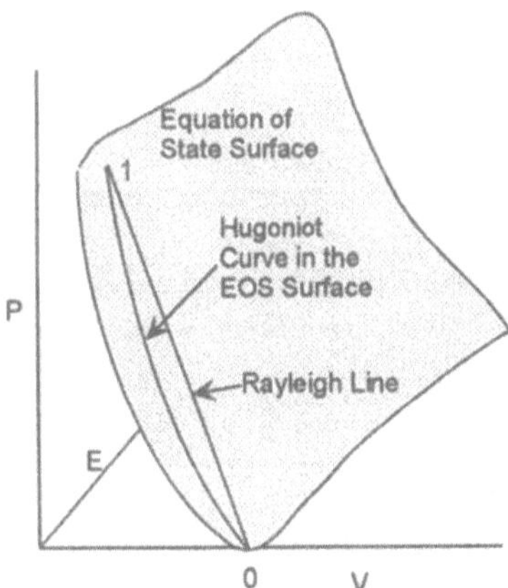

Figure 9.1. A schematic representation of an equation-of-state surface in pressure, specific volume and specific internal energy space. States 0 and 1 are the foot and head of a shock or steady wave. The Hugoniot curve is a locus of single shock end states that is on the EOS surface, the Rayleigh line is the line from the state 0 to the state 1. For a steady wave of finite thickness, the states on the Rayleigh line are defined in terms of stress. For a discontinuous shock, states on the Rayleigh line other than states 0 and 1 do not exist.

between the Rayleigh Line and the integral curve, a complete specification of a thermodynamic state at each stress point along the Rayleigh Line is possible; see Fig. 9.1. In the case of the discontinuous (catastrophic) shock, the notion of a continuum, on the scale of the shock width, is lost and attempts to specify a shock thickness by means of a dissipative function based in continuum theory are in error.

In the case of the catastrophic shock one envisions the possibility of direct mechanical excitation of a nonequilibrium distribution of energy (multiple temperatures not unlike the situation with laser excitation of electrons while leaving the lattice of ions cold or the case of heating a single component in a multicomponent material) within the modes of the system that relaxes on a (generally) fast time scale to a Boltzmann distribution. This will be irreversible, will increase the entropy, and will be dissipative. Such dissipation is certainly not viscous but is irreversible *heat production*. Although subsequent conduction of heat may occur, one supposes that it is the production of heat within the shock that is the initial dissipative contribution. If one can define the statistical distribution associated with the initial state and if one additionally knows the final equilibrium distribution the state goes to and the interactions by which this happens, then one can solve Boltzmann's transport equation, in principle, and get the rate of entropy production (see, for example, [15] or [3]). Also, if one knows the total energy deposited as heat and the modes excited initially, one can calculate a "temperature" for each particular mode. For example, one may view the electrons and ions in a metal as separate modes (with their own separate equations of state as in a continuum multicomponent system) for energy deposition. Then one has a case in which the excitation of one without excitation of the other is possible. In fact, in short pulse laser excitation of metals one finds that the free electrons absorb the laser energy and are raised to very high temperature while the lattice of ions remains essentially cold [16]. The ensuing approach to a single Boltzmann distribution occurs on a time scale mediated by the electron–phonon coupling parameter (the interaction) of the material. Until the electrons and ions come to the same temperature, this system is out of thermodynamic equilibrium. Again, in a thermodynamic sense this is a dissipative process in that the entropy increases both as the laser pulse heats the electrons and as the system approaches equilibrium. This picture of a nonequilibrium excitation of a molecular or atomic solid by a shock wave and the following dissipative relaxation to equilibrium is in complete accord with continuum mixture theory but it offers a great deal more. The possibility of lattice and/or molecular nonequilibrium set up by a catastrophic shock wave provides theories that are dependent on the nucleation of defects with a natural source term, the shock wave. Similarly, chemical reactivity induced by shock waves of sufficient strength may have roots in either an equilibrium population of excited states in the presence of a gradual compression or the direct, essentially instantaneous, population of a restricted set of modes in utter nonequilibrium established by a "catastrophic" shock (a nice classical calculation of a diatomic struck on one end—a beads on a string model—is presented in a paper by Rose and Martens [8]). At present, there are no definitive experimental answers to these questions of shock structure in condensed phase materials. There is ongoing work directed to this goal, namely that of Dlott et al., and the HERCULES team at Los Alamos.

Computational molecular-dynamics results obtained by Brenner [17] and more recently by Germann and Holian [18] suggest that, for a prototype reactive system with realistic potentials, one finds nonequilibrium over a wide range of input stresses but the character of the nonequilibrium depends on stress. At low stresses one sees preferential over-excitation of lattice modes (hot phonon modes) that decay into the molecules as they cool, whereas, at higher stresses, the molecules are directly excited by the shock wave. Both regions show non-equilibrium behavior.

9.3.1. Beads On A String: The Ultimate Catastrophic Shock

A particularly simple but illuminating example of the catastrophic shock and the resulting shock-velocity–particle-velocity relation may be found by examining the model of a set of identical beads on a string. The case of beads on a string is more complex than often depicted, athough it is the simplest "molecular-dynamics" calculation that can be done. The resulting expressions give the signaling velocity of the head of the disturbance in terms of the boundary velocity of the bead set in constant motion. The results are strongly dependent on the ratio of bead diameter to distance between beads. The diameter of each bead is taken to be d, and the distance between the contact surfaces of the beads is taken to be L. The ratio of bead separation to bead diameter will be set to N. Setting bead number 1 (this bead has very large mass relative to the other beads and is unperturbed by collision—it is a rigid wall) into uniform motion at a speed u_p causes it to strike bead number 2 in a time Δt. Bead 2 will leave the collision with a speed $2u_p$ (the collision is assumed to be perfectly elastic). In a time $\Delta t/2$ bead 2 will strike bead 3 and will come to rest only to be struck by bead 1 again at a time $2\Delta t$. This continues as long as bead 1 keeps moving (energy is input into the system) and as long as the number of beads on the string is not exhausted. For any N it may be shown that the relation between the speed at which the wave moves and the average speed with which the beads move is

$$U_S = \frac{2N+1}{N} u_p.$$

For N approaching zero (particles touching), the coefficient of u_p becomes arbitrarily large. The significant aspect of this expression is that the minimum value of the coefficient (particles small relative to the distance between) is 2 for elastic collisions. When the collisions are perfectly inelastic, one finds that

$$U_S = \frac{N+1}{N} u_p .$$

It is evident in the case of perfectly inelastic collisions that a wide range of values of coefficients of the particle velocity are available. For N large, that is, L much greater than d, we have the coefficient of the particle velocity approaching 1. If N becomes small, that is, L less than d, the value of the coefficient of the particle velocity may increase without apparent bound. If the particles are touching in this case, the "shock speed" becomes infinite so the least particle motion is greatly multiplied. Again, the important feature in the inelastic case is that values of the coefficient of the particle velocity have values as small as 1. Such values are found in shocks in gases and in highly porous materials. In general, these results are written as

$$U_S = C_0 + S u_p ,$$

where, experimentally, C_0 is approximately the bulk sound speed and S is a parameter between 1 and 3 to 4 that says something about how inelastic the interaction is in a shock process. It is comforting to note that the vast majority of data on shocked materials give values of S between 1 and 4 for inert materials. It is notable that many explosives have values of S the order of 2.5 to 3.5, for low-pressure shocks, that flattens to 1.5 to 2 as the pressure increases.

The model of beads on a string gives the ultimate in catastrophic shocks. Rose and Martens [8] use the model to do a classical calculation of the excitation of a diatomic in such a system with results that compare quite favorably to classical and quantum molecular-dynamics simulations at higher temperatures. The interaction that propagates the shock occurs only between two beads. The shock does not extend over tens or hundreds of beads – only two. Yet, within this entirely simple model, one recovers the experimental fact of the linear shock-velocity–particle-velocity relation with values ranging (mostly) over known experimental values. One should note the physical conditions under which the beads on a string model is appropriate—no interaction potential until the beads contact one another—then the potential is large. When the kinetic energy of the ions of a solid is much larger than the interaction potential holding the ions in place in the lattice, one may imagine that the beads on a string model of the shock wave process is becoming more and more physically reasonable.

9.3.2. Steady Waves

Finally, it is worth while to expend a few words on the issue of the realization of steady waves. Other than rare, well-thought-out, experimental situations, it is not likely that any wave measured in shock-wave experimentation is steady. Dissipative rates, as equilibrium is approached, often become linear in some thermodynamic variable and thus reach the steady state only asymptotically in time.

Bland [19] did an analysis for a particularly simple constitutive system that was carried through analytically. He found a dependence of the distance to "steady" state that goes as

$$\delta = \frac{8}{3}\frac{C_0}{S}\frac{1}{\dfrac{d\eta}{dt}},$$

where C_0 and S are the coefficients in the linear shock-velocity–particle- velocity relationship and $d\eta/dt$ is the maximum strain rate in the shock [12]. In this, Bland assumed a step profile relaxing to an appropriate dissipative shock thickness in the distance δ. As an example, look at aluminum and assume a strain rate of 10^8 s^{-1}. The parameters C_0 and S are 5.35 and 1.34, respectively, so δ is about 10^{-7} mm or half a nanometer. Thus, within the model of Bland, one expects a step input into aluminum with the parameters above to be steady instantly. In many systems of interest, we have multiple layers of materials in contact with explosives driven over a range of conditions. Wave-matching conditions, steady-shock thickness that differs from one material to another as Equations of State change, etc., argue that only very unexpectedly do we have steady waves in the sense of truly unchanging shock-wave profiles. A question that is perhaps better than whether or not we have a steady wave is what is the maximum error that we have in assuming that a wave is steady when, in fact, it is not?

Consider the case of a wave in the process of shocking up but not yet there. In this case, the loading is often assumed to be isentropic until the shock forms. Here one may compare the pressure achieved at a given specific volume on the isentrope through the initial state to that on the Hugoniot at the same specific volume. If one considers only weak shocks, then an approximate answer may be given directly for any Equation of State. Generally, one may show that

$$\Delta P_{H-S} = -\frac{1}{12 T_0}\left(\frac{\partial P}{\partial S}\right)_V \left(\frac{\partial^2 P}{\partial V^2}\right)_S (V_S - V_0)^3$$

in which ΔP_{H-S} is the pressure difference at volume V_S between the Hugoniot and the isentrope through the initial state with temperature T_0 and volume V_0. For compressive shocks in normal materials, the volume difference is negative. All other terms in the expression are positive, thus the pressure on the Hugoniot is larger than the pressure on the isentrope at the same specific volume. The actual amount being dependent on the details of the Equation of State.

As a particular example, in aluminum at a shock state particle velocity of 4 mm/μs, one has an Hugoniot pressure of 115.2 GPa and a temperature of 1036 K. On the isentrope at the same specific volume the values are 96.2 GPa and 959 K. The pressure error in this case is nearly 20% and the temperature error is 7%. In aluminum, at pressures to about 20 GPa, these differences are slight—no more than a few percent. In the case of high-pressure loading, however, the difference in being on the Hugoniot or elsewhere on the EOS surface may be profound. The pressure–volume plot of this example is shown in Fig. 9.2.

The case of excessively sharp loading, excess dissipation, is less certain. The experience most have with this is in plate impact calculations in hydrocodes. In this case, one finds the interface between the projectile and target to have the following characteristics. The stress and particle speed are continuous across the interface and satisfy the jump conditions, whereas the internal energy and density do not satisfy the jump conditions. At this early time the evolving wave is neither discontinuous nor steady. There is excess internal energy at the interface and the density is low. This occurs because the dissipative function that controls the shock thickness in the hydrocode depends, often quadratically, on the rate of displacement gradient (du/dx in one dimension). At the impact face, this dissipative function is converting work to heat at a very large rate—far in excess of the rate of dissipation required for a steady-wave profile. The degree to which a material fails to satisfy all of the jump conditions initially on impulsive loading is intimately connected to the dissipative mechanism that controls the structure of shocks in the material.

If ongoing kinetic processes induced by a wave are sensitive to the wave structure (risetime or maximum strain rate), then the rate of the structural evolution of the wave may be important to sustaining the kinetic process. It is noteworthy that, whereas dissipative (endothermic) processes like heat production or phase transitions will cause a discontinuous wave to spread and have finite thickness or even split into two waves, exothermic processes may cause a wave to steepen and become thinner.

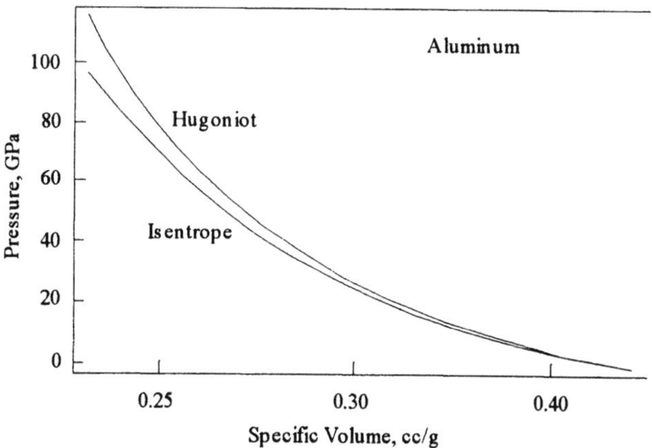

Figure 9.2. Hugoniot and isentrope for aluminum.

9.4. Some Experimental Notes

Experimental work in shock wave research has a robust and clever history. In general, one finds a way to strike a material impulsively and then to make a meaningful measurement of the resultant state (or sequence of states in some cases) as the shock passes through the sample. If one considers nuclear tests and the impact of comet Shoemaker–Levy on Jupiter at the largish end of the spectrum of observable shock-wave "experiments," then the small end might be the impact of molecules in molecular beams on inert targets with the subsequent generation of ionic fragments and their energies as the measured result. Similarly smallish experiments may also involve the "impact" of a "Gaussian plate" of photons in a very narrowly defined spatial and temporal pulse onto a target material that absorbs the photons in the electron structure with subsequent electron–phonon interactions generating a shock wave in some material of interest. Many other techniques have been used to generate shocks, including explosively driven experiments, gas guns firing flat projectiles to velocities of the order of 12 km/s, radiation drive in x-ray machines, etc. The limitations of a particular experiment in shock-wave physics have to do with the spatial and temporal resolution required by the phenomenon under study. If one is studying a problem with an intrinsic time constant of 1 ps, then measurements with time resolution of 1 ns will not elucidate the kinetics of the process—only the fact that such a process has occurred may be inferred from such measurements. For many applications, the knowledge that something has or has not happened at all may be sufficient. For the health and welfare of shock-wave physics research, such knowledge must be taken as no more than a starting point. The goal should be coming to an understanding of the underlying kinetics of the process of interest. Selection of an experimental approach appropriate to the spatial and temporal character of the process under study is essential to the possibility of success, although it certainly does not guarantee such success. However enchanting short (10–150 fs) pulse laser drive is as a well characterized and useful means of generating shocks, one must not lose track of the fact that plastic waves with steady-state risetimes of tens of nanoseconds are not going to be well studied by such means. Similarly, the detailed kinetics of the initial stages of shock-induced decomposition of complex molecules is not likely to benefit from VISAR or embedded gauging measurements until the time resolution of such probes gets down to the picosecond regime or faster. If one wishes to make a study of the localized onset of lattice destabilization, the formation of a dislocation for instance, and its subsequent motion, then spatial resolution of extraordinarily small size and corresponding temporal resolution will be required along with the ability to prepare simply exquisite samples—some guidance in this may be found in the recent work of Germann and Holian [18].

9.5. Recent Experimental Findings in Shock Structure

The remainder of this chapter discusses results obtained by the HERCULES Team at Los Alamos over the past few years. The HERCULES Project was started with the end goal being the experimental determination of the initial stages of condensed phase molecular excitation leading to reaction in strongly shocked high explosive molecules. In the course of planning an experimental program, it became quite clear that elucidation of the structure of the wave incident on the explosive sample would be of great importance. The samples used were similar in construction to those used by Dlott et al. [20]. A microscope coverslip was employed as the initial substrate onto which various shock-wave-driving materials such as aluminum, nickel, and gold were plated. Metal film thickness ranged from 50 nm to 2 μm. The experimental setup involved the generation of a high power, short duration laser pulse that was split into a pump and probe pulse [21]. Figure 9.3 shows the setup schematically.

The splitting of the laser pulse leaves the time registry of the two pulses unaltered. The pump pulse is focused to a spot size of approximately 75 μm on the converslip. The probe pulse is sent through an unbalanced Michelson interferometer to give a pair of pulses separated in time from each other by between 4 and 12 ps. These ultrafast probe laser pulses illuminate the sample surface over a 200-μm diameter where the shock wave will break out and are reflected from the surface onto the entrance slit of an imaging spectrograph. The frequency con-

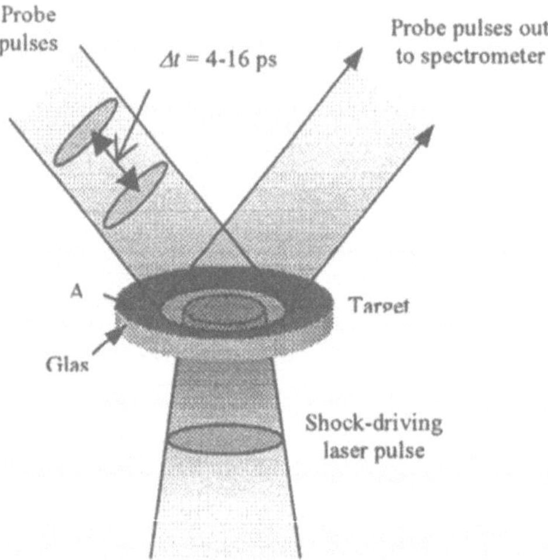

Figure 9.3. The experimental arrangement. The pump pulse enters from the bottom in this figure with the probes coming in at an angle and illuminating an area larger than the shocked area.

tent of the probe pulses allows them to be dispersed in the spectrograph and thus, be made to overlap one another spatially. A set of data is taken with the sample not moving prior to each shot. This serves as an optical fiducial interferogram. When the shot is fired, any motion of the surface between the arrival of the first and second probe pulses results in an altered interferogram. When these data are Fourier transformed, one may extract both the optical phase shift that has occurred and any changes in surface reflectivity. In the absence of optical effects, the free-surface displacement corresponding to the optical phase shift that occurs during the time between the two probe pulses is simply

$$\Delta x_{fs} = \Delta \phi \frac{\lambda}{4\pi \cos \theta}.$$

The angle θ is the angle of incidence of the probe pulses and λ is the wavelength. A target surface diameter of the elliptical region illuminated by the 200-μm-diameter probe pulses is imaged by the entrance slit of the spectrograph so phase shifts along this diameter are obtained with each shot. By continuously altering the delay between the pump pulse and the probes for a given probe to probe delay, one is able to follow the surface displacement of the free surface for tens or hundreds of picoseconds. In a typical experiment, a time history of the shock breakout event is built up from many repeated measurements.

The risetime of the free-surface velocity as the shock exits the thin film is recorded as the change in slope of the position versus time curve obtained as above, from the zeros recorded before shock arrival (no surface motion) to the nearly linear region after acceleration to the final free-surface velocity. The rise time we measure is the order of 2–7 ps for shocks of 5–15 GPa in aluminum, nickel, and gold [22]. There are, however, puzzling aspects to the data in both the aluminum and the gold that have been traced to different mechanisms.

Optical phase data from the shots on aluminum showed an unexpected decrease in the phase signal of the aluminum prior to the phase going positive and increasing as it should for positive surface displacements. The origin of this phase change was found to be an interband transition in aluminum at 800 nm. In the Brillouin zone of the fcc aluminum, one has a parallel band structure at one of the zone faces, (200), that originates just below the Fermi energy and terminates just above. The band below is nearly completely filled with electrons at room temperature, whereas that above is only slightly occupied. Because the change of band energy with wavevector in this region of the Brillouin zone is small, the joint density of states for the transition is quite large and absorption occurs. The energy jump between the bands is about 1.5 eV, which corresponds directly to the 800 nm laser light used in the experiment. The dependence of the position of this absorption on the pressure (shock state) in the aluminum provides a negative phase contribution to the signal [23]. When the pressure depend-

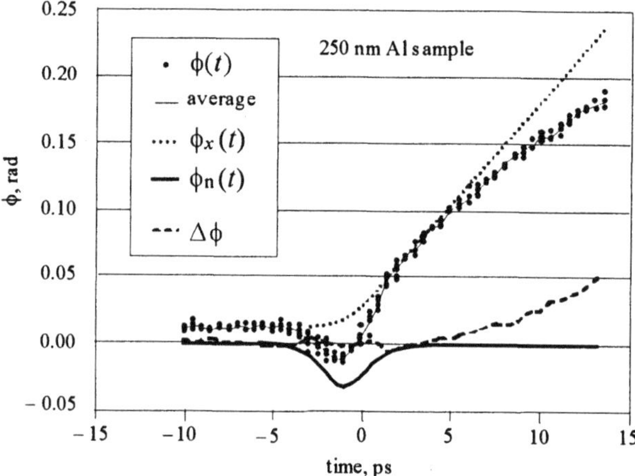

Figure 9.4. Phase vs. time of shock breakout from a thin Al film. Dots represent raw shot-to-shot data. $\phi_x(t)$ and $\phi_n(t)$ are the contributions to the phase of the surface displacement and refractive index respectively

ence of this interband absorption is taken into account and the data are modeled with near-free-surface pressure taken as proportional to surface acceleration (correct for a free surface), excellent agreement is found between theory and data [24]. Figure 9.4 shows a selection of phase data from a series of experiments on a thin aluminum film. Note that the time scale is in picoseconds.

It is evident from the record that the shock width in time is the order of 5 ps or less. Taking a value of 2 ps for the risetime and a shock speed of 6 nm/ps gives a shock thickness of 12 nm or about 25 lattice spacings in aluminium. If one identifies a mean free path with a lattice spacing, we have a shock thickness of about 25 mean free paths. In terms of mean free paths, this is about a factor of 2 greater than Gransdorf [8] calculates for gases. The flat portion of the record at negative times records the surface state prior to shock breakout. At about –5 ps the data trace begins to go negative and does not recover until near time 0. The analyzed trace is shown in the figure as the dark dotted curve where the effects of the interband transition have been removed. The shock rise is the region where the phase goes from flat to a steady linear increase. This "knee" is the shock risetime. Increasing the sample thickness from 250 nm to 2 µm does not alter this risetime.

There are some assumptions made in the analysis above that should be discussed. The raw optical phase signal is the bold series of dots. An assumption is made that the optical phase signal is composed of two parts—one due to the interband transition in aluminum and the other the free-surface displacement of the aluminum as the shock emerges. The surface displacement is taken to be a hyperbolic tangent

$$x_j(t) = \int_{t_i}^{t_t} u_p\left[1 + \tanh\left(\frac{t - t_j + \delta t_j}{\tau}\right)\right] dt$$

in which u_p is the shock state particle velocity, t is the time and the parameter, τ is the risetime given that the shock is well approximated by the hyperbolic tangent function. In addition to this term for the surface displacement with time, an additional term proportional to the shock state pressure is also included in the analysis. At a free surface, one may show that the shock state pressure is proportional to the surface acceleration. Given the form above for the displacement, two differentiations in time give the square of a hyperbolic secant. Thus, the pressure-induced phase contribution is modeled as

$$\Delta\phi_j(t) = \Delta\phi_{n,neff,k,keff} \operatorname{sech}^2\left(\frac{t - t_j + \delta t_j}{\tau}\right),$$

where

$$\Delta\phi_{n,neff,k,keff}$$

is a term that includes the optical phase shift due to a 4.9-nm-thick Al_2O_3 layer on the free surface (n, k) as well as a dependence on the effective complex index (n_{eff}, k_{eff}) of the underlying aluminum, which may be extracted from the Fresnel relations for s and p polarized light as a function of incident angle [24]. The total optical phase shift is then given as

$$\phi_j(t, \theta) = \Delta\phi_j(t) + \frac{4\pi\cos\theta}{\lambda_0} x_j(t),$$

where the angle θ is the incident angle of the probe pulses. It is evident that taking θ to be near $90°$ is advantageous in separating the state-dependent phase shift from the free-surface motion. Data were taken at angles of 32.6, 82.5, and 84.5. A non-linear (Levenberg–Marquardt) algorithm was used to simultaneously fit all data sets for the parameters: τ, u_p, t_j, n_{eff}, k_{eff}. The results of this technique for the three data sets taken at the angles specified above are shown in Fig. 9.5.

A number of important observations may be made as the result of these experiments. First, to accumulate such data sets vast numbers of shots must be done. The data represented in Fig. 9.5 alone would be the work of a lifetime if acquired on a gas gun. Second, when electronic effects may be anticipated in a shock-wave problem, time resolution appropriate to the problem is required. While a great number of optical free-surface-displacement measurements have been made on many metals, no evidence of state dependent complex index changes have been seen prior to the ones reported by Gahagan [22] and Funk [24]. The value of τ found from the fit to the data sets above is 2.34 ps. This rep-

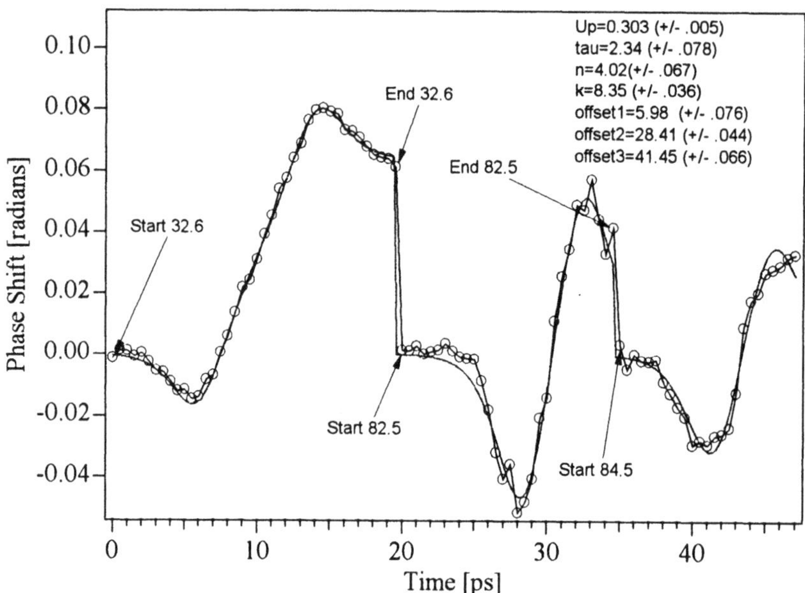

Figure 9.5. Data sets taken with a 750-nm thick aluminum sample and three different angles of incidence. Circles are averages of at least four shots at each time delay and the line is the fit to the functional form given above. The record represents nearly 500 individual shock wave experiments.

resents about the 60–70% risetime of the hyperbolic tangent, which may be a poor approximation to the actual wave shape. Thus, if one takes a risetime of the order of 2 ps for these shocks, the thickness is 12 nm (25 lattice spacings) for a pressure gradient of about 0.175 GPa per lattice spacing. The peak pressure in this experiment is only 4.6 GPa—well within the region where both an elastic and plastic wave will develop given sufficient time. In experiments on thicker samples at these pressures, an elastic–plastic structure is seen in the data. It is clear that the waves we are measuring at these pressures are not steady. In recent work at higher pressures (20 GPa in aluminum) we see no sign of an elastic precursor and the rise times are in close agreement with the values calculated for the case above. This is probably fortuitous.

In the case of gold, yet another odd result occurs upon short pulse laser irradiation. This particular experiment is not a shock-wave experiment; however, it is included as an example of the possibility of preferentially putting energy into a particular mode of a solid. In this case, the sample is viewed on the same side of the cover slip that is pumped. A small phase shift (and reflectivity change) is observed upon initial irradiation and about 1.5 ps later an additional "blip" in the phase is seen accompanied by a very large permanent (on the time scale of tens of picoseconds) change in surface reflectivity. The timing of the second "blip" corresponds to a round trip transit time of the twice reflected incident laser pulse in the thickness of the glass coverslip. Thus, it appears that the initial laser irradiation has formed a metastable state in the gold with optical properties that are not particularly different from the quiescent material. Gold has an interband transition from a valance d-band to the conduction band at the energy we are using. Thus, electrons moved from the d-band (a bonding band) to the conduction band may destabilize the lattice [25]. This state must be quite metastable, however, because it appears that the order of 3% of the incident laser energy is capable of causing the state to spontaneously nucleate to an equilibrium state [26]. Current belief is that this equilibrium state is melted gold. Calculations of the electron–phonon interaction and the resulting temperatures suggest that the timing is sufficiently close to allow this to happen. The optical data on the reflectivity and phase allow extraction of the complex index that, when compared to data for melted gold, is in quite good agreement. This experiment, and others similar to it, point out the possibility of using the optical property changes of metals undergoing phase transitions as a means of seeing the transitions and of gaining kinetic data on the transitions when they occur rapidly. The strong sensitivity of the technique to changes in the electronic structure of materials makes it a technique of choice where the time and spatial resolution are appropriate.

9.6. Conclusion

We set out in this chapter to provide some insight into the nature of the structure of shock waves. It is clear that our mathematical descriptions of such phenomena allow for the possibility of truly discontinuous changes of state when the material considered is taken to be a continuum. Real materials are, of course, not continua on a sufficiently small scale. The question naturally arises as to whether or not shock waves have structure of the order of atomic dimensions. We have measured shock-wave thickness at a maximum of 10–20 nm at modest pressures. This is just 20–40 lattice planes in most metals. For corresponding pressures of 10–20 GPa, we have stress gradients of 0.2–0.7 GPa per lattice spacing. We further expect these shock thicknesses to decrease with increasing stress, perhaps more slowly than the inverse fourth power, but still to decrease. It is well to note that the lateral spatial resolution of the HERCULES system is about 2 μm at 800-nm wavelength. Thus, there are the order of 1×10^5 aluminum atoms in this area and along a diameter there are the order of 1000. From Hayes and Grady [7], we may estimate the number of defects per unit length at 20 or so per micron (assuming a strain rate of 10^8 s^{-1}). Thus, our samples may have the order of 1000 defects within the lateral spatial resolution of our experiment. The possibility exists, therefore, that the measured risetimes we see are an average over the defect structure that exists beneath our current spatial resolution.

At the scale of these experiments, thousands of individual shock-wave measurements can be done daily. This is an unheard of data rate in studies of this kind. The incredible advantage of such a data rate is that poor signal to noise ratios in each measurement can be statistically eliminated. This is not feasible to do using any other shock-wave technique. For problems in the structure and dynamics of shock waves on space and time scales of nanometers and picoseconds, techniques of the kind discussed here are most appropriate at present. Experimental and theoretical work continues on this problem and we anticipate some truly novel discoveries in the area of shock structure and shock-induced molecular excitations in the years ahead. It is this sort of research and related work at the mesoscale that is the embodiment of the term "shock-wave physics".

Acknowledgments

In preparing this article I have benefited greatly from discussions with Yasuyuki Horie and Bob Graham. I want to thank them, but also caution the reader that the words are mine and the errors they may convey, mine also.

References

[1]   R. Becker, *Z. Physik* **8**, pp. 321–362 (1922)

[2]   Ya.B. Zel'dovich and Yu.P. Raizer, *Physics of Shock Waves and High-Temperature Hydrodynamic Phenomena*, Vol. II, Academic Press, New York (1967).

[3]   H.M. Mott-Smith, *Phys. Rev.* **82**, pp. 885–892 (1951).

[4]   A. Sakurai, *J. Fluid Mech.* **3**, pp. 255–260 (1957).

[5]   S. Ziering and F. Ek, *Phys. Fluids*, **4**, pp. 765–766 (1961).

[6]   P. Glansdorf, *Phys. Fluids*, **5**, pp. 371–379 (1962).

[7]   D.B. Hayes and D.E. Grady, *Shock Waves in Condensed Matter—1981* (eds. W.J. Nellis, L. Seaman, and R.A. Graham), American institute of Physics, New York, pp. 412–416 (1982).

[8]   D.A. Rose and C.C. Martens, *J. Phys. Chem. A*, **101**, pp. 4613–4620 (1997).

[9]   P. Embid and M. Baer, *Mathematical Analysis of a Two-Phase Model for Reactive Granular Flow*, Sandia National Laboratories report SAND88-3302, Dec 1989.

[10] A.C. Eringen, *Continuum Physics*, Vol III, pp 1–127, Academic Press, New York (1976).

[11] D.C. Wallace, *Phys. Rev. B.*, **22**, p. 4 (1980).

[12] J.W. Swegle and D.E. Grady, *J. Appl. Phys.* **58**, p. 692 (1985).

[13] D.C. Wallace, *Phys. Rev. B* **24**, pp. 5597–5606 (1981).

[14] R. A. Graham, *J. Chem. Phys.* **83**, p. 23 (1979).

[15] K. Huang, *Statistical Mechanics*, 2nd Edition, John Wiley and Sons, New York (1987)

[16] J. Hohlfeld, S.-S. Wellershoff, J. Gudde, U. Conrad, V. Jahnke, and E. Matthias, *Chemical Physics* **251**, pp. 237–258 (2000).

[17] D.W. Brenner, D.H. Robertson, M.L. Elert, and C.T. White, *Phys. Rev. Lett.* **70**, p. 2174 (1993); **76**, p. 2202(E) (1996).

[18] T.C. Germann, et al., in "Proceedings of the 12th Symposium (International) on Detonation," San Diego, CA, 11-16 Aug 2002 in press.

[19] D.R. Bland, *J. Inst. Maths. Applics.* **1**, pp. 56–75, (1964).

[20] G. Tas, J. Franken, S. A. Hambir, D. E. Hare, and D. D. Dlott, *Phys. Rev. Lett.* **78**, 4585, (1997).

[21] R. Evans, A.D. Badger, F. Fallies, M. Mahdieh, T.A. Hall, P. Audebert, J.-P. Geindre, J.–C. Gauthier, A. Mysyrowicz, G. Grillon, and A. Antonetti, *Phys. Rev. Lett.* **77**, p. 3359 (1996).

[22] K.T. Gahagan, D.S. Moore, D.J. Funk, R.L. Rabie, and S.J. Buelow, *Phys. Rev. Lett.* **85**, p. 15 (2000).

[23] H. Tups and K. Syassen, *J. Phys. F: Met. Phys.* **14**, p. 2753 (1984).

[24] D.J. Funk, D.S. Moore, K.T. Gahagan, S.J. Buelow, J.H. Reho, G..L. Fisher, and R.L. Rabie, "Ultrafast measurement of the optical properties of aluminium during shock-wave breakout," *Phys. Rev. B* **64**, p. 115114-1 (2001)

[25] H.O. Jeschke, M.E. Garcia, and K.H. Bennemann, *Phys. Rev. Lett.* **87**, p. 1 (2001).

[26] K. Lu and Y. Li, *Phys. Rev. Lett.* **80**, p. 20 (1998).

CHAPTER 10

What Is a Shock Wave to an Explosive Molecule?

Craig M. Tarver

An explosive molecule is a metastable chemical species that reacts exothermically given the correct stimulus. Impacting an explosive with a shock wave is a "wake-up call" or "trigger" that compresses and heats the molecule. The energy deposited by the shock wave must be distributed to the vibrational modes of the explosive molecule before chemical reaction can occur. If the shock pressure and temperature are high enough and last long enough, exothermic chemical decomposition can lead to the formation of a detonation wave. For gaseous, liquid, and perfect single-crystal solid explosives, after an induction time, chemical reaction begins at or near the rear boundary of the charge. This induction time can be calculated by high-pressure, high-temperature transition state theory. A "superdetonation" wave travels through the preshocked explosive until it overtakes the initial shock wave and then slows to the steady state Chapman-Jouguet (C–J) velocity. In heterogeneous solid explosives, initiation of reaction occurs at "hot spots" created by shock compression. If there is a sufficient number of large and energetic enough "hot spots," these ignition sites grow creating a pressure pulse that overtakes the leading shock front causing detonation. Because the chemical energy is released well behind the leading shock front of a detonation wave, a mechanism is required for this energy to reinforce the leading shock front and maintain its overall constant velocity. This mechanism is the amplification of pressure wavelets in the reaction zone by the process of de-excitation of the initially highly vibrationally excited reaction product molecules. This process leads to the development of the three-dimensional structure of detonation waves observed for all explosives. In a detonation wave, the leading shock wave front becomes a "burden" for the explosive molecule to sustain by its chemical energy release.

10.1. Introduction

What is a shock wave to an explosive molecule? There are several answers to this question depending upon the strength and time duration of the shock pulse. Because an explosive molecule (or a mixture of fuel and oxidizer molecules) is inherently metastable, it requires only an increase in its internal energy to overcome its activation energy barrier to reaction. This decomposition process may eventually become highly exothermic and cause deflagration (subsonic flow) or detonation (supersonic flow). Therefore, a shock wave is the "wake-up call" or the "trigger" that causes the molecule to release its chemical energy. The shock pulse must be of sufficient strength and time duration or self-sustaining exothermic chemical reaction does not occur. In a heterogeneous solid explosive, a weak shock wave can create a compressed material that does not react when subjected to subsequent shock waves. Stronger shock waves create reactive flows in their wake. These reactive flows can couple to, reinforce, and strengthen the shock front. The result is a detonation wave, in which the leading shock wave front is sustained by the chemical energy released behind it. Then the shock front is not only a "trigger" but also a "burden" to the explosive molecule since the shock is sustained by its exothermic chemical reaction.

Therefore, a shock wave can be many different things to an explosive molecule. In this chapter the current state of knowledge and future research directions for several compression regimes are briefly discussed in order of increasing pressure.

10.2. Non-Shock Impact Ignition

When a heterogeneous solid explosive charge is subjected to a low-velocity impact that produces only a few tenths of a GPa pressure, a two-stage compression wave is formed. This wave consists of an elastic wave that propagates through the explosive at longitudinal sound velocity followed by a plastic wave traveling at lower velocity [1,2]. Within the flow field produced by the plastic wave, regions of the explosive can be heated by void collapse, friction, shear, dislocation pile-up, and other dissipative mechanisms [3]. "Hot spots" are formed and can ignite and grow into an explosive energy release. Most of these ignitions result in subsonic deflagration waves driven by heat transfer from the hot reaction products into the surrounding explosive molecules. Impact ignition is one of the most important safety concerns, because it is caused by the smallest amount of energy delivered to the explosive molecules.

Several tests have been developed to study impact: drop hammers; drop weight impact machines; the Skid test; the Susan test; etc. In recent years, the Steven test at the Lawrence Livermore National Laboratory (LLNL) [4] and its modified version at the Los Alamos National Laboratory (LANL) [5] have been used to yield quantitative experimental data that can be simulated with reactive flow computer models. Figure 10.1 shows the LLNL version of the Steven test in which cylindrical high explosive charges are tightly confined by a Teflon ring and impacted by 6.01-cm-diameter steel projectiles of various radii of curvature. In the LANL version, this Teflon ring is not used, allowing the explosive charge to deform upon impact. Thus, the two configurations test two extremes of confinement typically used with high explosives. Several techniques, including embedded pressure gauges, a ballistic pendulum, and blast overpressure gauges, have been employed to measure the violence of the reactions produced in the Steven tests.

Thus far, several HMX-based plastic bonded explosives have yielded distinct threshold projectile impact velocities for reaction. The threshold velocities are slightly higher for the unconfined explosive charges in the LANL test. These threshold velocities fall in the usual order of impact sensitivity with PBX 9404 being the most sensitive and EDC-37 the least sensitive of the HMX-based materials tested. No TATB-based high explosives have yet reacted in either Steven test version. The measured violence of reaction increases with projectile velocity in both versions, but remains well below that of an intentional detonation of a Steven test charge. The embedded pressure gauges record input pressures of less than 0.1 GPa and pulse durations of approximately 60 μs. The measured times to reaction are 200 to 500 μs.

The measured pressures, pulse durations, and times to reaction, as well as the threshold velocities for ignition for several projectile types, have been calculated by the Ignition and Growth reactive flow model. As discussed in the next section, this model was originally developed for shock initiation and detonation predictions, but has been applied to other reactive flows, such as XDT (fracture-re-compaction-ignition-detonation transition), pressure-dependent deflagration, and deflagration-to-detonation transition (DDT). Normalizing the Ignition and Growth model to Steven test data has enabled, for the first time, predictions to be be made of impact ignition thresholds in accident scenarios that cannot be tested. Several ignition mechanisms (friction, shear, pore collapse, strain, etc.) have been used within the Ignition and Growth model framework to account for the onset of exothermic reaction in impact-induced "hot spots." Presently it is not clear which physical mechanism (or mechanisms) dominates the "hot spot" formation and ignition process under each set of impact conditions. Precise experiments in which one of more of these energy dissipation mechanisms is eliminated (or at least greatly reduced) are needed to better understand non-shock impact ignition.

10.3. The Ignition and Growth Reactive Flow Model

All reactive flow models require as a minimum: i) two equations of state, one for the unreacted explosive and one for its reaction products; ii) a reaction rate law for the conversion of explosive to products; and iii) a mixture rule to calculate partially reacted states in which both explosive and products are present. The Ignition and Growth reactive flow model [6] uses two temperature-dependent Jones–Wilkins–Lee (JWL) equations of state, one for the unreacted explosive and another one for the reaction products:

$$p = Ae^{-R_1 V} + Be^{-R_2 V} + \omega C_v T / V , \qquad (10.1)$$

where p is pressure in Megabars, V is relative volume, T is temperature, ω is the

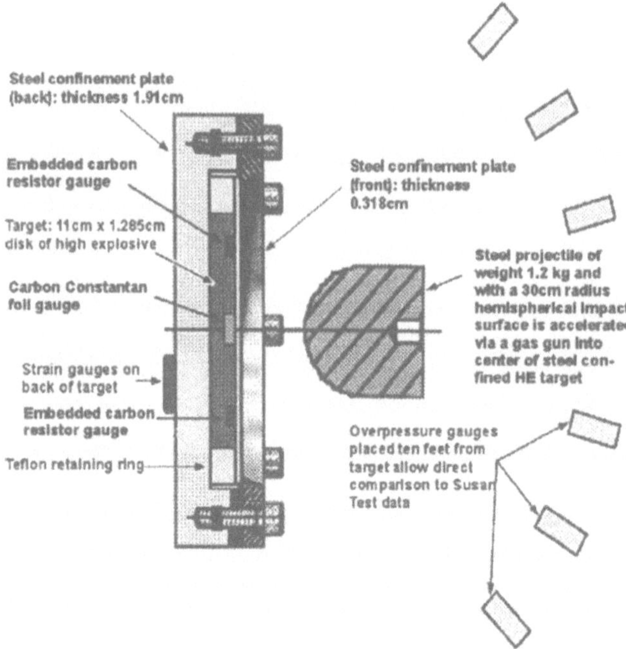

Figure 10.1. Geometry of the LLNL Steven impact test.

Grüneisen coefficient, C_v is the average heat capacity, and A, B, R_1, and R_2 are constants. The unreacted explosive equation of state is fitted to the available shock Hugoniot data, and the reaction product equation of state is fitted to cylinder test and other metal acceleration data. At the high pressures involved in shock initiation and detonation of solid and liquid explosives, the pressures of the two phases must be equilibrated, because interactions between the hot gases and the explosive molecules occur on nanosecond time scales depending on the sound velocities of the components. Various assumptions have been made about the temperatures in the explosive mixture, because heat transfer from the hot products to the cooler explosive is slower than the pressure equilibration process. In this version of the Ignition and Growth model, the temperatures of the unreacted explosive and its reaction products are equilibrated. Temperature equilibration is used, because heat transfer becomes increasingly efficient as the reacting "hot spots" grow and consume more explosive particles at high pressure and temperature. The reaction rate equation is

$$\frac{dF}{dt} = I\,(1-F)^b \left[\frac{\rho}{\rho_0} - 1 - a\right]^x + G_1\,(1-F)^c F^d p^y + G_2\,(1-F)^e F^g p^z \tag{10.2}$$

$$0 < F < F_{\text{igmax}}\,, \qquad 0 < F < F_{\text{G1max}}\,, \qquad F_{\text{G2min}} < F < 1,$$

where F is the fraction reacted, t is time in μs, ρ is the current density in g/cm^3, ρ_0 is the initial density, p is pressure in Mbars, and I, G_1, G_2, a, b, c, d, e, g, x, y, z, F_{igmax}, F_{G1max}, and F_{G2min} are constants. This three-term reaction rate law represents the three stages of reaction generally observed during shock initiation and detonation of pressed solid explosives [6]. The first stage of reaction is the formation and ignition of "hot spots" caused by the various possible mechanisms discussed for impact ignition as the initial shock or compression wave interacts with the unreacted explosive molecules. Generally, the fraction of solid explosive heated during compression is approximately equal to the original void volume. For shock initiation modeling, the second term in Eq. 10.2 describes the relatively slow process of the inward and/or outward growth of the isolated "hot spots" in a deflagration-type process. The third term represents the rapid completion of reaction as the "hot spots" coalesce at high pressures and temperatures, resulting in a transition to detonation. For detonation modeling, the first term again reacts a quantity of explosive less than or equal to the void volume after the explosive is compressed to the state at the von Neumann spike. The second term in Eq. 10.2 is used to model the fast decomposition of the solid into stable reaction product gases (CO_2, H_2O, N_2, CO, etc.). The third term is used to describe the relatively slow, diffusion-limited formation of solid carbon (amorphous, diamond, or graphite) as the equilibrium Chapman–Jouguet (C–J) state is approached. All of these reactive processes have been observed experimentally using embedded gauges and laser-interferometric techniques.

The Ignition and Growth reactive flow model has been applied to simulation of shock initiation and detonation experiments using many one-, two-, and three-dimensional hydrodynamic codes. In shock-initiation applications, it has successfully calculated many embedded gauge, run distance to detonation, short pulse duration, multiple shock, ramp-wave compression, and divergent-flow experiments on several high explosives. For detonation wave applications, the model has successfully calculated waveforms measured by embedded gauges and laser-interferometric methods, metal acceleration, failure diameter, corner turning, converging, diverging, and overdriven detonation experiments. Because the Ignition and Growth model and others of its generation use compression- and pressure-dependent reaction rate laws whereas chemical energy release actually depends upon the local temperatures everywhere in the compressed explosive, there are certain scenarios where this model must be used with care. One such scenario is discussed in the next section. Although pressure-dependent rate laws will still be used for the foreseeable future, efforts to develop next generation mesoscale and macroscale statistical "hot spot" models employing Arrhenius temperature-based chemical kinetics and heat transfer from the hot products to the cooler explosive have begun at several laboratories.

10.4. Weak Shock Compression

At slightly higher pressures than those produced in impact scenarios, the elastic and plastic waves merge into a relatively weak shock wave [1,2]. For homogeneous explosives, these shocks compress and heat the explosive molecules slightly, but little or no chemical decomposition occurs. For some heterogeneous solid explosives, there exists a narrow range of shock pressures in which all of the voids and other inhomogeneities can be compressed and perhaps react locally without creating growing "hot spots." The resulting fully dense explosive material cannot be shock initiated by subsequent strong shock waves or even detonation waves. This phenomenon is called "dead pressing" or "shock desensitization." Depending on the application, this can be a useful or a frustrating property of explosive charges. In the most thorough study of shock desensitization, Campbell and Travis [7] demonstrated that weak shock waves in the pressure range of 1–2.4 GPa could collapse the voids in PBX-9404 and Composition B-3, creating relatively homogeneous materials that failed to detonate when subjected to their own detonation waves. Interestingly, these detonation wave failures required times and run distances approximately equal to those measured for shock initiation by low-amplitude shocks. Thus, the detonation waves propagated until they reached material that had no more voids capable of forming new "hot spots." Tarver et al. [8] observed shock desensitization in the TATB-based explosive LX-17 in reflected shock experiments. They calculated the phenomenon using the Ignition and Growth model by assuming that a certain range of compressions [$h < (\rho/\rho_0 - 1) < a$] in the first term of Eq. 10.2 would result in no reaction ever occurring in the explosive. This assumption cannot address the observed times required for detonation failure in shock-desensitized explosives. The next generation of statistical "hot spot" reactive flow models with temperature-based reaction rate laws is needed to simulate the time dependent details of shock desensitization.

10.5. Homogeneous Explosives

Homogeneous explosives include gases, liquids without bubbles or suspended solids, and perfect crystals of solid explosives. In these materials, planar shock waves uniformly compress and heat the explosive molecules. There has long been some debate about the definition of the thickness of a shock wave. Zel'dovich and Raizer [9] define the width of a shock wave as the distance at which the viscosity and heat conduction become negligible. This occurs within a few molecular collisions in a gas. At the same time, the internal modes of gaseous explosive molecules are becoming excited: translational modes in a few collisions; rotational modes in tens of collisions; and vibrational modes in hundreds of collisions. These equilibration processes have long been studied in shock tubes [10]. Internal energy equilibration is now being studied in shocked liquid and solid explosives by Dlott et al. [11] and Fayer et al. [12]. In condensed phases, the phonon modes are initially excited followed by multi-phonon excitation of the lowest-frequency vibrational modes and then the higher-frequency modes by multi-phonon up-pumping and internal vibrational energy redistribution (IVR) [13]. Once the explosive molecules have attained vibrational equilibrium, chemical decomposition may start to occur.

For gaseous explosives, the nonequilibrium processes that precede chemical reaction are easily measured, since they can be lengthened to nanosecond or even microsecond time frames by dilution with inert gases or by the use of low initial gas pressures. The calculation of these states is also straightforward, because the perfect gas law applies. The initial reaction rates for the dissociation of the weakest chemical bond present in the explosive molecule/mixture are also easily measured in shock tube experiments and calculated using Arrhenius chemical kinetics. If the shock wave heats the explosive molecules to temperatures at which sufficient dissociation occurs before the shock compression ends and rarefaction wave cooling begins, the newly formed atoms react with surrounding molecules. An exothermic chain reaction process follows in which reaction product gases are formed in highly vibrationally excited states [14]. These excited products either undergo reactive collisions with the surrounding explosive molecules or nonreactive collisions with their neighbors in which one or more quanta of vibrational energy is transferred. Some collisions are "super-collisions" [15] in which several quanta of vibrational energy are transferred. Because reaction rates increase rapidly with each quantum of vibrational energy available to the colliding species, reactive collisions are very frequent and the main chemical reactions are extremely fast. Once the chain reaction process is completed, the remainder of the reaction zone is dominated by the de-excitation of highly vibrationally excited product molecules as chemical equilibrium is approached. This de-excitation process controls the length of the reaction zone and provides the chemical energy necessary for shock wave amplification during shock-to-detonation transition (SDT) and self-sustaining detonation.

The Non-Equilibrium Zel'dovich–von Neumann–Döring (NEZND) theory of detonation [14,16–19] was developed to explain the various nonequilibrium processes that precede and follow chemical energy release in self-sustaining detonation waves. As pressure wavelets pass through the subsonic reaction zone, they are amplified by chemical energy released in vibrational de-excitation processes. The opposite effect—shock-wave damping by a nonequilibrium gas that lacks vibrational energy after expansion through a nozzle—is a well-known phenomenon [9]. The pressure wavelets then interact with the main shock front and replace the energy lost during compression, acceleration, and heating of the explosive molecules. During shock initiation, this interaction process increases the shock front pressure and velocity. If the initial shock wave is amplified to a strength at which chemical reaction occurs close to the front, then self-sustaining detonation occurs. The pressure wavelet amplification process then provides the required chemical energy by developing and sustaining a three-dimensional Mach stem detonation wave structure. The complex three-dimensional leading shock wave is still a "trigger" for reaction, but it is also a "burden" for the explosive molecules to sustain at supersonic velocity with their energy release.

The three-dimensional structures of detonation waves have been observed for gaseous, liquid, and solid explosives [20]. In gaseous detonations, the details are very well known and several excellent reviews of the subject are available [21]. For liquid and perfect single-crystal solid explosives, the situation is much more complex and thus more difficult to observe and calculate than in gases. The high initial densities of the condensed phases make the measurement and calculation of the states attained behind a shock wave more difficult, because the processes now take tens or hundreds of picoseconds and the perfect gas law does not apply. The distribution of the shock compression energy between the potential (cold compression) energy of the unreacted liquid or solid and its thermal energy is a complex function of shock strength. The lack of voids, cracks, particle boundaries, etc., eliminates "hot spot" formation as an initiation mechanism. If the shock-compressed state lasts long enough for exothermic reaction to begin at the corresponding shock temperature, initiation of exothermic reaction occurs at or near the impacted boundary of the explosive charge. This "thermal explosion" creates a "superdetonation" wave that propagates through the precompressed explosive at a velocity in excess of its equilibrium Chapman-Jouguet (C–J) velocity. When this wave overtakes the leading shock wave, its velocity decreases rapidly until steady state C–J velocity is attained. This phenomenon has been measured and calculated for several detonating liquids [22] and solid perfect crystals [23]. Liquid explosives exhibit a wide range of shock sensitivity [24]. Perfect single crystals of sensitive solid explosives like PETN can be shock initiated [23], but single crystals of HMX cannot be initiated by a detonation wave from the 94% HMX plastic bonded explosive PBX 9404 [7].

The "induction" time for the initial "thermal explosion" can be calculated using the high pressure, high temperature transition state theory. Experimental unimolecular gas phase reaction rates under relatively low temperature ($<1000°$) shock conditions obey the usual Arrhenius law:

$$K = Ae^{-E/RT} ,$$
(10.3)

where K is the reaction rate constant, A is a frequency factor, E is the activation energy, and T is temperature, at low temperatures, but "fall-off" to less rapid rates of increase at high temperatures [25]. Nanosecond reaction zone measurements for solid explosives overdriven to pressures and temperatures exceeding those attained in self-sustaining detonation waves have shown that the reaction rates increase very slowly with shock temperature [26]. Eyring [27] attributed this "falloff" in unimolecular rates at the extreme temperature and density states attained in shock and detonation waves to the close proximity of vibrational states, which causes the high-frequency mode that becomes the transition state to rapidly equilibrate with the surrounding modes by IVR. These modes form a "pool" of vibrational energy in which the energy required for decomposition is shared. Any large quantity of vibrational energy that a specific mode receives from an excitation process is shared among the modes before reaction can occur. Conversely, sufficient vibrational energy from the entire pool of oscillators is statistically present in the transition state long enough to cause reaction. When the total energy in the vibrational modes equals the activation energy, the reaction rate constant K is given by

$$K = \frac{kT}{h} e^{-s} \sum_{i=0}^{s-1} \left(\frac{E}{RT} \right)^i \frac{1}{i!} e^{-E/RT} \qquad (10.4)$$

where k, h, and R are Boltzmann's, Planck's, and the gas constant, respectively, and s is the number of neighboring vibrational modes interacting with the transition state. The main effect of this rapid IVR among $s+1$ modes at high densities and temperatures is to decrease the rate constant dependence on temperature. Reasonable reaction rate constants were calculated for detonating solids and liquids using Eq. 10.4 with realistic equations of state and values of s [17]. For the lower temperatures attained in shock initiation of homogeneous liquid and solid explosives, the reaction rate constants calculated using Eq. 10.4 are larger than those predicted by Eq. 10.3. At detonation temperatures, the reaction rate constants from Eq. 10.4 are lower than those for regular Arrhenius kinetics in Eq. 10.3 and agree well with experimentally measured detonation reaction-zone length measurements. Reaction rate constants from Eqs. 10.3 and 10.4 are compared to induction time results for gaseous norbornene, liquid nitromethane, and single crystal PETN in Figs. 10.2–10.4, respectively [18]. Despite

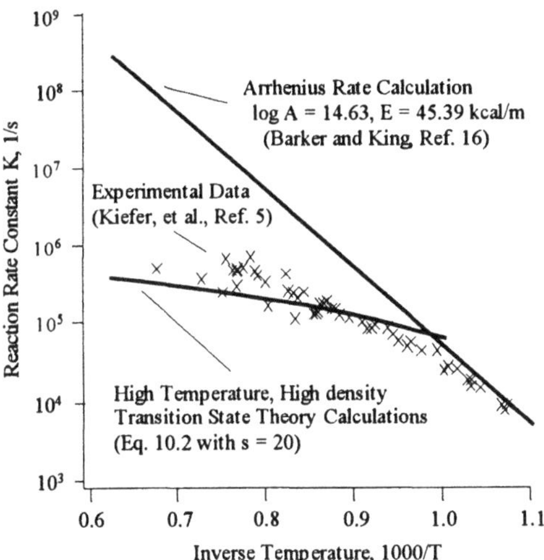

Figure 10.2. Reaction rate constant vs. inverse temperature for the unimolecular decomposition of norbornene

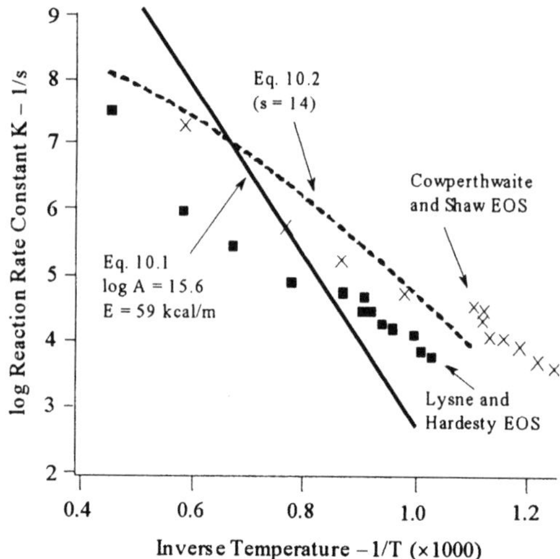

Figure 10.3. Reaction rate constants for nitromethane as functions of shock temperature

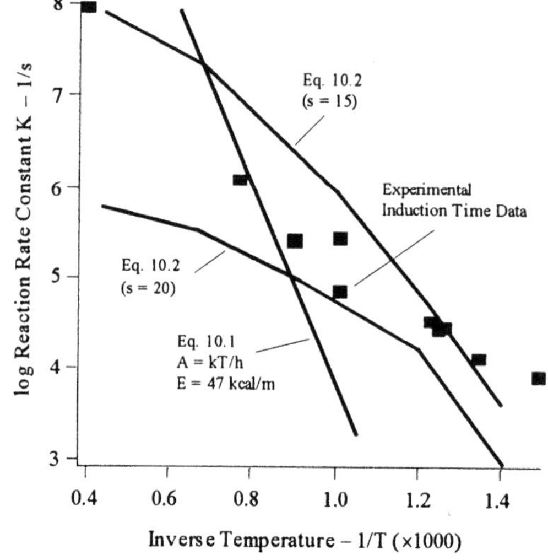

Figure 10.4. Reaction rate constants for single-crystal PETN as functions of shock temperature

uncertainties in the calculated shock temperatures for various equations of state, it is clear that Eq. 10.4 agrees well with all three sets of data using reasonable values of s of 15 to 20. This range of vibrational modes corresponds to the number of vibrations involving five or six neighboring closely packed atoms. Therefore, high-pressure, high-temperature transition state theory seems to accurately calculate induction times for shock-induced reactions during shock initiation and detonation of homogeneous gaseous, liquid, and solid explosives. More nanosecond time resolved experimental induction time data are clearly needed.

10.6. Heterogeneous Explosives

For heterogeneous explosives (liquids with bubbles or suspended solid particles and pressed or cast solids with voids, binders, metal particles, etc.), an initiating shock wave does not have to heat the entire material to the point of thermal explosion. Thermal energy is concentrated in local sites by the physical processes of void collapse, friction, shear, dislocation pile-up, etc. Liquid explosives containing bubbles can undergo partial reactions known as low velocity detonation (LVD) at hot spot sites created by collapsing voids. LVD can propagate long distances in metal pipes and is a major safety concern. LVD can fail to propagate or transition to full detonation in various scenarios [28].

It has long been known that shock initiation of solid explosives is controlled by ignition of hot spots [3]. How large and how hot does a hot spot have to be to react and begin to grow? Critical conditions for the growth or failure of hot spots in HMX- and TATB-based explosives have been calculated using multistep Arrhenius kinetic chemical decomposition models derived from thermal explosion experiments [29]. Figure 10.5 shows some calculated critical spherical hot spot temperatures in HMX and TATB. Once ignited, the growth rates of reacting hot spots into neighboring solid explosive particles and the interactions of several growing hot spots have been calculated for various geometries [30]. Figure 10.6 shows the times required for spherical HMX particles of various radii to completely deflagrate from the outside surface to the center under various boundary-temperature conditions. These relatively long reaction times imply that large explosive particles fragment during shock initiation, producing smaller particles with more reactive surface area for the hot gases to ignite. As growing hot spots coalesce at high pressures and temperatures, the transition from shock-induced reaction to detonation occurs very rapidly.

The buildup of pressure and particle velocity behind the shock wave front during shock initiation has been thoroughly studied using embedded gauges [31, 32] and laser-interferometric techniques [33]. These reactive flows have been modeled in multidimensional codes by the Ignition and Growth model of shock initiation and detonation [34]. Figure 10.7 shows measured and calculated particle velocity histories obtained for a shock-initiation experiment on the TATB-based explosive LX-17 [31]. The first term in the reaction rate law shown in Eq. 10.2 is used to account for the increase in shock front pressure recorded by the gauges in Fig. 10.7. The second or growth term in Eq. 10.2 is then used to simulate the growth of reaction behind the shock front, which causes the maximum particle velocities in Fig. 10.7. The third or completion term then models the rapid transition to detonation shown on the last three gauge records in Fig. 10.7.

This rapid transition occurs when growing hot spots are large enough to interact and rapidly transfer large amounts of heat to the remaining explosive particles.

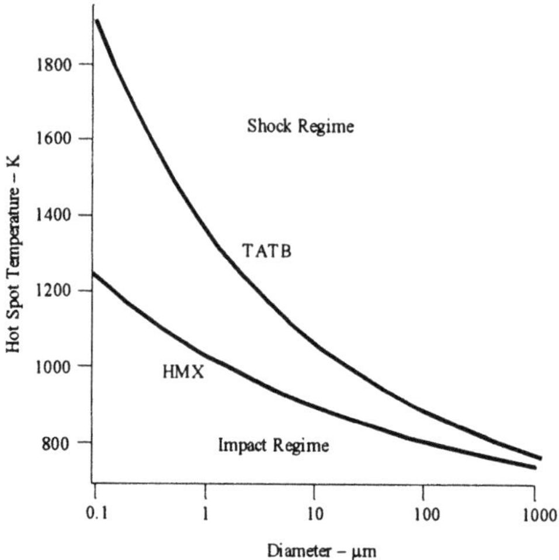

Figure 10.5. Critical spherical hot spot temperatures in HMX and TATB at various diameters.

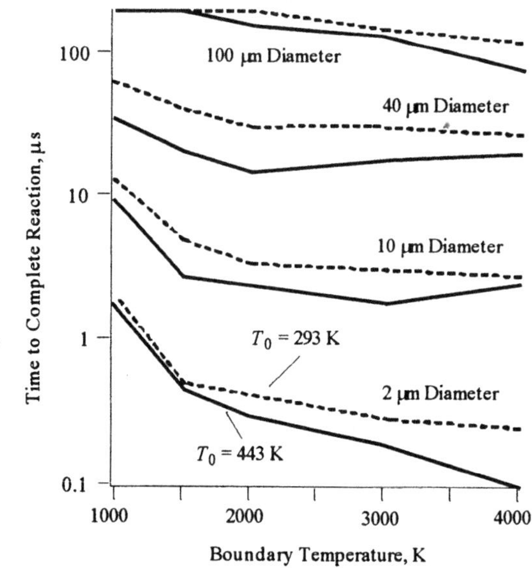

Figure 10.6. Reaction times for inward deflagrating HMX particles

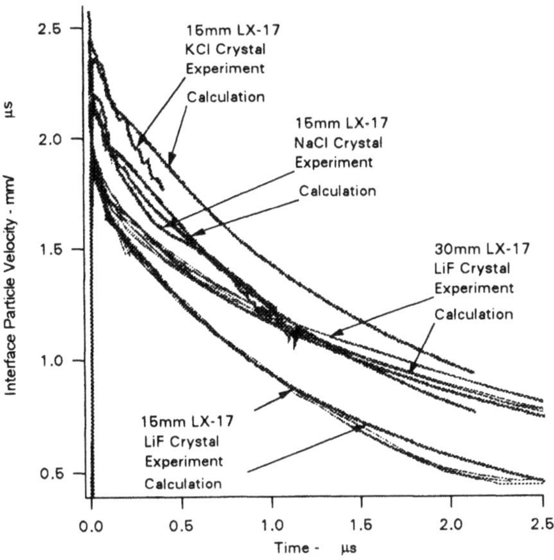

Figure 10.7. Particle velocity histories for LX-17 shock initiated by a Kel-F flyer plate at 2.951 mm/µs.

Detonation wave reaction zone structures in solid explosives and their metal acceleration properties have also been measured by embedded gauges and laser interferometry and calculated by the Ignition and Growth model [35]. Figure 10.8 shows the measured and calculated interface velocity histories for detonating LX-17 impacting various salt crystals [35]. The von Neumann spike state, a relatively fast reaction, a slower reaction, and finally the initial expansion of the products are clearly evident in Fig. 10.8. Figure 10.9 illustrates the measured and calculated free-surface velocities of 0.267-mm-thick tantalum discs driven by 19.871 mm of detonating LX-17. The momentum associated with the LX-17 reaction zone, which is approximately 3 mm long, and early product expansion are accurately measured and calculated in these small-scale experiments. The larger copper cylinder test is generally used to measure more of the reaction product expansion process. Because the main use of detonating solid explosives is to accelerate metals and other materials to high velocities, accurate measurements of the unreacted shock state (the "von Neumann spike"), the pressure profile of the chemical reaction zone, and the subsequent expansion of the reaction products as they deliver their momentum to the metal are essential. Currently these properties are known to within a few percent with nanosecond resolution [23,33,35,36]. Improved accuracy and time resolution are experimental goals.

Owing to solid particle interactions, one expects the detonation front structure to be more complex and less regular in heterogeneous explosives than in homogeneous ones. The subnanosecond techniques needed to resolve this wave

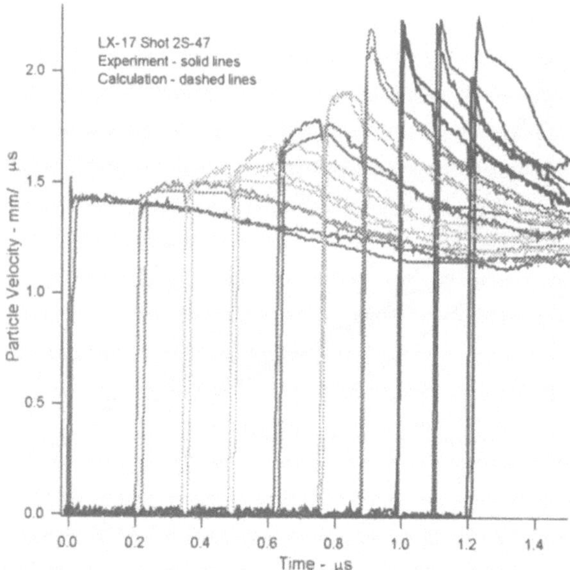

Figure 10.8. Interface particle velocity histories for detonating LX-17 and various salt crystals

front structure are becoming available. Figure 10.10 illustrates the various processes that precede and follow exothermic chemical reactions behind each wavelet of the three-dimensional structure comprising the reaction zone of a condensed phase detonation wave. Eventually all of these nonequilibrium physical and chemical mechanisms, plus perhaps others that have not yet been identified, will need to be measured experimentally and modeled in advanced multidimensional reactive flow models. Then the interactions of shock waves with explosive molecules and vice versa can be better understood. This understanding may lead to the production of safer, more energetic explosive molecules and formulations.

10.7. Future Research

Although a great deal has been learned in recent years about the interaction of shock waves with explosive molecules, greater spatial and time resolution is needed in shock-wave experiments and calculations. For understanding low-velocity-impact ignition mechanisms, the relative roles of void collapse, friction, shear, dislocation pile-up, strain, etc., need to be determined by clever experimentation. Many of these postulated "hot spot" formation mechanisms depend upon the magnitude of the viscosity in and behind shock-wave fronts, which has not yet been measured for shock waves in condensed phase explosives. If the dominant hot spot mechanism (or mechanisms) can be identified experimentally and successfully modeled, modifications to existing explosive formulations can be made. New processes and new materials (explosives, binders, additives, etc.) can be developed to produce safer products.

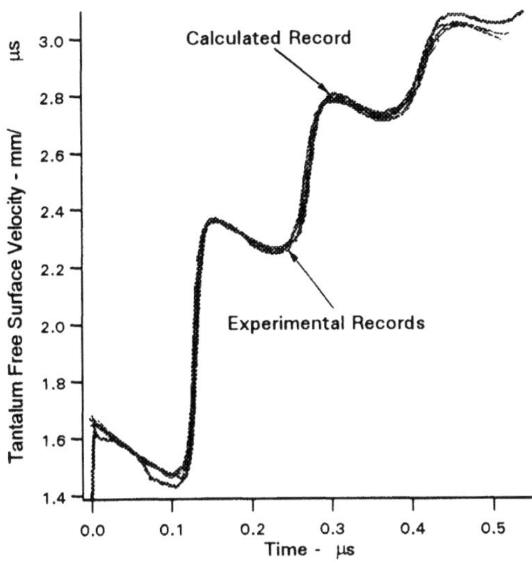

Figure 10.9. Free surface velocities for 0.267 mm thick tantalum disks driven by 19.871 mm of LX-17

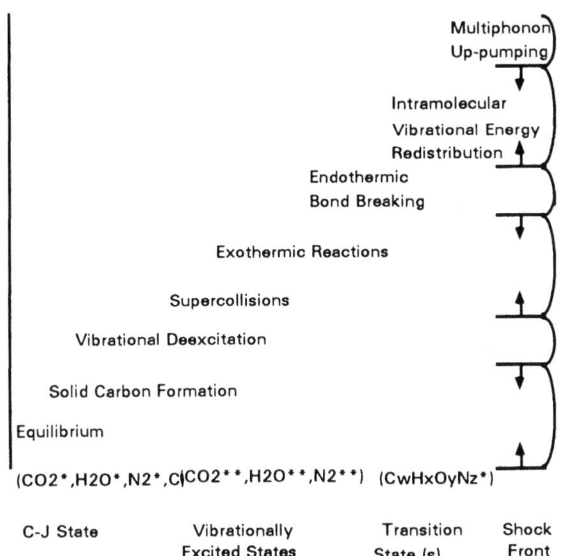

Figure 10.10. The Non-Equilibrium Zel'dovich–von Neumann–Döring (NEZND) model of detonation for condensed phase explosives

Because chemical reaction rates are controlled by the local temperature of a region of molecules, the most important research area in shock initiation is experimental measurement of local temperatures in all regions of shocked explosives: in and around "hot spots", in deflagration waves, in the reactive flows behind shock fronts, and in detonation waves. With these types of data, improved equations of state and temperature based, statistical "hot spot" reactive flow models can be developed to better predict the effects of shock waves on explosive molecules and vice versa [30]. Eventually it will become possible to model shock induced reactions as thermal decomposition mechanisms are modeled today by identifying intermediate reaction product species and following their concentration changes. To do this effectively, nanosecond or faster time resolved experimental data on the rates of consumption of the unreacted explosive, the concentrations of intermediate species, and the rates of production of the final stable products is needed. Accurate determination of the three-dimensional structures of detonation waves in condensed phase explosives is required to determine the level of detail required in reactive flow models to yield more realistic simulations in two- and three-dimensional hydrodynamic codes.

10.8. Summary

This chapter can only begin to address the complex question: What is a shock wave to an explosive molecule? Through several compression and heating mechanisms, a shock wave is the "wake-up call" or the "trigger" by which the exothermic power of the metastable explosive molecule is unleashed. Many possible outcomes of the initial hot spot formation process are possible: no reaction; shock desensitization; weak explosion; violent explosion; deflagration; shock-wave formation; decay or amplification of the shock front; and transition to detonation. Detonation is the desired result of an intentional shock initiation, but must be avoided at all costs during unintentional initiations (accidents). At the maximum rate of energy delivery in a detonation wave, the leading shock wave initiates the chemical reaction but then must be sustained by the chemical energy released. This chemical energy is initially released into highly vibrationally excited reaction products, whose relaxation to chemical equilibrium amplifies pressure wavelets propagating through the subsonic reaction zone. These pressure wavelets then overtake the shock-wave front, replacing its lost energy and creating the complex three-dimensional detonation wave front structure observed for all explosives. Understanding this intimate connection between nonequilibrium chemical kinetics and the three-dimensional detonation wave structure is one key to developing improved reactive flow models and safer, more powerful explosives. Another major key is to understand the mechanical processes of energy dissipation involved in "hot spot" formation, ignition, and growth induced by shock waves in explosive molecules.

Acknowledgments

The author would like to thank all of the excellent scientists that have taught him so much about explosives and life, especially Michael Cowperthwaite, Bob Woolfolk, Jerry Forbes, Paul Urtiew, Bud Hayes, LeRoy Green, Leroy Erickson, Jace Nunziato, Mel Baer, Ed Lee, Bill Davis, John Bdzil, and many others.

This work was performed under the auspices of the U.S. Department of Energy by Lawrence Livermore National Laboratory (Contract No. W-7405-ENG-48).

References

[1]   J.J. Dick, A.R. Martinez, and R.S. Hixson, in *Eleventh International Detonation Symposium*, Office of Naval Research, ONR 33300-5, Arlington, VA, pp. 317–324, (1998).

[2]   C.M. Tarver, P.A. Urtiew, S.K. Chidester, and L.G. Green, *Propellants, Explosives, Pyrotechnics* **18**, pp. 117–127 (1993).

[3]   J.E. Field, N.K.Bourne, S.J.P. Palmer, and S.M.Walley, *Phil. Trans. R. Soc. Lond.* **A 339**, pp. 269–299 (1992).

[4]   S.K. Chidester, C.M. Tarver, and R.G. Garza, in *Eleventh International Detonation Symposium*, Office of Naval Research, ONR 33300-5, Arlington VA, pp. 93–100 (1998).

[5]   D.J. Idar, R.A. Lucht, J.W. Straight, R.J. Scammon, R.V. Browning, J Middleditch, J.K. Dienes, C.B. Skidmore, and G.A. Buntain, in *Eleventh International Detonation Symposium*, Office of Naval Research, ONR 33300-5, Arlington, VA, pp. 101–110 (1998).

[6]   C.M. Tarver, J.O. Hallquist, and L.M. Erickson, in *Eighth Symposium (International) on Detonation* (ed. J.M. Short), Naval Surface Weapons Center NSWC MP86-194, Silver Spring, MD, pp. 951–961 (1985).

[7]   A.W. Campbell and J.R. Travis, in *Eighth Symposium (International) on Detonation* (ed. J.M. Short), Naval Surface Weapons Center NSWC MP86-194, Silver Spring, MD, pp. 1057–1068 (1985).

[8]   C.M. Tarver, T.M. Cook, P.A. Urtiew, and W.C. Tao, in *Tenth Symposium (International) on Detonation*, Office of Naval Research ONR 33395-12, Arlington, VA, pp. 696–703 (1993).

[9]   Y.B. Zel'dovich and Y.P. Raizer, *Physics of Shock Waves and High-Temperature Hydrodynamic Phenomena*, Academic Press, New York (1966).

[10] E.F. Greene and J.P. Toennis, *Chemical Reactions in Shock Waves*, Academic Press, New York (1964).

[11] X. Hong, S. Chen, and D.D. Dlott, *J. Phys. Chem.* **99**, pp. 9102–9109 (1995).

[12] W. Holmes, R.S. Francis, and M.D. Fayer, *J. Chem. Phys.* **110**, pp. 3576–3583 (1999).

[13] R.E. Weston, Jr. and G.W. Flynn, *Ann. Rev. Phys. Chem.* **43**, pp. 559–592 (1993).

[14] C.M. Tarver, *Comb. Flame* **46**, pp. 111–133 (1982).

[15] V. Bernshtein and I. Oref, *J. Phys. Chem.* **100**, pp. 9738–9758 (1996).

[16] C.M. Tarver, *Comb. Flame* **46**, pp. 135–155 (1982).

[17] C.M. Tarver, *Comb. Flame* **46**, pp. 157–179 (1982).

[18] C.M. Tarver, in *Shock Waves in Condensed Matter—1997* (eds. S.C. Schmidt, D.P. Dandekar, and J.W. Forbes), AIP Press, New York, pp. 301–304 (1998).

[19] C.M. Tarver, *J. Phys. Chem. A* **101**, pp. 4845–4851 (1997).

[20] W. Fickett and W.C. Davis, *Detonation*, University of California Press, Berkeley, (1979).

[21] J.H.S. Lee, *Detonation Waves in Gaseous Explosives, in Handbook of Shock Waves*, Volume 3 (eds. G. Ben-Dor, O. Igra, T. Elperin, and A. Lifshitz), Academic Press, New York, pp. 309–415 (2001).

[22] C.S. Yoo and N.C. Holmes, in *High-Pressure Science and Technology—1993* (eds. S.C. Schmidt, J.W. Shaner, G. Samara, and M. Ross), AIP Press, New York, pp. 1567–1570 (1994).

[23] C. M. Tarver, R. D. Breithaupt, and J.W.Kury, *J. Appl. Phys.* **81**, pp. 7193–7202 (1997).

[24] C.M. Tarver, R. Shaw, and M. Cowperthwaite, *J. Chem. Phys.* **64**, pp. 2665–2673 (1976).

[25] J.H. Kiefer and S.S. Kumaran, *J. Chem. Phys.* **99**, pp. 3531–3544 (1993).

[26] L.G. Green, C.M. Tarver, and D.J. Erskine, in *Ninth Symposium (International) on Detonation*, Office of the Chief of Naval Research OCNR 113291-7, Arlington, VA, pp. 670–682 (1989).

[27] H. Eyring, *Science* **199**, pp. 740–743 (1978).

[28] A.A. Schilperood, in *Seventh Symposium (International) on Detonation*, Naval Surface Warfare Center NSWC MP 82-334, Annapolis, MD, pp. 575–582 (1982).

[29] C.M. Tarver, S.K. Chidester, and A.L. Nichols, III, *J. Phys. Chem.* **100**, pp. 5795–5799 (1996).

[30] C.M. Tarver and A.L. Nichols, III, in *Eleventh International Detonation Symposium*, Office of Naval Research, ONR 33300-5, Arlington VA, pp. 599–605 (1998).

[31] R.L. Gustavsen, S.A. Sheffield, R.R. Alcon, C.M. Tarver, J.W. Forbes, and F. Garcia, in *Shock Compression of Condensed Matter—2001* (eds. M.D, Furnish, N.N. Thadhani and Y. Horie), AIP Press, New York, pp. 1019–1026(2002).

[32] S.A. Sheffield, R.L.Gustavsen, L.G. Hill, and R.R. Alcon, in *Eleventh International Detonation Symposium*, Office of Naval Research, ONR 33300-5, Arlington VA, pp. 451–458 (1998).

[33] R.L. Gustavsen, S.A. Sheffield, and R.R. Alcon, in *Eleventh International Detonation Symposium*, Office of Naval Research, ONR 33300-5, Arlington, VA, pp. 821–827 (1998).

[34] C.M. Tarver, J.W. Forbes, F. Garcia, and P.A. Urtiew, in *Shock Compression of Condensed Matter-2001,* (eds. M.D Furnish, N.N. Thadhani and Y. Horie), AIP Press, New York, pp. 1043–1046 (2002).

[35] C.M. Tarver, J.W. Kury, and R.D. Breithaupt, *J. Appl. Phys.* **82**, pp. 3771–3782 (1997).

[36] J.W. Kury, R.D. Breithaupt, and C.M. Tarver, *Shock Waves* **9**, pp. 227–237 (1999).

Author Index

Subject Index

MIX
Papier aus verantwortungsvollen Quellen
Paper from responsible sources
FSC® C105338
FSC
www.fsc.org

If you have any concerns about our products,
you can contact us on
ProductSafety@springernature.com

In case Publisher is established outside the EU,
the EU authorized representative is:
Springer Nature Customer Service Center GmbH
Europaplatz 3, 69115 Heidelberg, Germany

Printed by Libri Plureos GmbH
in Hamburg, Germany